Planning, Development and Management of Water Resources

Planning, Development and Management of Water Resources

Edited by **Raven Spoon**

SYRAWOOD
PUBLISHING HOUSE
New York

Published by Syrawood Publishing House,
750 Third Avenue, 9th Floor,
New York, NY 10017, USA
www.syrawoodpublishinghouse.com

Planning, Development and Management of Water Resources
Edited by Raven Spoon

International Standard Book Number: 978-1-68286-286-5 (Hardback)

Printed in China.

Contents

Preface

Water resources management is one of the prime concerns of our time. The inverse relationship between population growth and water availability has made planning and management of water resources even more crucial. This book discusses topics like impact of climate change on hydrology, planning, management, monitoring and protection of water resources, etc. This text will prove to be immensely beneficial to students and researchers in this field as it comprises discussions and case studies provided by internationally renowned experts.

This book is a comprehensive compilation of works of different researchers from varied parts of the world. It includes valuable experiences of the researchers with the sole objective of providing the readers (learners) with a proper knowledge of the concerned field. This book will be beneficial in evoking inspiration and enhancing the knowledge of the interested readers.

In the end, I would like to extend my heartiest thanks to the authors who worked with great determination on their chapters. I also appreciate the publisher's support in the course of the book. I would also like to deeply acknowledge my family who stood by me as a source of inspiration during the project.

Editor

Quantification of Water and Salt Exchanges in a Tidal Estuary

Janesh Devkota [1] and Xing Fang [2,*]

[1] Dynamic Solutions International LLC, Edmonds, WA 98020, USA; E-Mail: jdevkota@ds-intl.biz
[2] Department of Civil Engineering, Auburn University, Auburn, AL 36849-5337, USA

* Author to whom correspondence should be addressed; E-Mail: xing.fang@auburn.edu;

Academic Editor: Say-Leong Ong

Abstract: A calibrated three-dimensional hydrodynamic model was applied to study subtidal water and salt exchanges at various cross sections of the Perdido Bay and Wolf Bay system using the Eulerian decomposition method from 6 September 2008 to 13 July 2009. Salinity, velocity, and water levels at each cross section were extracted from the model output to compute flow rates and salt fluxes. Eulerian analysis concluded that salt fluxes (exchanges) at the Perdido Pass and Dolphin Pass cross sections were dominated by tidal oscillatory transport F_T, whereas shear dispersive transport F_E (shear dispersion due to vertical and lateral shear transport) was dominant at the Perdido Pass complex, the Wolf-Perdido canal, and the lower Perdido Bay cross sections. The flow rate Q_F and total salt transport rate F_S showed distinct variation in response to complex interactions between discharges from upstream rivers and tidal boundaries. Q_F and F_S ranged from −619 $m^3 \cdot s^{-1}$ (seaward) to 179 $m^3 \cdot s^{-1}$ (landward) and −13,480–6289 $kg \cdot s^{-1}$ at Perdido Pass when river discharges ranged 11.0–762.5 $m^3 \cdot s^{-1}$ in the 2008–2009 simulation period.

Keywords: Eulerian decomposition; salt exchange; numerical simulations; flow characteristics; estuary

1. Introduction

Estuaries are extremely productive regions due to the high flux of nutrients from the land and serve as breeding and nursery grounds for many species [1]. The impact of anthropogenic activity on the estuarine environment is a frequent concern because many major cities are located next to estuaries [2]. Neither estuaries nor oceans are capable of assimilating pollutants indefinitely; therefore, the increasing environmental concerns require the proper regulation and management of pollutant released in coastal zone [3]. To study the ecology of tidal estuaries, the exchanges of fresh and salt water are important not only in determining the characteristics of the environment [4] but also for water quality control and management. The circulation and salinity distribution patterns in an estuary depend on several factors such as river inflow pushing seaward, denser ocean water sliding landward, and tidal currents stirring and mixing the two [5]. The "exchange flow" or "gravitational circulation", which is characterized by deep inflow (denser salt intrusion in bottom layers) and shallow outflow (seaward freshwater flow in surface layers) through a cross section, dominates circulation in many estuaries [5]. Mean circulation and the circulation with frequencies lower than the semidiurnal and diurnal tides are often collectively called the residual circulation [6,7] because they are the residual of a time average over the principal periods.

The magnitude and direction of velocity and salinity determine the salt transport, which may affect inlet design of water supply in an estuary. The net outflow continually removes salt from the estuary whereas salt is brought back into the estuary by the "estuarine salt transport", *i.e.*, the vertical and lateral variations of tidally-averaged velocity and salinity, and "tidal dispersion", *i.e.*, the tidal correlations of velocity and salinity [8]. The magnitude of salt transport and the processes that contribute to it depend on the bathymetry of the estuary and on the strength of the physical forcing (e.g., tides, freshwater inflows, wind, *etc.*). In strongly stratified estuaries, salt transport is predominantly due to advection by the net landward flow and the estuarine salt transport [9]. However, in relatively well-mixed estuaries that are weakly stratified, tidal dispersion plays a larger role in the salt balance [10]. The stratification of partially stratified estuaries varies between these extremes. To understand characteristics of the time-variant estuary fluxes at different cross sections, these fluxes are further divided into two components: a subtidal component, which is a result of residual velocity and salinity; and an oscillatory tidal component, which is related to correlations in velocity and salinity at tidal time scales [11,12].

Several methods of determining the flow and salt exchange in estuaries have been formulated and verified successfully by various scientists [3,5,13–18]. Ketchum [4] developed an empirical method which computes the average distribution of fresh and salt water in an estuary and explains the exchange across various cross sections resulting from tidal oscillations. Bilgili *et al.* [3] used a Lagrangian particle tracking method embedded within a two-dimensional finite element model to study the transport and ocean-estuary exchange processes in the relatively well-mixed Great Bay Estuarine System in New Hampshire, USA. Using the Regional Ocean Modeling System, Chen *et al.* [18] used isohaline coordinate analysis to study the exchange flow characteristics in two different estuaries: the long (with respect to tidal excursion) Hudson River and the short Merrimack River.

This study used a previously calibrated Environmental Fluid Dynamic Code (EFDC) model [19] to investigate the flow and salt exchange across various cross sections to understand the underlying transport phenomena and mechanisms in the Perdido Bay and Wolf Bay (PBWB) system (Figure 1) including additional calibration on filtered water surface elevation (Figure 2). Xia *et al.* [20] used the EFDC model to understand distributions and dynamics of salinity and dissolved oxygen at Perdido Bay and the Gulf of Mexico. Xia *et al.* [21] studied the responses of estuarine plumes near the shoreline of the Gulf of Mexico under different wind conditions. They found that outflow from the estuary to the Gulf were strongest under northerly winds and could be stopped by southerly winds. Xia *et al.* [21] used the EFDC model to understand plume dynamics in Perdido Bay using idealized sensitivity experiments to examine the influence of wind stress on the three-dimensional plume signatures.

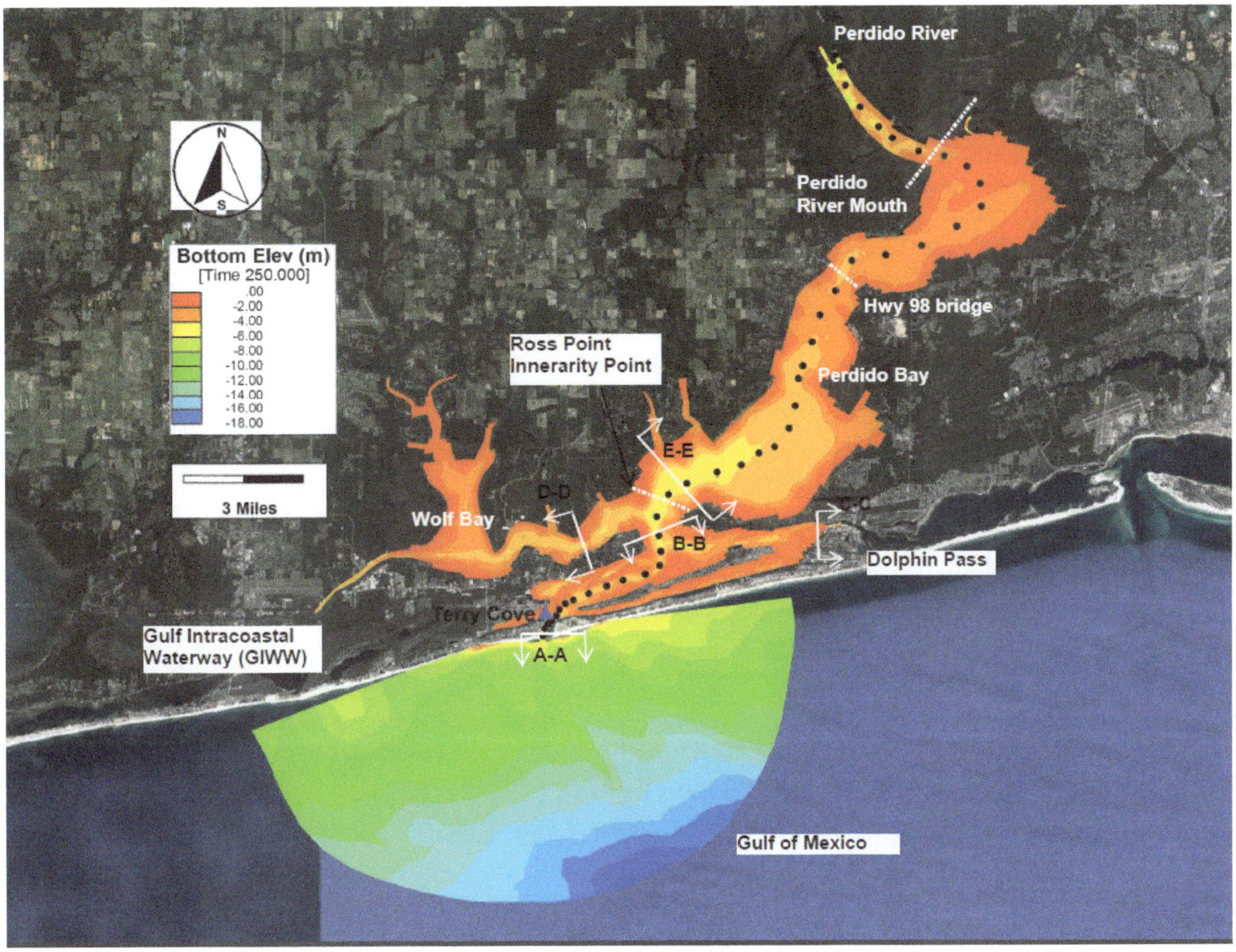

Figure 1. Environmental Fluid Dynamic Code (EFDC) model domain for Wolf Bay and Perdido Bay connected to the Gulf of Mexico through Perdido Pass, to eastern Big Lagoon through Dolphin Pass, and to western Mobile Bay through the Gulf Intracoastal Waterway (GIWW), showing the colored contours of bottom elevation and cross sections (A–A, B–B, C–C, D–D, and E–E) where water and salt fluxes were calculated. A centerline from Perdido Pass to Perdido River is represented using a series of dots for studying the salinity profile. White dashed lines refer to three cross sections used in Figure 3 and for discussion in the paper.

Devkota *et al.* [19] developed and calibrated an EFDC model for the PBWB system to understand the age of water in the system under different inflows from the rivers flowing into Wolf Bay and Perdido Bay. The model was calibrated from 6 September 2008 (Julian Day 250) to 13 July 2009 (Julian Day 560 starting from 1 January 2008). The PBWB system is connected to the Gulf of Mexico via three open boundaries (Figure 1): Perdido Pass on the south, Dolphin Pass—Connected to Big Lagoon and eventually to Pensacola Bay—On the East, and the Gulf Intracoastal Waterway (GIWW—Connected to the Mobile Bay and eventually to the Gulf of Mexico—To the West.

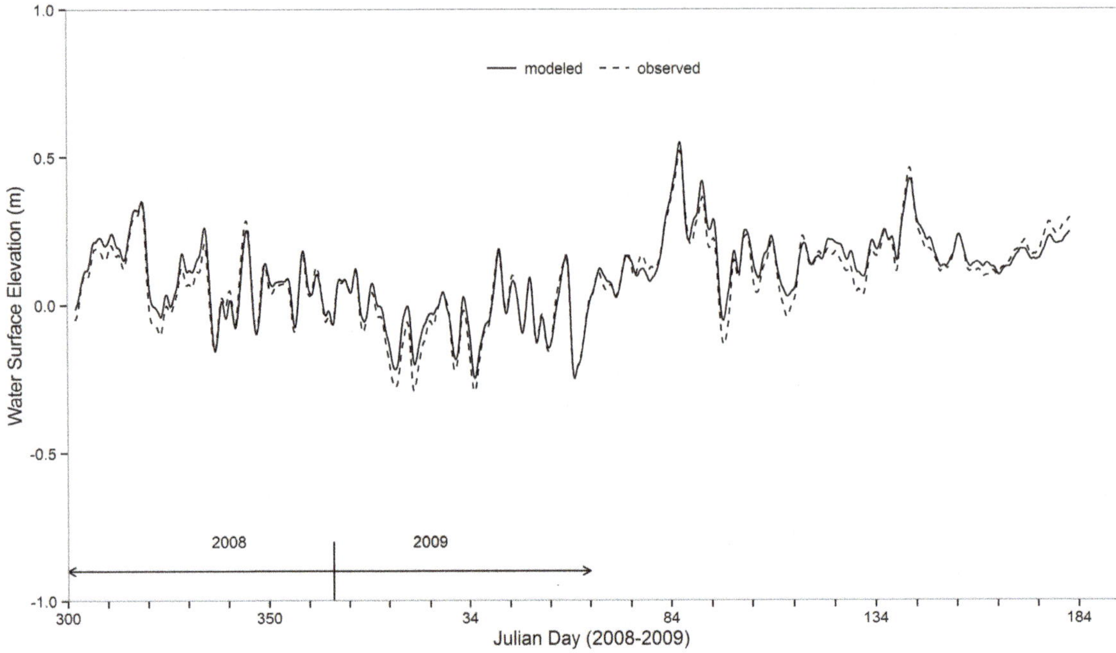

Figure 2. Modeled and observed filtered water surface elevations at Terry Cove from 27 October 2008 (Julian Day 300) to 3 July 2009 (Julian Day 184).

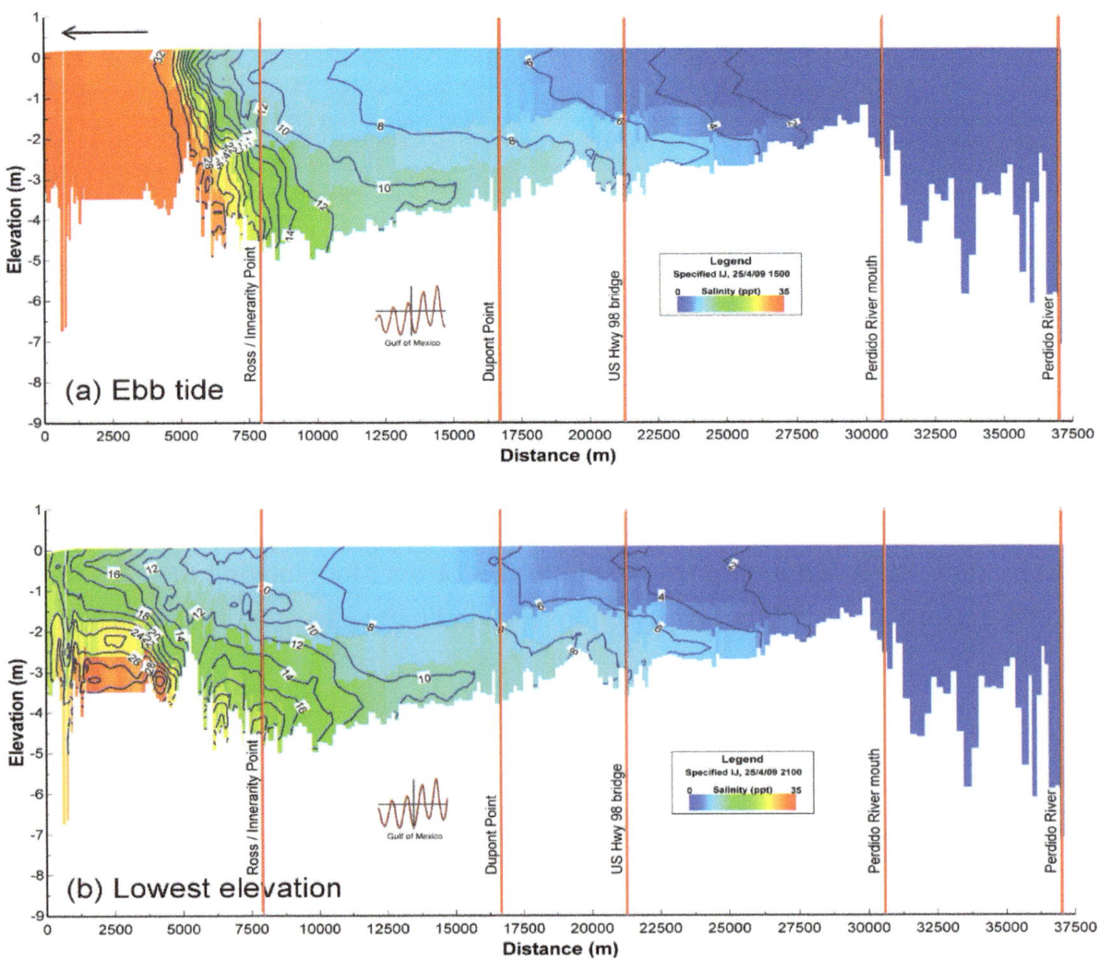

Figure 3. *Cont.*

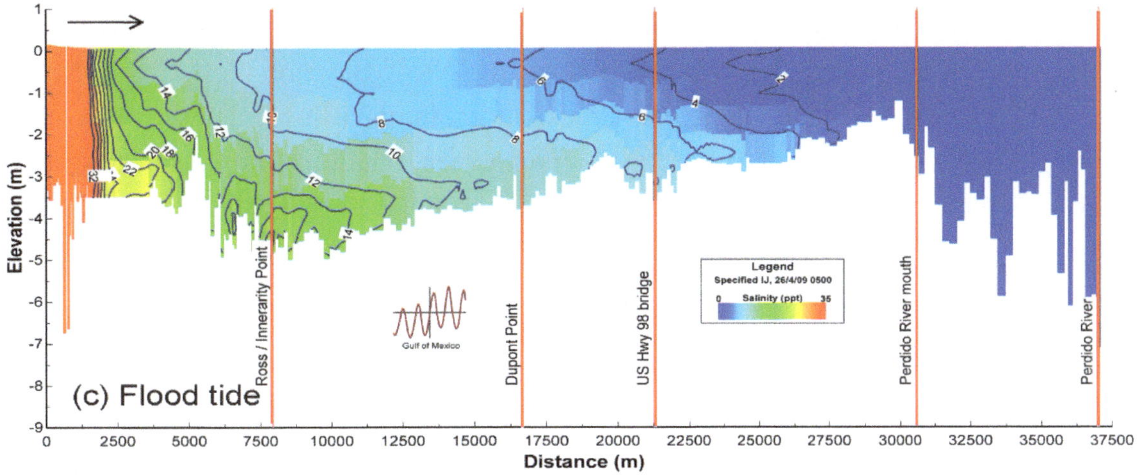

Figure 3. Simulated salinity profile contour plots through the centerline (Figure 1) from Perdido Pass to Perdido River during low flows from upstream under (**a**) ebb tide (15:00 25 April 2009) (**b**) the lowest water surface elevation (21:00 25 April 2009); and (**c**) flood tide (05:00 26 April 2009) at the Gulf of Mexico. Simulated salinity contour lines of 0–34 psu are shown in addition to colored contour maps.

In this study, we tried to answer the following three flow and salt exchange related questions: (1) How much salt is imported into Perdido Bay via Perdido Pass connection and how much fresh water is discharged from Perdido Bay to the Gulf of Mexico? (2) What factors are responsible for flow exchanges at different locations? (3) How much salt and water exchange takes place between Wolf Bay and Perdido Bay? In this paper we attempted to answer these questions by using the Eulerian salt decomposition method [5,22,23] with the calibrated EFDC model.

2. Materials and Methods

2.1. Study Area

The PBWB system (Figure 1) is a shallow to moderately deep inshore body of water [24] which has a length of 53.4 km and an average width of 4.2 km. Perdido Bay has an average depth of 2.6 m that increases gradually from the Perdido River mouth via lower Perdido Bay to the Gulf as indicated by elevation contours in Figure 1. The water depth in Perdido Bay tends to increase southward with the deepest parts of the estuary located at lower Perdido Bay, *i.e.*, the cross section connecting Ross Point and Innerarity Point (Figure 1). The major freshwater inflows into Perdido Bay are Perdido River, Styx River, Elevenmile Creek, and Bayou Marcus [24]. The Perdido River combined with the Styx River provides more than 70% of the freshwater input into the estuary, and mean river discharge recorded at the U.S. Geological Survey (USGS) station at Barrineau Park, Florida is 21 $m^3 \cdot s^{-1}$ [19,21]. Perdido Pass is the primary pathway of salt water to Perdido Bay and controls salinity distributions in Perdido Bay [24]. The width of Perdido Pass ranges from 200 to 500 m, and depth is around 5 m. The diurnal tidal range at Perdido Pass is about 0.2 m [19,20]. Using a 48-hour Lanczos low-pass filter [25], the filtered water surface elevations at three open boundaries (GIWW, the Gulf of Mexico, and Big Lagoon, Table 1) have similar statistical characteristics during the calibration period. Wolf Bay is a sub-estuary of Perdido Bay with a connection to the GIWW (Figure 1) and has a contributing watershed area of 126 km^2 [26]. The rivers that flow into Wolf Bay are Wolf Creek, Miflin Creek, Owens Bayou, Graham Bayou, and Hammock Creek, and combined flow into Wolf Bay ranged from 0.95–15.37 $m^3 \cdot s^{-1}$ with mean flow of 1.95 $m^3 \cdot s^{-1}$, which is very small in comparison to stream inflows to Perdido Bay (Table 1). Water in Wolf Bay flows through the GIWW into either Perdido Bay or Mobile Bay and ultimately into the Gulf of Mexico, and the flow directions depend on the magnitudes of wind and tide and the effect of the moon [27].

Table 1. Statistical summary of river inflows into Wolf Bay and Perdido Bay and filtered water surface elevations at GIWW (Gulf Intracoastal Waterway, west boundary), the Gulf of Mexico (south boundary), and Big Lagoon (east boundary).

	Discharges ($m^3 \cdot s^{-1}$)		**Water Surface Elevation (m)**		
Statistical Parameters	**Wolf-Bay Rivers**	**Perdido-Bay Rivers**	**GIWW**	**Gulf of Mexico**	**Big Lagoon**
Minimum	0.95	11.05	−0.29	−0.31	−0.19
Maximum	15.37	762.45	0.58	0.56	0.54
Mean	1.95	38.81	0.10	0.10	0.14
Standard Deviation	1.97	66.05	0.14	0.15	0.17
1st Quartile	1.11	14.61	0.019	0.01	0.00
Median	1.24	16.56	0.11	0.11	0.13
3rd Quartile	1.74	32.83	0.20	0.20	0.25

2.2. Hydrodynamic Model

The calibrated EFDC model used for this study was developed by Devkota *et al.* [19] for the PBWB system, called the Perdido EFDC model with site-specific model inputs. EFDC is a three-dimensional hydrodynamic model framework that solves continuity, momentum, salt and heat transport equations. EFDC uses hydrostatic and Boussinesq approximations [28]. EFDC uses orthogonal curvilinear or Cartesian horizontal coordinates and a stretched sigma vertical coordinate. EFDC uses an internal-external mode splitting to numerically solve the momentum equations and solves internal and external modes at the same time step, but it solves the external mode semi-implicitly with respect to barotropic pressure gradient term in depth-averaged momentum equations [28–30]. The Mellor and Yamada level 2.5 turbulence closure scheme is implemented in the model [31,32]. EFDC simulates density and topographically-induced circulation, and tidal and wind-driven flows in an estuary in order to predict spatial and temporal distributions of salinity and temperature [28,33]. The Perdido EFDC model has a total of 4878 curvilinear horizontal grids and 4 uniform spacing vertical sigma layers. The grid sizes for the model ranged from 27 to 368 m and layer thicknesses ranged from 0.04 to 1.78 m in the PBWP system (not including the small portion of the Gulf). The details of the model configurations are presented by Devkota *et al.* [19].

Upstream boundaries for the Perdido EFDC model include inflows from the four rivers into Perdido Bay and the five streams into Wolf Bay [19]. During the simulation period (2008–2009) of this study, the mean and maximum inflow rates from the four rivers to Perdido Bay are 38.8 $m^3 \cdot s^{-1}$ and 762.45 $m^3 \cdot s^{-1}$ (Table 1), respectively. Overall, the discharges from upstream rivers during the calibration period 2008–2009 represent the typical flows, which resulted from typical rainfall in 2008 and 2009 (49.84 inches in dry 2008 and 91.40 inches in wet 2009). Wind speed plays an important role in Wolf Bay and Perdido Bay and ranged from 0 to 11.5 $m \cdot s^{-1}$ with an average value of 2.9 $m \cdot s^{-1}$ during the simulation period (2008–2009). The dominant winds during simulation periods are northerly, southerly and northeasterly winds. Southerly winds tend to push the water into the system via Perdido Pass whereas northerly and northeasterly winds push water out of the Perdido Bay. There are three locations in the model where the flow exchange takes place between the Perdido Wolf system and the Gulf of Mexico, Mobile Bay, and Pensacola Bay. The southern boundary "the Gulf of Mexico" (Figure 1) is the measured water surface elevation at Dauphin Island, a NOAA's tides and currents station. Dolphin Pass is the location where the flow exchange between Perdido Bay and Pensacola Bay takes place via Big Lagoon. Measured time series of water surface elevation from the U.S. Army Corps of Engineers was used as the boundary condition at Dolphin Pass. The boundary condition at GIWW is the measured water surface elevation at the Gulf Shores NOAA's tides and currents station to represent the flow exchange taking place between Wolf Bay and Mobile Bay (Figure 1).

2.3. Theoretical Formulation for Eulerian Decomposition Method

Eulerian decomposition approach can be used to study the salt flux in tidal rivers and estuaries and has been successfully used to predict the estuarine exchange [5,18,22,23]. In this approach, the tidally averaged (subtidal) salt flux (F_S) through an estuarine cross section is decomposed into river flux (F_R), exchange flow flux (F_E), and tidal flux (F_T). The equation to compute the subtidal salt flux F_S (mass flow rate in $kg \cdot s^{-1}$) is given below based on Lerczak *et al.* [22]:

$$F_S = \left\langle \int usdA \right\rangle \tag{1}$$

where $\langle \, \rangle$ denotes a low-pass, subtidal temporal filter that produces tidally averaged/integrated component, u is the velocity in $m \cdot s^{-1}$ normal to the cross section as function of time and position on the cross section, s is the salinity in psu, and dA is the differential area of integration in m^2. The subtidal volumetric flow rate Q_F ($m^3 \cdot s^{-1}$) through a cross-sectional area (A) is determined as

$$Q_F = \left\langle \int udA \right\rangle \tag{2}$$

To simplify the calculation of salt fluxes through a cross section, the area is divided into a number of the cells along the section with constant differential areas dA along depth, but dA varies with time (expands and contracts tidally) and cell location (bottom elevation change). Velocity, salinity, and salt flux are also separated (decomposed) into three components. They are (a) tidally and cross-sectionally averaged; (b) tidally averaged and sectionally varying; and (c) tidally and sectionally varying remainder.

The tidally averaged properties are normalized using the tidally averaged, cross-sectional area $A_0 \equiv \left\langle \int dA \right\rangle$ [22]. Therefore, the tidally and sectionally averaged velocity (u_0) and salinity (s_0) are given by

$$u_0 \equiv \frac{\left\langle \int udA \right\rangle}{A_0} = \frac{Q_F}{A_0}, \quad s_0 \equiv \frac{\left\langle \int sdA \right\rangle}{A_0} \tag{3}$$

Both u_0 and Q_F are negative when they are towards the ocean (seaward) because flow coming from the open boundary towards the model domain is considered positive. The tidally averaged and sectionally varying velocity (u_E) and salinity (s_E) are defined as

$$u_E \equiv \frac{\langle udA \rangle}{dA_0} - u_0, \quad s_E \equiv \frac{\langle udA \rangle}{dA_0} - s_0 \tag{4}$$

where $dA_0 \equiv \langle dA \rangle$ is tidally averaged integration area. The exchange flow components (u_E and s_E) include the laterally and vertically varying part of the gravitational circulation [5]. The tidally varying and sectionally varying velocity (u_T) and salinity (s_T), which vary with time and over the cross section, are defined by Lerczak *et al.* [22] as

$$u_T \equiv u - u_0 - u_E, \quad s_T \equiv s - s_0 - s_E \tag{5}$$

The components u_T and s_T vary predominantly on tidal scales, while u_0, s_0, u_E, and s_E vary only on subtidal scales.

Finally, the subtidal salt flux can be decomposed into three components (river, exchange, and tidal) as follows:

$$\begin{aligned} F_S &= \left\langle \int (u_0 + u_E + u_T)(s_0 + s_E + s_T) dA \right\rangle \approx \left\langle \int (u_0 s_0 + u_E s_E + u_T s_T) dA \right\rangle \\ &\equiv Q_F s_0 + F_E + F_T \end{aligned} \tag{6}$$

In the above simplification, Lerczak *et al.* [22] have made use of the following properties to eliminate six of the nine terms;

$$\int u_E dA_0 = 0, \int s_E dA_0 = 0, \langle u_T dA \rangle = 0, \langle s_T dA \rangle = 0 \tag{7}$$

The three fluxes in the right-hand side of Equation (6) represent the subtidal salt fluxes due to cross-sectional average advective transport, shear dispersion due to vertical and lateral shear transport, and tidal oscillatory salt transport due to temporal correlations between *u* and *s* [22].

3. Results and Discussions

3.1. Comparison of Subtidal Water Surface Elevations

The Perdido EFDC model was run from 6 September 2008 (Julian Day 250) to 13 July 2009 (Julian Day 560) during the calibration. The model was calibrated against measured water surface elevation, water temperature, and salinity at various monitoring stations. The details of the calibration are given by Devkota *et al.* [19]. In this paper, the observed and modeled time series of water surface elevations at Terry Cove and Blue Angels Park were revisited and subtidal components were studied using simulation outputs at each hour.

The tides in the PBWB system are dominated by diurnal tides: O1 and P1 tidal constituents [19]. It is essential to remove the tidal effects in the time series of water surface elevation and salinity to understand the underlying residual subtidal components and the processes governing the flow. To begin with, to remove the tidal effects, a 24-hour low-pass filter was employed, but it was unable to remove the tidal effects completely. A 48-hour Godin low-pass filtering program [34] was used by the USGS to remove tidal signals from observed discharges in the Mobile River that links to the Gulf of Mexico [35]. Similarly, a 48-hour Lanczos low-pass filter was used by Kim and Park [23] to study water and salt exchange for a micro-tidal, stratified northern Gulf of Mexico estuary. Therefore, a 48-hour Lanczos low-pass filter [25] was used to eliminate the tidal effects from the time series of water surface elevation at Terry Cove. Figure 2 shows the time series plot of subtidal observed and modeled water surface elevations at Terry Cove from Julian Day 300 (26 October 2014) to Julian Day 560 (13 July 2009). The observed and modeled subtidal water surface elevations match reasonably well. The mean absolute differences between observed and modeled subtidal water surface elevations at Terry Cove from 300–560 days was 0.04 m (Table 2). During the calibration period, there were no continuous time series data for other transport parameters such as temperature and salinity; therefore, time series comparison between observed and modeled subtidal temperature and salinity could not be made.

Table 2. Statistical summary of the differences (Observed − Modeled) and absolute differences (|Observed − Modeled|) between observed and modeled subtidal water surface elevations (m) at the monitoring stations Blue Angeles Park and Terry Cove.

Statistical Parameters	Blue Angels Park		Terry Cove	
	Observed − Modeled	**|Observed − Modeled|**	**Observed − Modeled**	**|Observed − Modeled|**
Minimum	−0.13	0.00	−0.1	0.00
Maximum	0.18	0.18	0.05	0.10
Mean	0.00	0.03	−0.02	0.02
Standard Deviation	0.04	0.03	0.03	0.02
1st Quartile	−0.02	0.01	−0.04	0.01
Median	0.00	0.02	−0.02	0.02
3rd Quartile	0.02	0.04	0.00	0.04

3.2. Salinity Distributions in Perdido Bay

The temporal and spatial salinity distributions in Perdido Bay are first illustrated using salinity profile distribution plots. Three examples of salinity distribution in Perdido Bay along the centerline (Figure 1) from Perdido Pass to Perdido River are plotted for low inflows (Figure 3) on 25–26 April 2009. The inflows on 25–26 April 2009 (Julian Days 481–482 in Figure 4) were low (about 19.5 $m^3 \cdot s^{-1}$) and chosen to represent low inflow conditions. Three water surface elevations at Perdido Pass were chosen to represent ebb tide (seaward flow), the lowest elevation in the tidal cycle, and flood tide (landward flow).

Figure 3 demonstrates that salinity fields vary throughout Perdido Bay as a result of dynamic inflows, tides, and wind forcing. The *x*-axis in Figure 3 represents the horizontal distance from the mouth of Perdido Pass (0 m) to the mouth of Perdido River (37,500 m, Figure 1) and the *y*-axis represents the depth elevations (m) of the computational grids along the centerline shown in Figure 1. Salinity (psu) distributions along the centerline are represented using color contours with blue being the lowest salinity (0 psu) and red being the highest salinity (35 psu). Isohalines from 0–35 psu were also plotted in Figure 3 using 2 psu increment. Various locations along the centerline such as Perdido River, Perdido River mouth, US Hwy 98 Bridge, DuPont Point, and Ross Point and Innerarity Point are labeled using vertical lines that allow readers to visualize the temporal and spatial salinity distributions. Time series of water level at the Gulf of Mexico is displayed using small windows with the vertical line representing the corresponding time of the snapshot of salinity distribution.

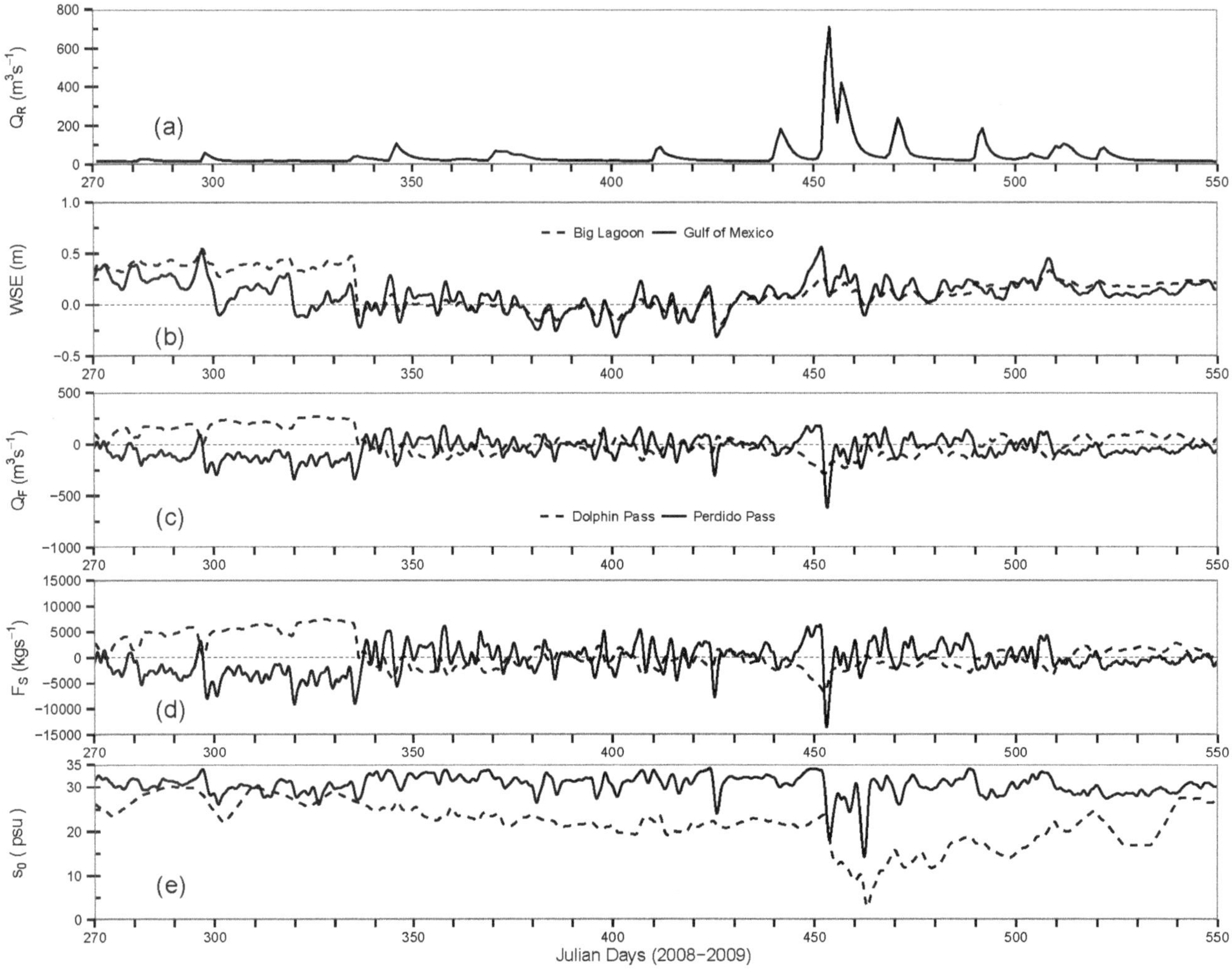

Figure 4. Time series of (**a**) discharge Q_R ($m^3 \cdot s^{-1}$) from Perdido River and Styx River; (**b**) water surface elevation (WSE, m) at the Gulf of Mexico (solid line) and Big Lagoon (dashed line); (**c**) volumetric flow rate Q_F ($m^3 \cdot s^{-1}$); (**d**) salt transport rate F_S ($kg \cdot s^{-1}$); and (**e**) cross-sectional average salinity s_0 (psu) at Perdido Pass and Dolphin Pass. Positive and negative fluxes in (c) and (d) indicate flow into and out of the Perdido Bay.

Figure 3a shows that salt water was advancing towards the Gulf of Mexico due to the ebb tide (an arrow was used to indicate the flow direction) at the downstream boundary at 15:00 on 25 April 2009 under low inflows from upstream. The Perdido Pass complex includes the area from Perdido Pass (Section A-A) to the Section B-B in Figure 1. The more or less well-mixed condition happened in the Perdido Pass complex with salinity about 34 psu throughout the depth. However, at the intersection of Ross Point and Innerarity Point (~7.5 km upstream from Perdido Pass), the vertical salinity stratification was strong and about 10 psu (11 and 21 psu at the surface and bottom layers, respectively). The salinity stratification decreased rapidly from Ross/Innerarity Point to DuPont Point and US Hwy 98 bridge due to the dilution created from the upstream freshwater inflows. In Figure 3b, at 21:00 on 25 April 2009, reduced salinity towards Perdido Pass indicated that the salinity was greatly reduced from ebb tide to the lowest water surface elevation at the downstream boundary. The Perdido Pass complex during the lowest water surface elevation at downstream had a stronger stratification compared to the stratification under ebb tide (Figure 3a). The strongest stratification occurred in the middle portion of the Perdido Pass complex where the surface salinity ranged from 14–18 psu and the bottom salinity from 28–30 psu (red region in Figure 3b). A snapshot of salinity distribution at 05:00 on 26 April 2009 (Figure 3c) shows that salinity was being introduced into the Perdido Bay by the flood tide from the Gulf of Mexico. During the flood tide, the stratification near Perdido Pass occurred at the lowest elevation (Figure 3b) and was destroyed because the water depth was relatively small and flow momentum from tides was large, and flood tides from downstream eventually pushed high saline water into the system. Because the inflows were small on 25–26 April 2009, salt water passed the US Hwy 98 Bridge (5–8 psu) and reached upper Perdido Bay, but Perdido River mouth still had freshwater under both ebb and flood tides (Figure 3) at Perdido Pass.

During the large inflows from upstream, the salinity at the Perdido Pass complex was relatively small regardless of ebb or flood tides at the Gulf of Mexico [36]. Therefore, complex temporal and spatial stratification patterns in Perdido Bay were the result of interactions between different inflows and tidal variations.

3.3. Eulerian Flux Decomposition in the PBWB System

To explore the salt fluxes into the PBWB system, the Eulerian method of salt flux decomposition was applied to hourly time series data of simulated velocity, water surface elevation, and salinity from calibrated Perdido EFDC model. The flux calculations were performed from 26 September 2008 (Julian Day 270)–3 July 2009 (Julian Day 550). To avoid any effect of initial conditions, the first 20 simulation days were treated as model spin up period. The model forcing includes observed river inflows, ocean tides, and meteorological inputs as boundary conditions. Flow exchanges were calculated at five different cross sections (Figure 1) in the PBWB system. For Eulerian flux method, Equations (1)–(7) were implemented to compute various flux components such as Q_F, F_S, s_0, $Q_F s_0$, F_E, F_T, u_E, and s_E. To be consistent about the sign convention, Perdido Bay is treated as the control area or reference area where the flow exchange takes place between Perdido Pass, Dolphin Pass, and Wolf Bay. All flows and fluxes moving from the external (open) boundary, or Wolf Bay into Perdido Bay, have positive values, and all flows and fluxes moving out of Perdido Bay to the external boundary, or Perdido Bay to Wolf Bay, have negative values (Figures 4–6 and in Tables 3–5).

The 2008–2009 simulation period had a wide range of freshwater river discharges Q_R (Figure 4a) flowing into Perdido Bay, which ranged from 14.2–760 $m^3 \cdot s^{-1}$ with an average inflow of 42.0 $m^3 \cdot s^{-1}$. The time series of filtered water surface elevation (WSE) at the Gulf of Mexico during the simulation period plotted in Figure 4b exhibited the variation from −0.32 m up to 0.77 m with an average value of 0.15 m. Filtered WSE at the Big Lagoon (east boundary, Figure 1) varied from −0.19 –0.54 m with an average value of 0.14 m, and was mostly higher than filtered WSE at the Gulf before Julian Day 335 (31 November 2008, Figure 4b). Average values of subtidal volumetric flow rate Q_F (Figure 4c) calculated from Eulerian flux method (Equation (2)) over the simulation period were −42.5 $m^3 \cdot s^{-1}$ and 16 $m^3 \cdot s^{-1}$ (Table 3) for Perdido Pass (Section A-A, Figure 1) and Dolphin Pass (Section C-C, Figure 1), respectively. The negative values of discharge Q_F and salt flux F_S indicate seaward flux from Perdido

Bay, *i.e.*, southward through Perdido Pass and eastward through Dolphin Pass. Both Q_F and F_S showed large temporal variations. Q_F through Perdido Pass ranged from −619 $m^3 \cdot s^{-1}$–180 $m^3 \cdot s^{-1}$ (Table 3). The subtidal salt flux F_S through Perdido Pass (Figure 4d and Table 3) ranged from −13,480–6289 $kg \cdot s^{-1}$ (average of −527 $kg \cdot s^{-1}$) and −6789–7393 $kg \cdot s^{-1}$ (average of 648 $kg \cdot s^{-1}$) through Dolphin Pass. With the range of variations 1–2 orders of magnitude larger than the corresponding means, the mean flux values are not representative of the water and salt exchange through Perdido Pass and Dolphin Pass.

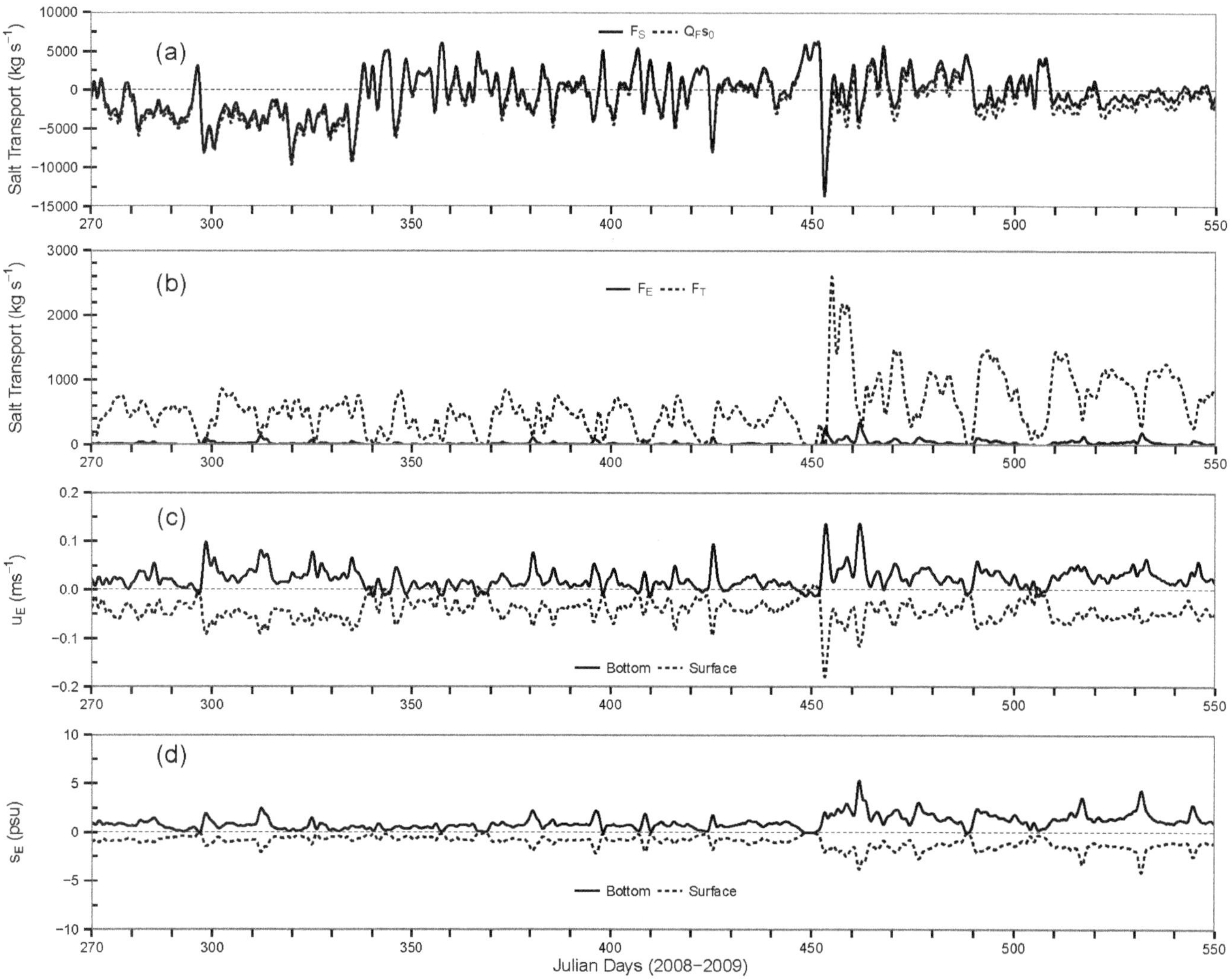

Figure 5. (**a**) Salt transport rate F_S ($kg \cdot s^{-1}$) and river flow flux component $Q_F s_O$; (**b**) salt transport components F_E and F_T through Section A-A (Perdido Pass); (**c**) subtidal estuarine exchange flow u_E ($m \cdot s^{-1}$); and (**d**) salinity s_E (psu) in the surface and bottom layers at the deepest channel cell of Perdido Pass. Positive (negative) flux in (a) and (b) and positive (negative) velocity in (c) indicate the flow into (out of) Perdido Bay. Note that *y*-axis scales are different in (a) and (b).

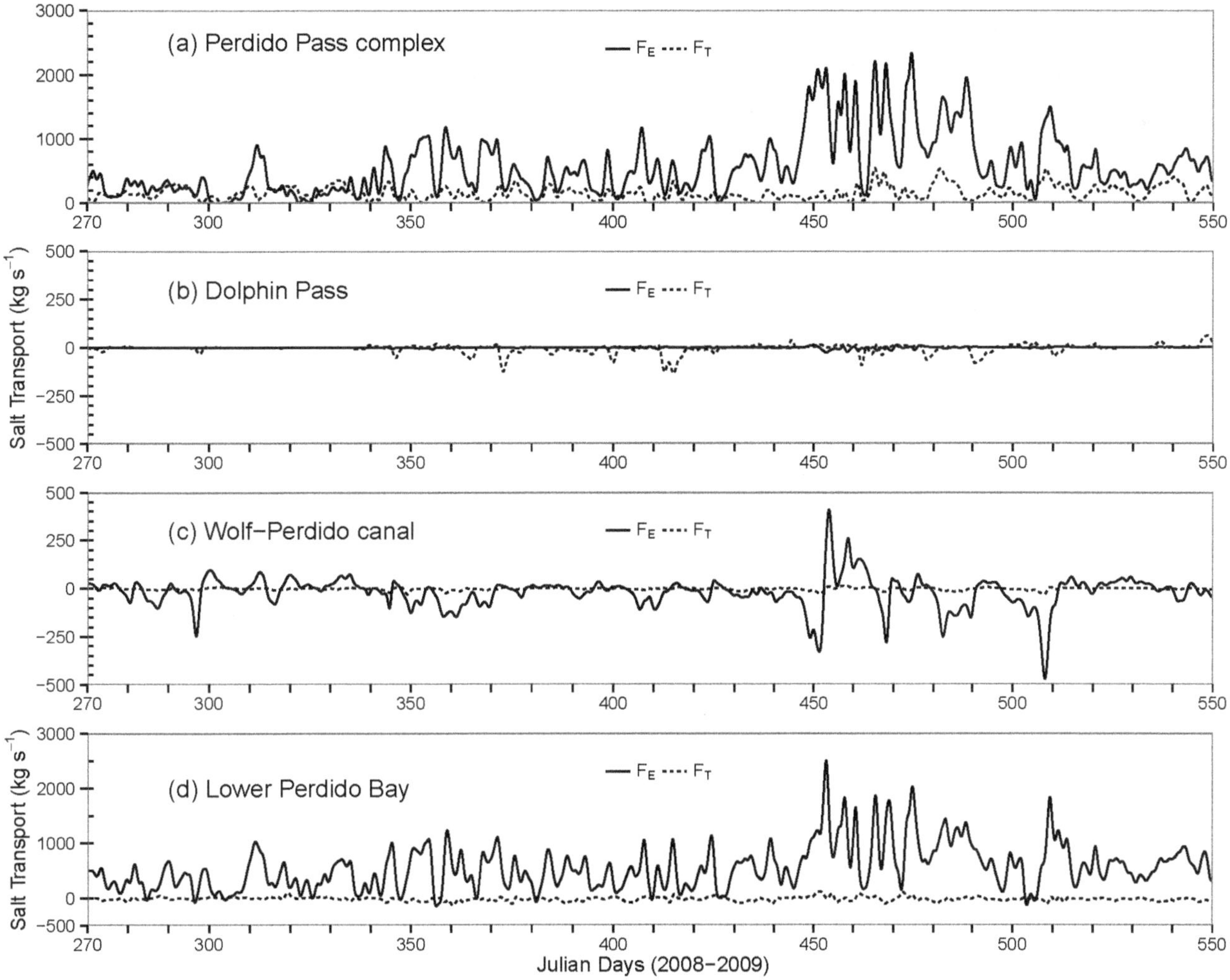

Figure 6. Shear dispersion (F_E) and tidal oscillatory (F_T) salt transport through four cross sections (Figure 1): (**a**) Perdido Pass complex (Section B-B); (**b**) Dolphin Pass (Section C-C); (**c**) Wolf-Perdido canal (Section D-D); and (**d**) Lower Perdido Bay (Section E-E).

Table 3. Statistical summary of Q_F ($m^3 \cdot s^{-1}$) and F_S ($kg \cdot s^{-1}$) computed using Eulerian decomposition method through five cross sections (Figure 1): Perdido Pass (Section A-A), the Perdido Pass complex (Section B-B), Dolphin Pass (Section C-C), the Wolf-Perdido canal (Section D-D), and lower Perdido Bay (Section E-E).

Statistical Parameters	Perdido Pass		Perdido Pass Complex		Dolphin Pass		Wolf-Perdido Canal		Lower Perdido Bay	
	Q_F	F_S	Q_F	F_S	Q_F	F_S	Q_F	F_S	Q_F	F_S
Minimum	−619	−13,480	−873	−14,810	−286	−6,789	−127	−2,181	−777	−11,430
Maximum	180	6,289	139	4,020	272	7,393	51	804	94	2,198
Mean	−42	−527	−34	113	16	648	−14	−309	−50	−360
Standard Deviation	98	2787	95	1,862	121	2,924	25	472	76	1,302
1st Quartile	−101	−2,114	−66	−713	−75	−1,474	−29	−570	−71	−921
Median	−40	−506	−15	430	3	44	−12	−231	−36	−125
3rd Quartile	18	1,118	21	1,153	89	1,852	2	11	−10	327

Table 4. Correlation coefficients of F_E (kg·s^{-1}), F_S (kg·s^{-1}), and Q_F (m^3·s^{-1}) through different cross sections (Figure 1).

Cross Sections\Variables	F_E	F_S	Q_F
Combined through Sections A-A and C-C *versus* through Section B-B	0.06	0.95	0.97
Combined through Sections B-B and D-D *versus* through Section E-E	0.91	0.98	0.99

Note: The Section A-A is Perdido Pass, B-B is the Perdido Pass complex, C-C is Dolphin Pass, D-D is the Perdido-Wolf canal, and E-E is lower Perdido Bay (Figure 1).

Table 5. Statistical summary of average percent changes (%) and standard deviations of the differences (numbers in the brackets) of exchange parameters between sensitivity model runs and the calibration run at Perdido Pass, Perdido Pass complex, and Wolf-Perdido canal. The "+50% Q" and "−50% Q" represent the 50% increase and decrease of freshwater inflows from all upstream rivers form the calibration run, respectively.

Parameter/Location	Perdido Pass		Perdido Pass Complex		Wolf-Perdido Canal	
	+50% Q	−50% Q	+50% Q	−50% Q	+50% Q	−50% Q
F_E	−34% (20.2 [1])	−10% (18.6)	41% (180.4)	−6% (121.2)	−288% (42.6)	3576% (87.6)
F_T	20% (130.7)	−22% (175.8)	4% (39.5)	−19% (51.6)	1560% (4.0)	−1317% (6.1)
F_S	−25% (223.5)	22% (301.8)	−1% (218.8)	19% (282.8)	−4% (61.6)	10% (95.5)
Q_F	−44% (18.3 [2])	38% (19.0)	4% (27.0)	−8% (27.2)	−2% (6.9)	13% (6.5)

Notes: [1] in *kg s^{-1}* for all numbers in the brackets for F_E, F_T, and F_S; [2] in *m^3·s^{-1}* for all numbers in the brackets for Q_F.

Subtidal flow rate Q_F at Perdido Pass had relatively strong correlation with river discharge Q_R and filtered WSE at the Gulf (W_{gulf}) and Big Lagoon boundaries (W_{bl}). A multi-linear regression equation was developed including above variables, and the regression equation explains 76% variations of Q_F:

$$Q_{F\,(PerdidoPass)} = 6.92 - 0.77\,Q_R + 648\,W_{gulf} - 614\,W_{bl} \quad (R^2 = 0.76) \tag{8}$$

Because Q_F is negative for seaward flux and Q_R is positive for flows from upstream, the regression coefficient for Q_R is negative. The regression equation shows that the seaward flux at Perdido Pass is proportional to upstream river discharge and WSE at Big Lagoon but inversely proportional to WSE at the Gulf. The regression equation approximately indicates that a 100 m^3·s^{-1} of river discharge from upstream is balanced with a 0.12 m WSE increase at the Gulf. The equation shows complex interactions between river discharges from upstream and WSEs directly or indirectly influenced by tides in the Gulf of Mexico. From Equation (8), it is evident that the water surface elevation boundary posed at Big Lagoon and the Gulf of Mexico have a large impact on overall fate and transport of salinity in the estuary.

Figure 4e shows the time series of tidally and sectionally averaged salinity (s_0) for Perdido Pass (solid line) and Dolphin Pass (dashed line). From Julian Day 270–330 s_0 at Dolphin Pass was in similar order of magnitude compared to the s_0 at Perdido Pass. From Julian Days 330–550 s_0 at Perdido Pass was much higher. The salinity s_0 at Perdido Pass ranged from 14–34 psu with a mean value of 30.4 psu, whereas s_0 at Dolphin Pass ranged from 3–30 psu with a mean value of 22 psu. The reason why s_0 is higher in Perdido Pass is because Perdido Pass is directly connected to the Gulf of Mexico where a constant salinity of 34 psu was applied, whereas Dolphin Pass is connected to the Gulf through Big Lagoon and Pensacola Bay that have a relatively lower salinity boundary condition applied.

3.4. Exchange through Perdido Pass

The relative contribution of three components ($Q_F s_0$, F_E, and F_T in Equation (6)) to F_S varies as function of water column stratification [23]. For Julian Days 270–450, F_S was almost entirely determined by $Q_F s_0$ (Figure 5a), which indicates the stratification at Perdido Pass was relatively weak. However, for Julian Days 450–550, the shear dispersive salt transport F_E and the tidal oscillatory salt transport F_T (Figure 5b, different y-axis scale from Figure 5a) were significant and were equally important as $Q_F s_0$, which indicates a relatively strong stratification at Perdido Pass. The F_T at Perdido Pass (Section A-A in Figure 1) was a dominant or important component of F_S in addition to $Q_F s_0$, because velocity and salinity components u_T and s_T were dominant or relatively large, as it is close to the Gulf of Mexico and within a tidal excursion. Dronkers and van de Kreeke [37] suggested that the larger magnitude of oscillatory flux, also called "nonlocal" salt transport, whose magnitude is a function of a variation of topography within a tidal excursion, plays a significant role rather than processes representative of the overall salt transport in an estuary. Many theoretical and numerical studies of salt transport mechanisms have considered estuaries of a uniform cross section [10,38,39]. The Section A-A at Perdido Pass where the salt flux was studied doesn't have uniform-depth cross section and is characterized by narrow width with deep channel and shallow overbank areas connecting Perdido Bay and the Gulf of Mexico. The sudden change in the area of flow and bathymetry between the Gulf of Mexico and Perdido Pass and between Perdido Pass and Perdido Bay can result in large horizontal dispersion (see more information in the Section 3.7) and the domination of F_T component at Perdido Pass. For Julian Days 450–560, F_T increased largely compared to F_T from 250–450 (Figure 5b); this is due to the interaction of large inflows from upstream and variations of downstream tides. F_E at Perdido Pass was up to 346 kg·s^{-1} (Figure 5b) with an average value of 31 kg·s^{-1}, but F_T was up to 2596 kg·s^{-1} with an average value of 582 kg·s^{-1}. F_T at Perdido Pass was on the average 19 times larger than F_E, and they are weakly correlated (correlation coefficient $R = 0.34$).

In Figure 5c,d exchange flow velocity and salinity (u_E and s_E) are displayed for surface and bottom layers at the deepest cell in the Perdido Pass cross section. These tidally averaged and sectionally varying velocities were dominantly positive values in the bottom layer but negative values in the surface layer. It means that the exchange flow velocity was dominantly seaward at the surface and landward at the bottom. The exchange flow velocity at the surface was strongly correlated to the velocity at the bottom and the correlation coefficient between them was −0.88. The data analysis shows that the exchange flow salinity s_E at the bottom layer was always positive because the bottom salinity was greater than the tidally and sectionally averaged salinity s_0, but s_E at the surface layer was always negative because the surface salinity was less than s_0 because the interaction with freshwater from upstream rivers. The magnitude of exchange flow salinity at the bottom layer was similar to the salinity in the surface layer with a correlation coefficient of −0.96 (Figure 5d). The estuarine salinity at the deep channel of Perdido Pass varied with moderate variation in stratification. Maximum stratification at Perdido Pass occurred on Julian Days 450–560 with the bottom-surface salinity difference (Δs_E) as large as 9 psu, whereas weak stratifications existed from Julian Days 270–450, with (Δs_E) ranging from 0–5.2 psu. Variation in stratification was largely determined by complex interactions of river discharges and tidal WSEs. The increase in river inflows from Julian Days 450–560 increased the stratification in Perdido Pass. The exchange flow, (Δu_E) defined as the difference between the surface and bottom u_E, had an average value of 0.06 $m \cdot s^{-1}$ and a maximum value up to 0.19 $m \cdot s^{-1}$.

In summary, from the Perdido Pass connection, the average salt flux going out of Perdido Bay is −527 $kg \cdot s^{-1}$ and the average salt flux coming into the estuary from the Gulf of Mexico is 648 $kg \cdot s^{-1}$. The average tidally and cross-sectionally averaged salinity at the Perdido Pass connection was 30.4 psu. The above provides the information to the first question posed in the end of the introduction.

3.5. Exchange through the Perdido Pass Complex and Dolphin Pass

Eulerian flow decomposition method was also applied to determine the flow exchange components through Sections B-B (Perdido Pass complex), C-C (Dolphin Pass), D-D (the Wolf-Perdido canal), and E-E (the lower Perdido Bay) in Figure 1, that would allow us to investigate the underlying factors that govern salt flux exchange through these cross sections. The distributions of salt flux components F_E and F_T are different in the Perdido Pass complex (Figure 6a) compared to the components in Perdido Pass (Figure 5b). The dominance of tidal oscillatory flux F_T was greatly reduced and exchange flux F_E became dominant in the Perdido Pass complex. The salinity and momentum differences between the incoming water flux from the Gulf and outgoing water flux from upstream controlled the amount of flow exchange through Section B-B. The water and salt exchange at the Perdido Pass complex was also affected by western/eastern fluxes from Dolphin Pass (Figure 1) in addition to seaward/landward fluxes from Perdido Pass.

Subtidal averaged Q_F through Section B-B varied from 139 $m^3 \cdot s^{-1}$ (incoming flux) to −873 $m^3 \cdot s^{-1}$ (outgoing flux, Table 3). The average value of Q_F from Julian Days 270–550 through Section B-B was −34 $m^3 \cdot s^{-1}$, which was slightly smaller than average Q_F at Perdido Pass (Figure 5c). The salt flux F_S through Section B-B ranged from −14,810–4020 $kg \cdot s^{-1}$ (average of 113 $kg \cdot s^{-1}$). In comparison to F_E (ranged from −27.9–14.0 $kg \cdot s^{-1}$ with an average value of −0.5 $kg \cdot s^{-1}$) through Section C-C (Dolphin Pass, Figure 6b), F_T was dominant (ranged from −134.7–61.1 $kg \cdot s^{-1}$ with an average value of −5.6) and flow exchange occurred in the east-west direction. F_T was dominant because the salinities of incoming and outgoing water through Dolphin Pass were not much different, and were directly influenced by the Big Lagoon boundary which is far away from the Gulf of Mexico. Subtidal Q_F through Section C-C (Dolphin Pass) varied from −286 $m^3 \cdot s^{-1}$ (western flux) to 272 $m^3 \cdot s^{-1}$ (eastern flux) with a small average value of 16 $m^3 \cdot s^{-1}$ (Table 3) in 2008–2009. It means that there were relatively large flows back and forth in the east-west directions through Dolphin Pass but the net flow was not significant.

3.6. Exchange through the Wolf-Perdido Canal and Lower Perdido Bay

The flow exchanges at the Wolf-Perdido canal (Section D-D in Figure 1), which connects Perdido Bay and Wolf Bay, and in lower Perdido Bay (Section E-E) take place in the east-west direction. At the

Wolf-Perdido canal, F_E was dominant (ranged from −472.2–407.7 kg·s^{-1} with an average value of −20.5 kg·s^{-1}) and the negative value of F_E indicated that the flow was moving from Perdido Bay to Wolf Bay (Figure 6c) due to inflows from upstream rivers into Perdido Bay. Q_F at the Wolf-Perdido canal varied from −127.2 m^3·s^{-1} (towards Wolf Bay) to 51.1 m^3·s^{-1} (towards Perdido Bay). It means that a small portion of river discharges Q_R from Perdido River and Styx River could flow through Section D-D into Wolf Bay. On average, there were more days for flows from Perdido Bay to Wolf Bay through Section D-D because average Q_F at D-D was −14 m^3·s^{-1}. Q_F at the Wolf-Perdido canal was typically smaller than Q_F at Perdido Pass and Dolphin Pass. The salt-water intrusion from Perdido Bay towards Wolf Bay mostly occurred when there were large inflows from upstream and large incoming flows via the Dolphin Pass and Perdido Pass boundaries. The amount of flow entering Wolf Bay from Perdido Bay was of small magnitude, which might be one of the reasons that Wolf Bay is pristine and less polluted than Perdido Bay. At lower Perdido Bay, F_E was dominant with mean value of 556.3 kg·s^{-1} and mean F_T was −15.7 kg·s^{-1}. Q_F at lower Perdido Bay ranged from −776.5–93.5 m^3·s^{-1} with an average value of −49.5 m^3·s^{-1} (75% of Q_F is less than −10.0 m^3·s^{-1} or seaward outflows). The salt flux F_S at lower Perdido Bay ranged from −11,430 kg s^{-1} –2198 kg·s^{-1} with an average value of −360 kg·s^{-1} (Table 3).

A data analysis was performed to examine possible correlations of salt and flow fluxes F_E, F_S, and Q_F through different cross sections. Figure 1 shows that the sum or combined salt and flow fluxes through the Perdido Pass complex (Section B-B) and the Wolf-Perdido canal (Section D-D) would possibly correlate with corresponding fluxes through lower Perdido Bay (Section E-E). It is also possible that the sum or combined salt and flow fluxes through Perdido Pass (Section A-A) and Dolphin Pass (Section C-C) would correlate with corresponding fluxes through the Perdido Pass complex (Figure 1). Derived correlation coefficients of F_E, F_S, and Q_F between above-mentioned cross sections are summarized in Table 4. The interaction among exchange salt fluxes F_E through the Perdido complex, Dolphin Pass, and Perdido Pass was complex. Therefore, the correlation coefficient between the sum of F_E through Perdido Pass and Dolphin Pass and F_E through the Perdido Pass complex was very small (only 0.06) because tidal oscillatory flux F_T was dominant at Perdido Pass. Other correlation coefficients ranged from 0.91–0.99 (Table 4) indicating that strong correlations of F_E, F_S, and Q_F indeed exist between these cross sections.

Thus, we explored the salt exchange between Wolf Bay and Perdido Bay by analyzing the salt flux at the Wolf-Perdido canal. On average, the flow moved from Perdido Bay towards Wolf Bay with a magnitude of 14 m^3·s^{-1} and the average salt flux towards Wolf Bay was 309 kg·s^{-1}. However, these numbers might change when there is a large river inflow from upstream of Perdido Bay and when there is a large flow from Big Lagoon and the Gulf of Mexico. Understanding the overall flow pattern in such navigational systems would allow the water quality manager to make remedial measures if Perdido Bay is polluted.

3.7. Horizontal Dispersion

Dispersion is an important mixing characteristic in an estuary and causes pollutants to spread. Dispersion is the result of velocity differences in space. Following Lerczak *et al.* [22], the one dimensional along-estuary (*x* coordinate in equations below) salt conservation equation can be written in the following form:

$$A_0(x)\frac{\partial S_0}{\partial t} = \frac{\partial}{\partial x}[Q_f S_0 + A_0(x)K_x(x)\frac{\partial S_0}{\partial x}] \quad (9)$$

Integrating the above equation and simplifying it, we get,

$$-A_0(x)K_x(x)\frac{\partial S_0}{\partial x} = F_E + F_T \quad (10)$$

Horizontal dispersion coefficient K_x was calculated using Equation (10) at three locations: Perdido Pass, Perdido Pass complex, and Wolf-Perdido canal. Three time series of K_x were plotted in Figure 7 with different y axis scales for three locations. The magnitude of horizontal dispersion along Perdido Pass was significantly higher compared to the horizontal dispersion at Perdido Pass complex and Wolf-Perdido canal. During the calibration period, the mean horizontal dispersion at Perdido Pass was 810.2 m^2/s whereas the mean horizontal dispersion at Perdido Pass complex and Wolf-Perdido canal were 40.2 m^2/s and 48.12 m^2/s, respectively. When the rate of change of the cross sectionally and time averaged salinity S_o with respect to the distance does not exactly correspond to the sign change for $F_E + F_T$ (Equation (10)), horizontal dispersion coefficients at Wolf-Perdido canal during certain time periods were negative (Figure 5c) because of the complex exchange characteristics at the Wolf-Perdido canal (Figure 6c).

The fact that tidal oscillatory flux F_T at Perdido Pass (Figure 5b) is dominant and significantly larger explains the higher magnitude of horizontal dispersion at Perdido Pass compared to other cross sections. The magnitude of horizontal dispersion at Perdido Pass was approximately 10 times larger than horizontal dispersion at Perdido Pass complex, which is because of its narrow width and long channel with the direct connection to the Gulf compared to the relatively wide cross section at Perdido Pass complex. The magnitude of K_x at Perdido Pass shows positive correlation with spring (S on Figure 7) and neap (N) tides.

3.8. Sensitivity Experiments

Two sensitivity numerical experiments were performed to evaluate the impact of freshwater into the system. The sensitivity experiments include two model runs with 50% increase and 50% decrease of freshwater inflows (Q) from all rivers into Wolf Bay and Perdido Bay. The water surface elevation boundaries (tidal influences) at the Gulf, Dolphin Pass and GIWW remained unchanged for both the sensitivity runs. The simulation results from the sensitivity model runs were then compared with results from the calibration run. The exchange parameters F_E and F_T for three cross sections: Perdido Pass, Perdido Pass complex, and Wolf-Perdido canal are plotted in Figure 8. The summary of statistics showing the average percent changes (%) and standard deviations of the differences (numbers in the brackets) of exchange parameters from the calibration run for F_E, F_T, F_S and Q_F are given in Table 5. The exchange parameters at different locations have different orders of magnitude and different directions. Therefore, to provide more insights in the interactions due to increase and decrease of freshwater inflows, both percent changes and standard deviations in $kg \cdot s^{-1}$ or $m^3 \cdot s^{-1}$ of the differences are presented in Table 5. When the magnitudes of parameters such as Q_F, F_E, F_T and F_S in the calibration run are small (near zero) in certain periods, the percent changes from the calibration run would be larger numbers, which would eventually impact the average percent change. It should be noted that since the interactions among various forcing factors are complex in an estuary, both graphical and tabular results only provide partial information.

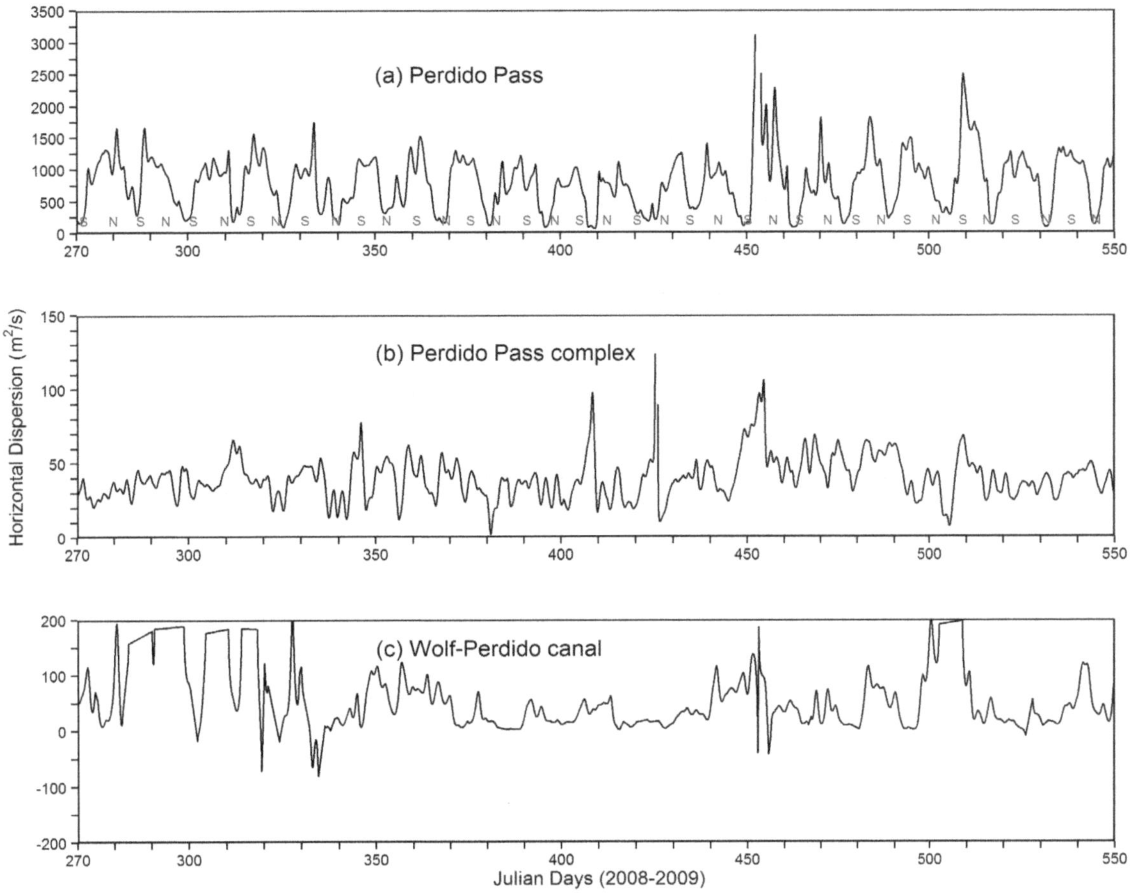

Figure 7. Horizontal dispersion (K_x) at three cross sections (Figure 1): (**a**) Perdido Pass; (**b**) Perdido Pass complex; and (**c**) Wolf-Perdido canal. The "S" and "N" in the top panel stand for spring and neap tides.

At Perdido Pass, total salt flux F_S would decrease on average by 25% for 50% Q increase, while it would increase by 22% for 50% Q decrease. This is because these results are more dominated by the tidal boundary at the Gulf, which is the same for sensitivity runs and the calibration run. The oscillatory flux F_T depends on characteristics of tidal excursion and local bathymetry. F_T exchange component, which is dominant in Perdido Pass, would increase on average by 20% with 50% Q increase, while F_T would decrease by 22% with 50% Q decrease. On average (Table 5), the shear dispersive transport component F_E at Perdido Pass would have smaller magnitude for both 50% increase and decrease of freshwater inflows.

At Perdido Pass complex, the salt flux component F_E with 50% Q increase would have consistent increase compared to the calibration run. This is because higher upstream inflow from Perdido Bay would cause the flow on upper layers to rapidly propagate towards the Gulf and increase overall stratification and dispersion. F_E with 50% Q decrease would have on average a small drop compared to the calibration run, except during Julian days 450–460, where there was a large inflow coming into the system from the rivers upstream of Perdido Bay.

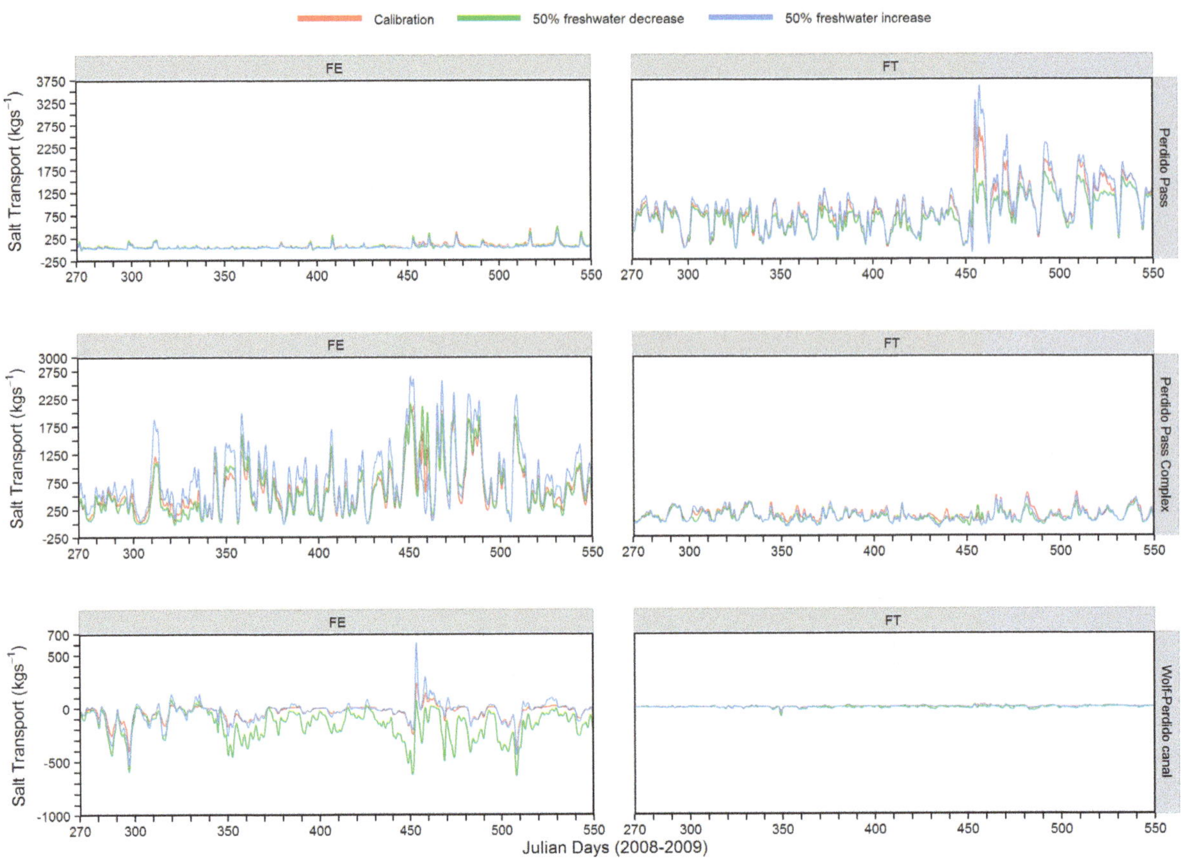

Figure 8. Time series of shear dispersive transport (F_E) and oscillatory tidal transport (F_E) at Perdido Pass, Perdido Pass complex, and Wolf-Perdido canal for the calibration run, 50% freshwater increase and 50% freshwater decrease sensitivity runs.

4. Summary and Conclusions

A previously calibrated three-dimensional hydrodynamic EFDC model was used to simulate flow and salinity distributions in the Perdido Bay and Wolf Bay system under unsteady flows from rivers, tides from three open boundaries, and atmospheric forcing in 2008–2009. The calibrated EFDC model provided simulated hourly velocities and salinity at different layers (depths) for all grids in five cross sections. Eulerian flux method was applied to determine the water and salt fluxes through these cross sections using hourly model outputs. The summaries of key findings from the study are as follows:

a. The salinity at Perdido Bay varied largely with upstream inflows (low and high inflows) and tides at open boundaries (ebb and flood tides). Inflow of highly saline water at the bottom layers and outflow of relative low salinity water at the surface layers resulted from the interaction of flood tide at the Gulf of Mexico and inflows from upstream. Salinity was always small from the US Hwy 98 Bridge towards Perdido River. During low inflows from upstream and flood tides from the Gulf of Mexico, maximum salinity up to 32 psu reached Ross/Innerarity Point (lower Perdido Bay).
b. Eulerian analysis concluded that tidal oscillatory fluxes (F_T) at Perdido Pass (Section A-A) and Dolphin Pass (Section C-C) were dominant compared to exchange flow component F_E, whereas

at Sections B-B (Perdido Pass complex), D-D (Wolf-Perdido canal), and E-E (lower Perdido Bay) the exchange flows F_E were dominant.

c. During the simulation period, it was found that a small amount of flow exchange occurred between Wolf Bay and Perdido Bay. During the high inflows from rivers in Perdido Bay and higher tides from the Gulf and Big Lagoon, the water from Perdido Bay moved towards Wolf Bay, however, during normal flows and gentle tides the water from Wolf Bay moved towards Perdido Bay. It means the influences from Perdido Bay and the Gulf of Mexico was relatively small in Wolf Bay under normal inflows.
d. Horizontal dispersion coefficient was computed at three locations for the calibration run and it was found that the magnitude of horizontal dispersion at Perdido Pass (a relatively narrow channel which directly connects to the Gulf) was approximately 10 times larger than horizontal dispersion coefficient at Perdido Pass complex (wide cross section).
e. Sensitivity runs with 50% increase and 50% decrease in freshwater inflow relative to the calibration run were performed. At Perdido Pass, total salt flux F_S would decrease on average by 25% for 50% Q increase while it would increase by 22% for 50% Q decrease, because tidal boundary at the Gulf is fixed for the calibration and sensitivity runs.

Acknowledgments

This study was partially funded through the project "Impacts of human activities and climate change on water resources and ecosystem health in the Wolf Bay Basin: A Coastal Diagnostic and Forecast System (CDFS) for integrated assessment" by Auburn University Water Resources Center. The authors would like to thank the reviewers for providing valuable comments that helped us to improve the manuscript.

Author Contributions

Authors collaborated together for the completion of this work. Janesh Devkota designed the study, developed and calibrated the Perdido EFDC model, ran the model for all cases, performed data analyses of model results and wrote the first draft of the manuscript. Xing Fang provided valuable instructions on study design, supervised the data analyses, and reviewed and revised the manuscript. Both authors contributed significantly in writing the manuscript.

Conflicts of Interest

The authors declare no conflict of interest.

References

1. Neilson, B.J.; Cronin, L.E. *Estuaries and Nutrients*; Humana Press: Clifton, NJ, USA, 1981.
2. Kennish, M.J. *Ecology of Estuaries, Volume i: Physical and Chemical Aspects*; CRC Press, Inc.: Boca Raton, FL, USA, 1986; p. 254.

3. Bilgili, A.; Proehl, J.A.; Lynch, D.R.; Smith, K.W.; Swift, M.R. Estuary/ocean exchange and tidal mixing in a gulf of maine estuary: A lagrangian modeling study. *Estuar. Coast. Shelf Sci.* **2005**, *65*, 607–624.
4. Ketchum, B.H. The exchanges of fresh and salt waters in tidal estuaries. *J. Marine Res.* **1951**, *10*, 18–38.
5. MacCready, P. Calculating estuarine exchange flow using isohaline coordinates. *J. Phys. Oceanogr.* **2011**, *41*, 1116–1124.
6. Lewis, R.E.; Lewis, J.O. The principal factors contributing to the flux of salt in a narrow, partially stratified estuary. *Estuar. Coast. Shelf Sci.* **1983**, *16*, 599–626.
7. Jay, D.A.; Musiak, J.D. Internal tidal asymmetry in channel flows: Origins and consequences. *Coast. Estuar. Stud.* **1996**, 211–249, doi:10.1029/CE050p0211.
8. Fischer, H.B. Mixing and dispersion in estuaries. *Annu. Rev. Fluid Mech.* **1976**, *8*, 107–133.
9. Hansen, D.V.; Rattray, M., Jr. New dimensions in estuary classification. *Limnol. Oceanogr.* **1966**, 319–326.
10. Smith, R. Buoyancy effects upon longitudinal dispersion in wide well-mixed estuaries. *Philos. Trans. R. Soc. Lond. Ser. A Math. Phys. Sci.* **1980**, *296*, 467–496.
11. Uncles, R.J.; Stephens, J.A. Salt intrusion in the tweed estuary. *Estuar. Coast. Shelf Sci.* **1996**, *43*, 271–293.
12. Fischer, H.B.; List, E.J.; Koh, R.C.Y.; Imberger, J.; Brooks, N.H. *Mixing in Inland and Coastal Waters*; Academic Press: San Diego, CA, USA, 1979.
13. MacDonald, D.G. Estimating an estuarine mixing and exchange ratio from boundary data with application to mt. Hope bay (massachusetts/rhode island). *Estuar. Coast. Shelf Sci.* **2006**, *70*, 326–332.
14. Gilcoto, M.; Pardo, P.C.; Alvarez-Salgado, X.A.; Perez, F.F. Exchange fluxes between the ria de vigo and the shelf: A bidirectional flow forced by remote wind. *J. Geophys. Res.-Oceans* **2007**, *112*, doi:10.1029/2005JC003140.
15. Traynum, S.; Styles, R. Exchange flow between two estuaries connected by a shallow tidal channel. *J. Coast. Res.* **2008**, *24*, 1260–1268.
16. Engqvist, A.; Stenstrom, P. Flow regimes and long-term water exchange of the himmerfjarden estuary. *Estuar. Coast. Shelf Sci.* **2009**, *83*, 159–174.
17. Valle-Levinson, A.; Gutierrez de Velasco, G.; Trasviña, A.; Souza, A.; Durazo, R.; Mehta, A. Residual exchange flows in subtropical estuaries. *Estuar. Coasts* **2009**, *32*, 54–67.
18. Chen, S.N.; Geyer, W.R.; Ralston, D.K.; Lerczak, J.A. Estuarine exchange flow quantified with isohaline coordinates: Contrasting long and short estuaries. *J. Phys. Oceanogr.* **2012**, *42*, 748–763.
19. Devkota, J.; Fang, X.; Fang, V.Z. Response characteristics of perdido and wolf bay system to inflows and sea level rise. *Br. J. Environ. Clim. Chang.* **2013**, *3*, 229–256.
20. Xia, M.; Craig, P.M.; Wallen, C.M.; Stoddard, A.; Mandrup-Poulsen, J.; Peng, M.; Schaeffer, B.; Liu, Z. Numerical simulation of salinity and dissolved oxygen at perdido bay and adjacent coastal ocean. *J. Coast. Res.* **2011**, *27*, 73–86.
21. Xia, M.; Xie, L.; Pietrafesa, L.J.; Whitney, M.M. The ideal response of a gulf of mexico estuary plume to wind forcing: Its connection with salt flux and a lagrangian view. *J. Geophys. Res.* **2011**, *116*, C08035.

22. Lerczak, J.A.; Geyer, W.R.; Chant, R.J. Mechanisms driving the time-dependent salt flux in a partially stratified estuary. *J. Phys. Oceanogr.* **2006**, *36*, 2296–2311.
23. Kim, C.-K.; Park, K. A modeling study of water and salt exchange for a micro-tidal, stratified northern gulf of mexico estuary. *J. Marine Syst.* **2012**, *96–97*, 103–115.
24. Livingston, R.J. *Trophic Organization in Coastal Systems*; CRC Press: Boca Raton, FL, USA, 2003.
25. Duchon, C.E. Lanczos filtering in one and two dimensions. *J. Appl. Meteorol.* **1979**, *18*, 1016–1022.
26. Wang, R.; Kalin, L. Modelling effects of land use/cover changes under limited data. *Ecohydrology* **2011**, *4*, 265–276.
27. Beasley, L.R. *Interaction of Groundwater, Surface Water and Seawater in Wolf Bay, Weeks Bay, and Dauphin Island Coastal Watersheds, Alabama*; Auburn University: Auburn, AL, USA, 2010.
28. Hamrick, J.M. *A Three-Dimensional Environmental Fluid Dynamics Computer Code: Theoretical and Computational Aspects*; Special Report 317; Virginia Institute of Marine Science, College of William and Mary: Gloucester Point, VA, USA, 1992.
29. Hamrick, J.M.; Mills, W.B. Analysis of water temperatures in conowingo pond as influenced by the peach bottom atomic power plant thermal discharge. *Environ. Sci. Policy* **2000**, *3*, 197–209.
30. Park, K.; Jung, H.-S.; Kim, H.-S.; Ahn, S.-M. Three-dimensional hydrodynamic-eutrophication model (hem-3d): Application to kwang-yang bay, korea. *Marine Environ. Res.* **2005**, *60*, 171–193.
31. Mellor, G.L.; Yamada, T. Development of a turbulence closure model for geophysical fluid problems. *Rev. Geophys. Space Phys.* **1982**, *20*, 851–875.
32. Galperin, B.; Kantha, L.H.; Hassid, S.; Rosati, A. A quasi-equilibrium turbulent energy model for geophysical flows. *J. Atmos. Sci.* **1988**, *45*, 55–62.
33. Gong, W.; Shen, J.; Hong, B. The influence of wind on the water age in the tidal rappahannock river. *Marine Environ. Res.* **2009**, *68*, 203–216.
34. Godin, G. *The Analysis of Tides*; University of Toronto Press: Toronto, ON, Canada, 1972; p. 264.
35. Processing and Publication of Discharge and State Data Collected in Tidally-Influenced Areas. Available online: http://water.usgs.gov/admin/memo/SW/sw10.08-final_tidal_policy_memo.pdf (accessed on 1 March 2014).
36. Sterling, M.; Beaman, F.; Morvan, H.; Wright, N. Bed-shear stress characteristics of a simple, prismatic, rectangular channel. *J. Eng. Mech.* **2008**, *134*, 1085–1094.
37. Dronkers, J.; van de Kreeke, J. Experimental determination of salt intrusion mechanisms in the volkerak estuary. *Neth. J. Sea Res.* **1986**, *20*, 1–19.
38. Fischer, H.B. Mass transport mechanisms in partially stratified estuaries. *J. Fluid Mech.* **1972**, *53*, 671–687.
39. Scott, C.F. A numerical study of the interaction of tidal oscillations and non-linearities in an estuary. *Estuar. Coast. Shelf Sci.* **1994**, *39*, 477–496.

2

Impact of Rural Domestic Wastewater Irrigation on the Physicochemical and Microbiological Properties of Pakchoi and Soil

Bo Yang [1,†], Xiao Kong [1,2,†], Bingjian Cui [2,3,†], Decai Jin [2,3,†], Ye Deng [2,3], Xuliang Zhuang [2,3], Guoqiang Zhuang [2,3] and Zhihui Bai [2,3,*]

[1] College of Environment and Safety Engineering, Qingdao University of Science and Technology, Qingdao 266042, China; E-Mails: yangbo16@126.com (B.Y.); kongxiaozaikeda@163.com (X.K.)

[2] Key Laboratory of Environmental Biotechnology, Research Center for Eco-Environmental Sciences, Chinese Academy of Sciences, Beijing 100085, China;
E-Mails: cuibingjian12b@mails.ucas.ac.cn (B.C.); jindecai19841019@gmail.com (D.J.); yedeng@rcees.ac.cn (Y.D.); xlzhuang@rcees.ac.cn (X.Z.); gqzhuang@rcees.ac.cn (G.Z.)

[3] College of Resources and Environment, University of Chinese Academy of Sciences, Beijing 100049, China

† These authors contributed equally to this work.

* Author to whom correspondence should be addressed; E-Mail: zhbai@rcees.ac.cn;

Academic Editor: Miklas Scholz

Abstract: Great attention has been paid to the potential of wastewater irrigation as a sustainable water source, particularly due to water scarcity and water pollution issues. However, few studies have focused on its adverse effects and on the health risks it may pose. In this study, the physicochemical properties of soils and plants irrigated with rural domestic wastewater and associated microbiological risks were investigated. The results showed that sewage irrigation could increase the production of vegetables and improve soil fertility. While the nitrate content of plants increased significantly, pathogens on plants and in soils increased after irrigation with raw wastewater. In particular, there was a wide range of pathogenic bacteria in the phyllosphere, which may indicate risks if contaminated vegetables are consumed directly. Treated wastewater irrigation was not significantly different from controls, which were irrigated by tap water; consequently, it can be used as an alternative water resource for agricultural irrigation. The presence of a wide spectrum pathogens in wastewater shows the necessity of long-term monitoring and further evaluation.

Keywords: rural domestic wastewater; pathogens; pakchoi; real-time quantitative PCR; phyllosphere

---

1. Introduction

The geographic distribution of water resources in China is very non-uniform, with 90% of surface water and 70% of groundwater distributed in various southern provinces. North China, which is situated in an arid and semi-arid climatic belt, receives less rainfall and experiences higher evaporation; it has consequently become one of the main water-deficient areas, particularly as the local demand for water has increased so rapidly with economic development and population growth, that it already surpasses local water supply capacity. Moreover, growing water pollution renders the situation even more serious. Water shortages and poor water quality are interacting with each other and seriously constraining economic and social development [1–3].

The scarcity of water resources has increasingly rendered recycled wastewater a valid substitute resource [4]. Treated wastewater is used directly for irrigation in many countries, especially in arid and semiarid areas. It contains essential plant nutrients, such as Ca, Mg, K, N, P and Fe, that are important for plant growth [5]. The reuse of treated wastewater avoids this being discharged into sensitive environments [6,7], helps conserve water resources, recycles nutrients (N and P) and minimizes pollution loads in receiving water bodies [8]. However, wastewater also contains various inorganic substances, including potential toxic elements and heavy metals (Cr, Cd, Pb, Ni and Hg) [9], which may be at phytotoxic levels or create health risks [10,11]. Safe and efficient management of this resource therefore cannot be achieved without appropriate monitoring [12], given that wastewater reuse also involves many potential risks to soil quality and to the growth and quality of crops [13–15].

Considerable research has been focused on the effect of wastewater irrigation on soil properties and on the quality of vegetables. Lado *et al.* [6] studied the effect of long-term secondary treated wastewater irrigation on soil chemical properties. Lv *et al.* [16] reported the effect of sewage irrigation on growth and development, yield and quality of potato. Similarly, Fayyad *et al.* [17] conducted a field experiment to investigate the effect of different treatments of potable and treated wastewater on the quality of tomato fruit. Singh *et al.* [18] described a year-long experiment to observe the effect of sewage wastewater irrigation on soil properties, crop yield and the environment. Maldonado *et al.* [19] conducted research on heavy metal content in soils under different wastewater irrigation patterns. However, little attention has been concentrated on the microbiological risk of wastewater irrigation.

The consumption of contaminated vegetables presents a potential risk to human health, with these containing large numbers of microbial pathogens in addition to heavy metals and other potential toxic elements. The presence of pathogenic bacteria may pose a health hazard for consumers of raw fruits and vegetables [20]. For example, *Arcobacter* spp. can cause persistent diarrhea, accompanied by abdominal pain, stomach cramps, nausea, vomiting, fever and other symptoms, and the infection rate is difficult to assess.

Given the significance of these risks, in this study, we tested the content of pathogenic microorganisms in rural domestic sewage and conducted research on pathogen residues on the surfaces of leaves and in soil of plant roots using real-time quantitative PCR. Additionally, the effects of wastewater irrigation on growth and nutrition plant indices were also investigated.

2. Experimental Section

2.1. Experimental Design

The experiment was performed in a greenhouse at Beijing Forestry University Forestry Science Co. Ltd. (40°00'96" N, 116°34'69" E, Beijing, China). The original experimental soil was collected from suburban farmland of Beijing. Table 1 lists the physicochemical characteristics of this soil.

Table 1. Physicochemical properties of the experimental soil sample.

Parameters	Soil		
	>0.05 mm	31.25	sand
	0.05–0.02 mm	22.20	
Soil particle proportion (%)	0.02–0.005 mm	13.40	lime
	0.005–0.002 mm	6.30	
	<0.002 mm	26.85	clay
pH	7.96		
EC (μs/cm)	571		
Organic matter (g/kg)	1.58		
NO_3-N (mg/kg)	8.56		
NH_3-N (mg/kg)	0.063		
Available P (mg/kg)	3.1		
Available K (mg/kg)	107.7		
Cu (mg/kg)	20.3		
Zn (mg/kg)	67.3		
Cd (mg/kg)	0.023		
Pb (mg/kg)	36.3		

Wastewaters were obtained from a rural domestic wastewater treatment plant, with a membrane bioreactor (MBR) processes (Figure 1), located in Huairou District of Beijing; the water samples' quality parameters are shown in Table 2.

Pakchoi seeds were purchased from the Chinese Academy of Agricultural Sciences (CAAS). The seeds were surface sterilized using 3% sodium hypochlorite solution disinfection for 15 min, followed by several cycles of rinsing with sterilized distilled water. The sterile seeds were then placed in a petri dish to sprout at room temperature and appropriate humidity.

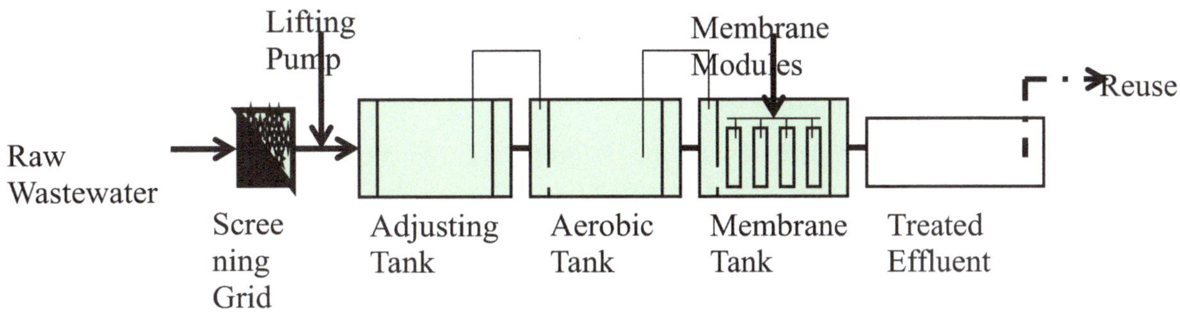

Figure 1. Process flow diagram of the wastewater treatment system with a membrane bioreactor.

Table 2. Physical and chemical properties of experimental water samples.

Parameters	Raw Wastewater	Treated Wastewater	Criteria of Tap Water [21]	Criteria of Irrigation Water [22]
pH	7.60	7.59	6.5–8.5	5.5–8.5
EC (μs/cm)	754	376	--	≤1,000
Salinity (mg/L)	415–566	207–282	≤1,000 (TDS)	-
Temperature (°C)	12.2	12.1	--	≤35
COD (mg/L)	346	35	--	≤100 [a] or 60 [b]
SS (mg/L)	26	21	--	≤60 [a] or 15 [b]
anionic surfactant (mg/L)	2.00	no detectable (<0.05)	≤0.3	≤5
TN (mg/L)	54.6	26.2	--	--
TP (mg/L)	4.97	7.66	--	--
Zn (mg/L)	0.153	no detectable (<0.006)	≤1.0	≤2
Cu (mg/L)	no detectable (<0.01)	no detectable (<0.01)	≤1.0	≤1
Pb (mg/L)	0.003	no detectable (<0.001)	≤0.01	≤0.2
Cd (mg/L)	0.0002	0.0002	≤0.005	≤0.01
Cl^- (mg/L)	44.2	56.1	≤250	≤350
NH_3 -N (mg/L)	45.7	8.59	≤0.5	--
NO_3- N (mg/L)	no detectable (<0.2)	16.0	≤20	--
Total bacteria count/mL	$(1.24 \pm 0.05) \times 10^6$	$(2.80 \pm 0.20) \times 10^5$	≤100	--
Total coliforms/mL	$6.05 \times 10^5 \pm 1.75 \times 10^5$	no detectable (<30)	no detectable	--
Fecal coliforms/mL	240,000	no detectable (<3)	no detectable	≤20 [a] or 10 [b]
Escherichia coli	8.9×10^4	no detectable (<30)	no detectable	--
Ascaris Lumbricoides (eggs/L)	34	no detectable	no detectable	≤2[a] or 1[b]

Notes: [a] Vegetables need processing, cooking or peeling; [b] rabbit food, melons and fruit; --, no data; TDS, total dissolved solids.

The soil was air-dried and sieved (20 mesh). Each pot (23 cm in top diameter, 16 cm bottom and 20 cm high) contained 2.5 kg dry soil and 20 seeds of pakchoi. After germination, the seedlings were thinned out, and 10 seedlings were left for growth. Irrigation was carried out from the seedling stage based on actual water requirement (irrigated approximately every 3 days, 100 mL at a time in one pot, 12 times and 1200 mL in total), with a specified amount of conventional fertilizer (0.15 g urea, 0.06 g calcium superphosphate, 0.15 g potassium sulfate per pot) added to the water. After the plants matured (about 6 weeks), the vegetables were harvested in summer 2013. Then, the second round of planting was performed with the same method in autumn 2013. The third round of the experiments began on 19 March 2014 and extended until 5 May 2014 with the same method for the long-term investigation of the effect of wastewater irrigation.

Experimental treatments involved surface irrigation with three types of water: tap-water (DW), raw wastewater (WW) and treated wastewater (TW). At the same time, controls with nothing sowed and irrigated with tap water (CK-DW) and raw wastewater (CK-WW) were set up. Each treatment was applied in four replicates with a randomized block design.

2.2. Collection of Microbial Samples and DNA Extraction

After the third round of planting, vegetables were harvested at maturity (about 6 weeks) using sterile scissors, which separated the aboveground parts and roots. Harvested plants were placed in valve bags and transported immediately to the laboratory for analysis.

Leaves (10 g) were stuffed in an aseptic polypropylene tube containing 100 mL sodium phosphate buffer (0.2 M, pH 7.0), shaken for 30 min in a vapor-bathing constant temperature vibrator and then treated by ultrasonic waves (40 kHz) for 30 min using ultrasonic agitation (Ningbo Xingzhi Biological Technology Co. Ltd., Ningbo, China). The buffer solution was then filtered through a 0.22-μM cellulose nitrate microporous composite membrane. The microbes from the leaves were accumulated on the membrane, which was stored at −80 °C subsequently.

Surface soil and rhizosphere soil samples of each pot were air-dried in a sterile chamber at room temperature, then sieved through a 20 mesh sieve and stored at −80 °C.

Water samples were filtered through 0.22-μm membranes to collect microbes, then the membrane was put in a sterile centrifuge tube and stored at −80 °C.

Genomic DNA of the above microbes' samples was extracted using the Fast DNA SPIN Kit for Soil (MP Biomedicals, Santa Ana, CA, USA) following the manufacturer's instructions.

2.3. Physical and Chemical Properties Detection of Plants and Soil

The pH and electrical conductivity (EC) of soil samples were detected in aqueous extract (soil:deionized water = 1:5) using a multi-parameter water quality monitoring instrument [23]. Organic matter was measured using the potassium bichromate titrimetric method. The activity of urease and phosphatase was assayed based on the method of Tabatabai [24].

Fresh and dry weights of mature plants were measured by the gravimetric method, and the plants' height was measured by a ruler. The salicylic acid method was used to measure the vegetables' nitrate content [25], while soluble sugar was determined by the anthrone colorimetry method [26]. The Coomassie Brilliant Blue G250 staining method was used to determine the concentration of soluble protein in vegetable samples [27]. The vegetables' vitamin C content was determined by the 2,6-diohloroindophenol titration method [28].

2.4. Molecular Assays

This study focused on common pathogens, including pathogens present in the environment and several pathogens that have emerged as a focus of research interest in recent years. Specific pathogens considered were: *Aeromonas hydrophila*, *Arcobacter* spp., *Bacillus cereus*, *Clostridium difficile*, *Clostridium perfringens*, *Escherichia coli*, *Legionella* spp. and *Mycobacterium* spp.

Traditional PCR was conducted to confirm the presence of pathogens. Based on the PCR results, positive pathogens were investigated further, with the corresponding primer pairs for each target and expected product size shown in Table 3.

Table 3. Primers of pathogenic bacteria for real-time PCR assays.

Pathogenic Bacteria	Targeted Gene	Primer Sequences	Amplicon Size (bp)	Ref.
Aeromonas hydrophila	Cytolytic enterotoxin	AHCF1: GAGAAGGTGACCACCAAGAACA AHCR1: AACTGACATCGGCCTTGAACTC	232	[29]
Arcobacter spp.	23S rRNA	ARCO1: GTCGTGCCAAGAAAAGCCA ARCO2: TTCGCTTGCGCTGACAT	331	[30]
Bacillus cereus	16S rRNA	F: TCGAAATTGAAAGGCGGC R: GGTGCCAGCTTATTCAAC	288	[31]
Clostridium difficile	16S rRNA	Clo-16F: TTGAGCGATTTACTTCGGTAAAGA Clo-16R: CCATCCTGTACTGGCTCACCT	157	[32]
Clostridium perfringens	16S rRNA	Clp-F: ATGCAAGTCGAGCGA(G/T)G Clp-R: TATGCGGTATTAATCT(C/T)CCTTT	120	[32]
Escherichia coli	*uidA*	Eco-F: CTGCTGCTGTCGGCTTTA Eco-R: CCTTGCGGACGGGTAT	205	[33]
Legionella spp.	16S rRNA	LEG448:GAGGGTTGATAGGTTAAGAGC LEG858:GTCAACTTATCGCGTTTGCT	430	[34]
Mycobacterium spp.	16S rRNA	Myco F: ATGCACCACCTGCACACAGG Myco R: GGTGGTTTGTCGCGTTGTTC	470	[35]

Positive PCR products were purified with the E.Z.N.A.™ Gel Extraction Kit (Omega Bio-tek, Norcross, GA, USA) and ligated to a pGEM-T Easy vector system I (Promega, Madison, WI, USA). Recombinant plasmids were transferred into *E. coli* DH5α-competent cells (Biomed, Beijing, China) and coated on LB agar plates containing ampicillin, X-gal and IPTG, as recommended by the manufacturer. PCR was performed to confirm the sizes of insert fragments and sequenced by Ruibio BioTech Co., Ltd. (Beijing, China). The gene sequences were aligned by BLAST [36] on NCBI. Plasmid DNA was extracted using the E.Z.N.A.™ Plasmid Mini Kit I (Omega Bio-tek, Norcross, GA, USA). The gene copy number was calculated according to concentrations of plasmid DNA determined by a Nanodrop2000 UV-VIS Spectrophotometer (Thermo Scientific, Waltham, MA, USA) [37]. The plasmid was diluted using the ten-fold dilution method and used as the standard curve for the quantitative PCR.

Real-time PCR amplification was performed in a 20-μL reaction volume using GoTaq qPCR Master Mix (Promega, Madison, WI, USA) on a Stratagene Mx3005P instrument (Agilent Technology, Santa Clara, CA, USA). The PCR mixture contained 10 μL of Master Mix, 0.5 μL of each primer

(10 μM·L^{-1}), 2 μL of template DNA and 7 μL nuclease-free water. The reaction procedure was as follows: 95 °C for 2 min, 45 cycles at 95 °C for 15 s, 58–62 °C for 45 s, 72 °C for 45 s. All experiments were performed in triplicate. For each run, template DNA was replaced with double-distilled water as a negative control.

2.5. Statistical Analysis

Statistical analyses were performed with Microsoft Office Excel 2013 and SPSS Statistics 19.0. In all cases, differences were considered to be significant if the *p*-value for the χ^2 test was less than 0.05.

3. Results and Discussion

3.1. Effects of Sewage Irrigation on Soil Properties

In order to investigate the influence of rural domestic wastewater on the soil and plant properties, we compared the basic soil chemical properties after irrigating with tap and wastewater for three seasons. The results showed that the soil pH decreased after planting and irrigation, mainly because of the neutralization. There was no significant difference in soil pH between non-planted soils irrigated with tap water and wastewater, while soil pH was significantly higher in non-planted than in planted soil (Figure 2), because the pH of plant root exudates is lower than soil. On the other hand, it can be seen from Figure 1 that soil pH declined significantly after planting and irrigation, regardless of whether this took place with wastewater or tap water. The result indicated that wastewater irrigation may not significantly affect soil pH, but that vegetables may indicate the significant influence of this parameter. The pH of all of the soil samples meets the standards for farmland soil [38].

Electrical conductivity(EC) and salinity (positively correlated with each other) of soil irrigated with wastewater were significantly lower than in soil irrigated with tap water and treated wastewater, due to the wastewater irrigated plants growth being better, so that more mineral salts were absorbed by the plant. EC and salinity of soil increased regardless of the source of irrigation water, due to the mineral salts from irrigation water being concentrated in soil with water evaporation. Soil salinities of all soil samples are lower than the standard limit (1.0 g/kg) of the environmental quality evaluation standards for farmland for edible agricultural products [38].

The difference in organic matter (OM) among different treatments is not significant ($p > 0.05$). However, we can see from Figure 1 that OM increased after the irrigation. This might be because the irrigation promoted lichen and autotrophic bacteria growth. Soil organic matter contains a large number of major elements and trace elements that are essential for plant growth, and it is therefore one of the important indicators of soil fertility [39]. On the other hand, growth-promoting substances, such as vitamins, amino acids, plant hormone and gibberellin, are released when soil organic matter degrades, stimulating the growth of higher plants and microorganisms [40].

The urease activities of soil irrigated with wastewater have no significant difference with the control and those of soil irrigated with treated water. This activity increased in all treatments after the experiment, possibly because N and P fertilizers were added during the course of the experiment. The phosphatase activity of soil irrigated by wastewater was not significantly different than that of soil irrigated with tap water or treated wastewater. However, values were lower in planted treatments than in unplanted ones. The results indicate that changes in N and P fertilizers led to an increase of phosphatase activity, and P was greatly in demand by the plant. The conversion speed of organic and inorganic nutrients in soil mainly depends on the enzymatic reaction of redox enzymes. Soil enzyme activity could reflect the strength and direction of biochemical processes and is an important index for evaluating soil fertility and self-purification ability. It is therefore important to improve soil enzyme activity in order to improve the soil ecological environment and soil fertility.

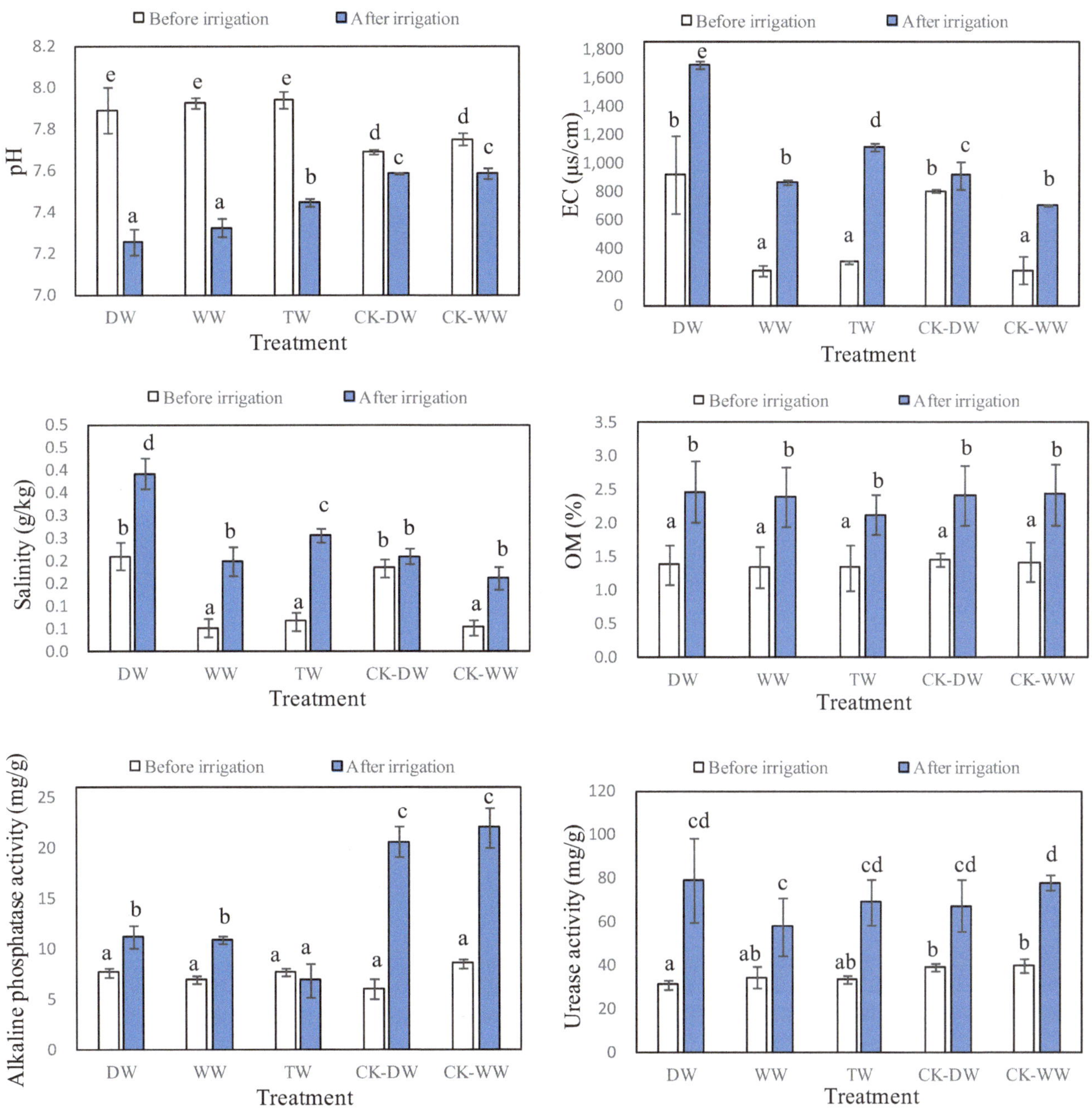

Figure 2. Properties of soil before and after irrigation of the third season planting. Abbreviations: DW, treatment irrigated with tap water and plant; WW, treatment irrigated with wastewater and plant; TW, treatment irrigated with treated wastewater and plant; CK-DW, treatment with no plant and irrigated with tap water; CK-WW, treatment with no plant and irrigated with wastewater. The standard deviation of the mean (n = 3) is shown; different letters demonstrate a significant difference at $p < 0.05$ based on ANOVA test analysis.

3.2. Effects of Sewage Irrigation on Plant Properties

Table 4 shows the quality parameters of pakchoi after the experiment. The height and fresh weight of vegetables irrigated with wastewater were significantly higher than those of vegetables irrigated with tap water and slightly higher than those of vegetables irrigated with treated wastewater. The height and fresh weight of vegetables irrigated with treated wastewater were higher than those of vegetables irrigated with tap water, but the difference was not significant ($p > 0.05$). The result indicates that wastewater irrigation could greatly increase the output of pakchoi and that irrigation with treated wastewater led to a slight increase in output.

Table 4. Quality parameters of pakchoi following irrigation with DW (control), WW and TW.

Samples	Height (cm)	Fresh Weight (g)	Soluble Sugar (%)	Soluble Protein (mg/g)	Nitrate (g/kg)	Vc (mg/g)
DW	15.25 ± 0.76^{a}	5.06 ± 0.57^{a}	0.18 ± 0.04^{a}	9.74 ± 0.33^{a}	1.19 ± 0.10^{a}	1.02 ± 0.11^{a}
WW	19.18 ± 0.76^{b}	10.83 ± 1.22^{c}	0.16 ± 0.01^{a}	10.16 ± 0.42^{a}	$1.53 \pm 0.09^{a,b}$	1.06 ± 0.19^{a}
TW	$17.75 \pm 0.89^{a,b}$	$7.31 \pm 1.05^{a,b}$	0.15 ± 0.01^{a}	9.26 ± 0.11^{a}	1.58 ± 0.07^{b}	0.89 ± 0.08^{a}

Notes: The standard deviation of the mean ($n = 3$) is shown; ^{a,b,c} different letters demonstrate a significant difference at $p < 0.05$ based on ANOVA test analysis.

Wastewater irrigation did not make significant changes in the content of soluble sugar, soluble protein and vitamin C (Vc), which means that wastewater irrigation had no clear influence on the nutritional quality of pakchoi. However, nitrate content significantly increased after irrigation with wastewater. Human consumption of vegetables in which nitrates exceed relevant standards may be seriously harmful to health; for this reason, even if the nitrate content of vegetables irrigated with wastewater conformed to safety standards (agricultural product-safety requirements for non-environmental pollution vegetables: GB18406.1-2001), further studies should be carried out on this aspect.

3.3. Quantitative PCR Detection of Pathogenic Bacteria

Real-time PCR detection of pathogenic microorganisms was conducted to monitor the amount of pathogens in wastewater, plant phyllosphere, rhizosphere and non-rhizosphere soil. The number of pathogenic bacteria in each sample is shown in Table 5 and Figure 3.

Table 5. Number of pathogens in wastewater and treated wastewater.

Pathogens	WW (Copies/L) ^{a}	TW (Copies/L) ^{b}
Aeromonas hydrophila	$(5.48 \pm 0.22) \times 10^{10}$	$(1.96 \pm 0.02) \times 10^{7}$
Arcobacter spp.	$(1.01 \pm 0.04) \times 10^{11}$	$(1.45 \pm 0.01) \times 10^{8}$
Bacillus cereus	$(2.01 \pm 0.10) \times 10^{8}$	$(2.02 \pm 0.06) \times 10^{7}$
Clostridium difficile	$(4.20 \pm 0.10) \times 10^{6}$	$(6.35 \pm 0.11) \times 10^{4}$
Clostridium perfringens	$(2.85 \pm 0.09) \times 10^{9}$	$(5.00 \pm 0.25) \times 10^{4}$
E. coli	$(1.02 \pm 0.06) \times 10^{9}$	$(2.40 \pm 0.36) \times 10^{5}$
Legionella spp.	$(1.27 \pm 0.09) \times 10^{6}$	$(6.10 \pm 0.77) \times 10^{5}$
Mycobacterium spp.	$(5.71 \pm 0.40) \times 10^{7}$	$(3.17 \pm 0.35) \times 10^{7}$
Total bacteria	$(3.13 \pm 0.38) \times 10^{11}$	$(1.56 \pm 0.08) \times 10^{10}$

Notes: The standard deviation of the mean ($n = 3$) is shown; different letters ($^{a,\ b}$) demonstrate a significant difference at $p < 0.05$ based on ANOVA test analysis.

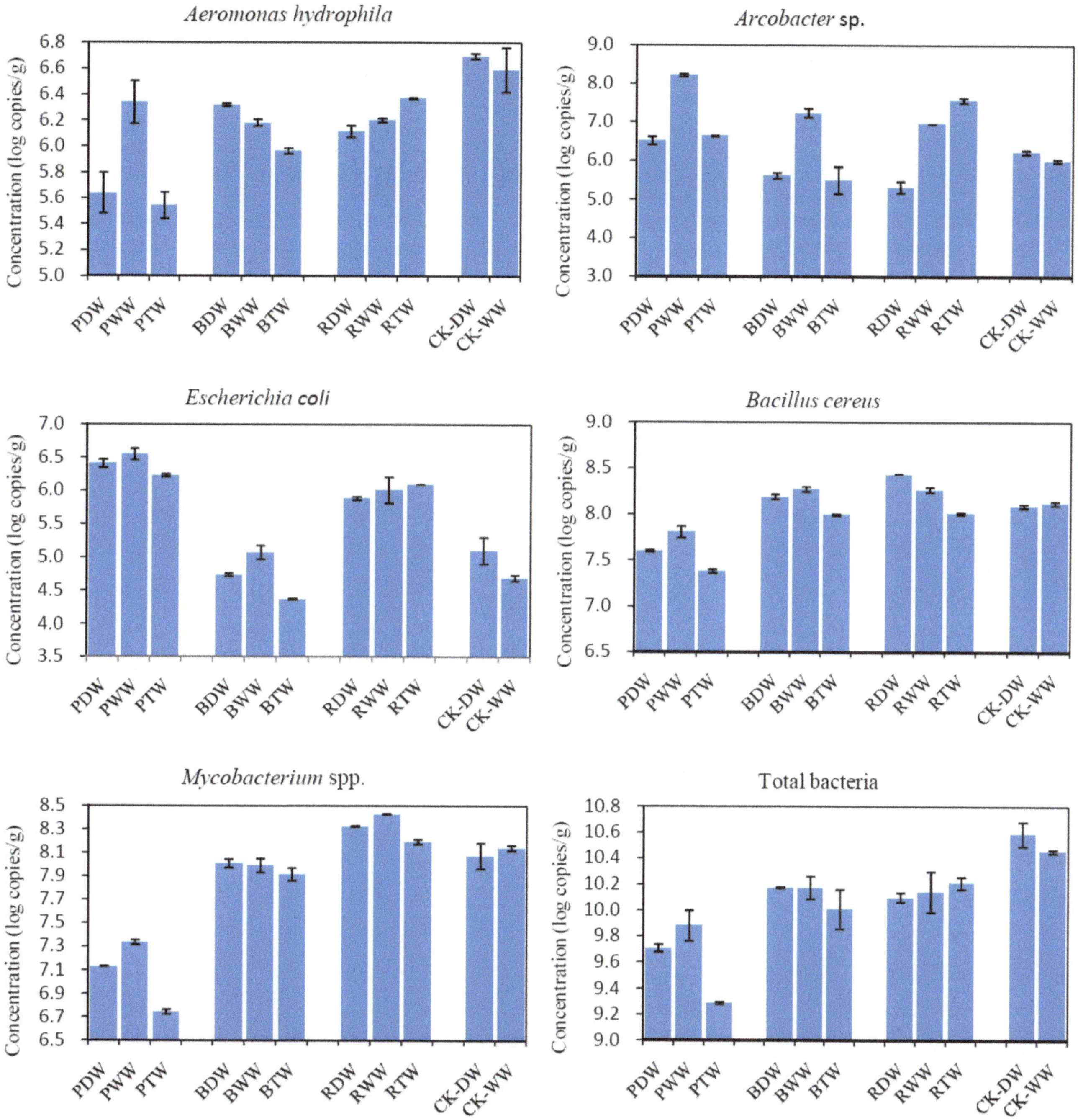

Figure 3. Pathogenic bacteria in each sample. Abbreviations: PDW, phyllosphere irrigated with tap water; PWW, phyllosphere irrigated with wastewater; PTW, phyllosphere irrigated with treated wastewater; BDW, non-rhizosphere irrigated with tap water; BWW, non-rhizosphere irrigated with wastewater; BTW, non-rhizosphere irrigated with treated wastewater; RDW, rhizosphere irrigated with tap water; RWW, rhizosphere irrigated with wastewater; RTW, rhizosphere irrigated with treated wastewater.

In raw wastewater and treated wastewater, Aeromonas hydrophila, Arcobacter spp., Bacillus cereus, Clostridium difficile, Clostridium perfringens, *E. coli*, *Legionella* spp. and Mycobacterium spp. were detected. Of these, Clostridium difficile, Clostridium perfringens and *Legionella* spp. were not found in other samples. Most of the pathogens decreased after treatment. The result shows that although the sewage treatment could remove pathogenic microorganisms significantly, treated wastewater would still contain numerous pathogens, the risks to human health of which were not clear till now. Further study should be conducted for the standard limit of these pathogens in treated wastewater.

Aeromonas hydrophila was very common in raw wastewater, and its numbers were still very high after treatment. There were no significant differences between the frequency of *A. hydrophila* in the phyllosphere, rhizosphere soil and non-rhizosphere soil. There were also no significant differences between treatments irrigated with raw wastewater and treated wastewater, as compared to the control. The quantity of *A. hydrophila* was still up to seven orders of magnitude. *A. hydrophila* is widely distributed in different aquatic environments in nature; it is an opportunistic pathogen, which could lead to food poisoning, waterborne diseases, infectious diarrhea, secondary infections and sepsis. It is therefore very dangerous to come into contact with *A. hydrophila* directly or by eating vegetables infected with the bacteria.

The number of *Arcobacter* spp. in wastewater was very high and could reach the same order of magnitude as the total bacteria. Quantities of *Arcobacter* spp. in the phyllosphere were higher than in soil; similarly, the bacterium was more frequent in plants irrigated with wastewater than in the control, while plants irrigated with treated wastewater were not significantly different from the control. *Arcobacter* spp. is a foodborne and waterborne pathogen. Of all of the species, *A. cryaerophilus*, *A. butzleri* and *A. skirrowii* are pathogens associated with human gastroenteritis and bacteremia [41,42]. So far, the pathogenesis of *Arcobacter* spp. is not clear. The possible infection pathways of *Arcobacter* spp. may include direct contact or eating contaminated food or drinking water.

Human pathogenic bacteria can be colonized both in phyllosphere and rhizosphere, and accordingly, some outbreaks of foodborne illnesses have been reported, which were associated with human consumption [43]. Contamination with *E. coli* in all treatments followed the order of phyllosphere > rhizosphere > non-rhizosphere. This was quite different from other bacteria mentioned above; *E. coli* more easily to survived in the phyllosphere, which might be more easily colonized than other pathogens. Some studies have proven the ability of intestinal pathogens, such as *E. coli*, *Salmonella* spp. and *Bacteroides*, to internalize and colonize into plant leaves through wounds, stomata or root uptake and migration [44,45], which could be contaminated via rainwater and irrigation water. *E. coli* is a foodborne pathogen used as a health standard for drinking water and food. Pathogenic *E. coli* could cause a disease outbreak through contaminated drinking water, food and aquatic environment, which could potentially be life-threatening.

In summary, there were no significant differences between the control and plants irrigated with treated wastewater, suggesting that treated wastewater can be used as a valid alternative for agricultural reuse. This result is coincident with the report by Jang *et al.* [46], which concluded that reclaimed wastewater irrigation in rice paddies presents no increased human health or eco-environmental risks. However, the pathogens in both phyllosphere and soils irrigated by raw wastewater showed higher contents than the control. Pathogens in sewage are directed into the soil through irrigation, and soil can provide more suitable conditions (such as temperature, moisture, nutrients, and so on) for the survival of microorganisms. The potential risks of pathogens in the wastewater irrigation process need further assessment.

4. Conclusions

Although rural domestic wastewater irrigation could improve soil fertility and plant yields, pathogens might be transported into soil through irrigation with raw wastewater and then transmitted to the plant in greenhouse conditions. Since vegetables contaminated with pathogenic bacteria are an important pathway of intestinal infectious diseases, more attention should be paid to the potential risks of wastewater irrigation in greenhouses for vegetable planting. The pathogens in the wastewater irrigation process should be further investigated in field conditions.

Acknowledgments

This work was supported by the Key Project of Chinese Academy of Sciences (Nos. KZZD-EW-09-3 and KZZD-EW-11-3), the National Science and Technology Major Project of China (Nos. 2012ZX07203-006 and 2014ZX07204-005) and the Special Funds of the National Academy Alliance (2012-1).

Author Contributions

Bo Yang, Xiao Kong, Bingjian Cui, Decai Jin and Zhihui Bai designed and carried out the experiments and analyzed the data. Xiao Kong wrote the main manuscript text. All authors reviewed and improved the manuscript. Zhihui Bai supervised the project.

Conflicts of Interest

The authors declare no conflict of interest.

References

1. Jiang, Y. China's water scarcity. *J. Environ. Manag.* **2009**, *90*, 3185–3196.
2. Li, L.J.; Li, B.Q. The problems of water resources in north China and the strategy. *Chin. Geogr. Sci.* **1992**, *2*, 183–196.
3. Zhang, J.Y.; He, R.M.; Qi, J.; Liu, C.S.; Wang, G.Q.; Jin, J.L. A new perspective on water issues in North China. *Adv. Water Sci.* **2013**, *24*, 303–310.
4. Srinivasan, J.T.; Reddy, V.R. Impact of irrigation water quality on human health: A case study in India. *Ecol. Econ.* **2009**, *68*, 2800–2807.
5. Sharma, R.K.; Agarwal, M.; Marshall, F. Heavy metal contamination in vegetables grown in wastewater irrigated areas of Varanasi, India. *Bull. Environ. Contam. Toxicol.* **2006**, *77*, 312–318.
6. Lado, M.; Bar-Tal, A.; Azenkot, A.; Assouline, S.; Ravina, I.; Erner, Y.; Fine, P.; Dasberg, S.; Ben-Hur, M. Changes in Chemical Properties of Semiarid Soils under Long-Term Secondary Treated Wastewater Irrigation. *Soil Sci. Soc. Am. J.* **2012**, *76*, 1358–1369.
7. Toze, S. Reuse of effluent water- benefits and risks. *Agric. Water Manag.* **2006**, *80*, 147–159.
8. Al-Sa'ed, R. Pathogens assessment in reclaimed effluent used for industrial crops irrigation. *Int. J. Environ. Res. Public Health* **2007**, *4*, 68–75.

9. Rehman, K.; Ashraf, S.; Rashid, U.; Ibrahim, M.; Hina, S.; Iftikhar, T.; Ramzan, S. Comparison of proximate and heavy metal contents of vegetables grown with fresh and wastewater. *Pak. J. Bot.* **2013**, *45*, 391–400.
10. Pereira, L.S.; Oweis, T.; Zairi, A. Irrigation management under water scarcity. *Agric. Water Manag.* **2002**, *57*, 175–206.
11. Murtaza, G.; Ghafoor, A.; Qadir, M.; Owens, G.; Aziz, M.A.; Zia, M.H.; Saifullah. Disposal and use of sewage on agricultural lands in Pakistan: A review. *Pedosphere* **2010**, *20*, 23–34.
12. Peasey, A.; Blumenthal, U.; Mara, D.; Ruiz-palacios, P.G. A Review of Policy and Standards for Wastewater reuse in Agriculture: A Latin American Perspective. *Well Study, Task No. 68 Part II,* 2000. Available online: http://www.lboro.ac.uk/well (accessed on 13 January 2014).
13. Elifantz, H.; Kautsky, L.; Mor-Yosef, M.; Tarchitzky, J.; Bar-Tal, A.; Chen, Y.N.; Minz, D. Microbial activity and organic matter dynamics during 4 years of irrigation with treated wastewater. *Microb. Ecol.* **2011**, *62*, 973–981.
14. Meli, S.; Porto, M.; Belligno, A.; Bufo, S.A.; Mazzatura, A.; Scopa, A. Influence of irrigation with lagooned urban wastewater on chemical and microbiological soil parameters in a citrus orchard under Mediterranean condition. *Sci. Total. Environ.* **2002**, *285*, 69–77.
15. Speir, T.W. Soil biochemical properties as indices of performance and sustainability of effluent irrigation systems in New Zealand—A review. *J. Roy. Soc. N. Z.* **2002**, *32*, 535–553.
16. Lv, M.C.; Cai, H.J.; Chen, X.M. Effects of sewage irrigation on physiological characteristics of potato and soil environment. *J. Irrig. Drain.* **2007**, *26*, 26–29.
17. Al-Lahham, O.; el Assi, N.M.; Fayyad, M. Impact of treated wastewater irrigation on quality attributes and contamination of tomato fruit. *Agric. Water Manag.* **2003**, *61*, 51–62.
18. Singh, P.K.; Deshbhratar, P.B.; Ramteke, D.S. Effect of sewage wastewater irrigation on soil properties, crop yield and environment. *Agric. Water Manag.* **2012**, *103*, 100–104.
19. Maldonado, V.M.; Rubio Arias, H.O.; Quintana, R.; Saucedo, R.A.; Gutierrez, M.; Ortega, J.A.; Nevarez, G.V. Heavy Metal Content in Soils under Different Wastewater Irrigation Patterns in Chihuahua, Mexico. *Int. J. Environ. Res. Public Health* **2008**, *5*, 441–449.
20. Tsado, E.K.; Adesina, O.A.; Oyeleke, S.B. A Survey on the Bacterial Load of Selected Fruits and Leafy Vegetables in Minna Metropolis of Niger State. *Niger. J. Anim. Prod. Adv.* **2013**, *3*, 6–11.
21. Ministry of Health of P.R. China; Standardization Administration of P.R. China. *Standards for Drinking Water Quality*; GB 5749–2006; The National Standard of the People's Republic of China; Ministry of Health of P.R.: Beijing, China, 2006.
22. Ministry of Environmental Protection of P.R. China. *Standards for Irrigation Water Quality*; GB 5084–92; The National Standard of the People's Republic of China; Ministry of Environmental Protection of P.R. China: Beijing, China, 1992;.
23. Morugán-Coronado, A.; García-Orenes, F.; Mataix-Solera, J.; Arcenegui, V.; Mataix-Beneyto, J. Short-term effects of treated waste water irrigation on Mediterranean calcareous soil. *Soil Tillage Res.* **2011**, *112*, 18–26.
24. Tabatabai, M.A. Chemical and microbiological properties. In *Methods of Soil Analysis*; Soil Science Society of America: Madison, WI, USA, 1982; pp. 903–947.
25. Cataldo, D.A.; Haroon, M.; Schrader, L.E.; Young, V.L. Rapid colorimetric determination of nitrate in plant tissues by nitration of salicylic acid. *Commun. Soil Sci. Plant Anal.* **1975**, *6*, 71–80.

26. Dubois, M.; Gilles, K.A.; Hamilton, J.K.; Rebers, P.A.; Smith, F. Colorimetric method for determination of sugars and related substances. *Anal. Chem.* **1956**, *28*, 350–356.
27. Lowry, O.H.; Rosebrough, N.J.; Farr, A.L.; Randall, R.J. Protein measurement with the folin phenol reagent. *J. Biol. Chem.* **1951**, *193*, 265–275.
28. Tee, E.S.; Young, S.; Young, S.I.; Ho, S.K.; Mizura, S.S. Determination of vitamin C in fresh fruits and vegetables using the dye-titration and microfluorometric methods. *Pertanika* **1988**, *11*, 39–44.
29. Kingombe, C.I.; Huys, G.; Tonolla, M.; Albert, M.J.; Swings, J.; Peduzzi, R.; Jemmi, T. PCR detection, characterization, and distribution of virulence genes in *Aeromonas* spp. *Appl. Environ. Microbiol.* **1999**, *65*, 5293–5302.
30. Bastyns, K.; Cartuyvels, D.; Chapelle, S.; Vandamme, P.; Goossens, H.; DeWachter, R. A variable 23S rDNA region is a useful discriminating target for genus-specific and species-specific PCR amplification in *Arcobacter* species. *Syst. Appl. Microbiol.* **1995**, *18*, 353–356.
31. Priha, O.; Hallamaa, K.; Saarela, M.; Raaska, L. Detection of *Bacillus cereus* group bacteria from cardboard and paper with real-time PCR. *J. Ind. Microbiol. Biot.* **2004**, *31*, 161–169.
32. Rinttila, T.; Kassinen, A.; Malinen, E.; Krogius, L.; Palva, A. Development of an extensive set of 16S rDNA-targeted primers for quantification of pathogenic and indigenous bacteria in faecal samples by real-time PCR. *J. Appl. Microbiol.* **2004**, *97*, 1166–1177.
33. Kaushik, R.; Balasubramanian, R.; de la Cruz, A.A. Influence of air quality on the composition of microbial pathogens in fresh rainwater. *Appl. Environ. Microbiol.* **2012**, *78*, 2813–2818.
34. Miyamoto, H.; Yamamoto, H.; Arima, K.; Fujii, J.; Maruta, K.; Izu, K.; Shiomori, T.; Yoshida, S. Development of a new seminested PCR method for detection of *Legionella* species and its application to surveillance of legionellae in hospital cooling tower water. *Appl. Environ. Microbiol.* **1997**, *63*, 2489–2494.
35. Mendum, T.A.; Chilima, B.Z.; Hirsch, P.R. The PCR amplification of non-tuberculous mycobacterial 16S rRNA Sequences from soil. *FEMS Microbiol. Lett.* **2000**, *185*, 189–192.
36. Basic Local Alignment Search Tool. Available online: http://blast.ncbi.nlm.nih.gov/Blast.cgi (accessed on 7 December 2013).
37. Whelan, J.A.; Russell, N.B.; Whelan, M.A. A Method for the absolute quantification of cDNA using real-time PCR. *J. Immunol. Methods* **2003**, *278*, 261–269.
38. Ministry of Environmental Protection of P.R. China. *Environmental Quality Evaluation Standards for Farmland of Edible Agricultural Products*; The National Standard of the People's Republic of China; Ministry of Environmental Protection of P.R. China: Beijing, China, 2006.
39. Song, C.-Y.; Zhang, X.-Y.; Liu, X.-B.; Gao, C.-S. Effect of soil organic matter on soil fertility and crop productivity. *Syst. Sci. Comp. Stud. Agric.* **2008**, *24*, 357–362.
40. Brady, N.C.; Weil, R.R. *The Nature and Properties of Soil*, 13th ed.; Prentice Hall: Upper Saddle River, NJ, USA, 2002.
41. Ho, H.T.K.; Lipman, L.J.A.; Gaastra, W. *Arcobacter*, what is known and unknown about a potential foodborne zoonotic agent. *Vet. Microbiol.* **2006**, *115*, 1–13.
42. Bi, S.L.; Meng, H.C. Biological characteristics and diagnoses of pathogenic *Arcobacter* species. *Mod. Food Sci. Technol.* **2013**, *29*, 211–214.

43. Quilliam, R.S.; Williams, A.P.; Jones, D.L. Lettuce cultivar mediates both phyllosphere and rhizosphere activity of *Escherichia coli* O157:H7. *PLoS ONE* **2012**, *7*, doi:10.1371/journal.pone.0033842.
44. Kroupitski, Y.; Golberg, D.; Belausov, E.; Pinto, R.; Swartzberg, D.; Granot, D.; Sela, S. Internalization of Salmonella enterica in leaves is induced by light and involves chemotaxis and penetration through open stomata. *Appl. Environ. Microbial.* **2009**, *75*, 6076–6086.
45. Solomon, E.B.; Yaron, S.; Matthews, K.R. Transmission of *Escherichia coli* O157:H7 from contaminated manure and irrigation water to lettuce plant tissue and its subsequent internalization. *Appl. Environ. Microbiol.* **2002**, *68*, 397–400.
46. Jang, T.; Jung, M.; Lee, E. Assessing environmental impacts of reclaimed wastewater irrigation in paddy fields using bioindicator. *Irrig. Sci.* **2013**, *31*, 1225–1236.

3

Assessing Potential Algal Blooms in a Shallow Fluvial Lake by Combining Hydrodynamic Modelling and Remote-Sensed Images

Monica Pinardi [1,2,*], Andrea Fenocchi [3], Claudia Giardino [1], Stefano Sibilla [3], Marco Bartoli [2] and Mariano Bresciani [1]

[1] Institute for Electromagnetic Sensing of the Environment, National Research Council, Via Bassini 15, Milan 20133, Italy; E-Mails: giardino.c@irea.cnr.it (C.G.); bresciani.m@irea.cnr.it (M.B.)

[2] Department of Life Sciences, University of Parma, Viale Usberti 11/a, Parma 43124, Italy; E-Mail: marco.bartoli@unipr.it

[3] Department of Civil Engineering and Architecture, University of Pavia, Via Ferrata 3, Pavia 27100, Italy; E-Mails: andrea.fenocchi@unipv.it (A.F.); stefano.sibilla@unipv.it (S.S.)

* Author to whom correspondence should be addressed; E-Mail: pinardi.m@irea.cnr.it;

Academic Editor: Say-Leong Ong

Abstract: Shallow fluvial lakes are dynamic ecosystems shaped by physical and biological factors and characterized by the coexistence of phytoplankton and macrophytes. Due to multiple interplaying factors, understanding the distribution of phytoplankton in fluvial lakes is a complex but fundamental issue, in the context of increasing eutrophication, climate change, and multiple water uses. We analyze the distribution of phytoplankton by combining remotely sensed maps of chlorophyll-*a* with a hydrodynamic model in a dammed fluvial lake (Mantua Superior Lake, Northern Italy). The numerical simulation of different conditions shows that the main hydrodynamic effects which influence algal distribution are related to the combined effect of advection due to wind forces and local currents, as well as to the presence of large gyres which induce recirculation and stagnation regions, favoring phytoplankton accumulation. Therefore, the general characters of the phytoplankton horizontal patchiness can be inferred from the results of the hydrodynamic model. Conversely, hyperspectral remote-sensing products can be used to validate this model, as they provide chlorophyll-*a* distribution maps. The integration of ecological, hydraulic, and remote-sensing techniques may therefore help the monitoring and protection of inland water quality, with important improvements in management actions by policy makers.

Keywords: fluvial lake; chlorophyll-*a* distribution; water circulation; hydrodynamic model; remote sensing; hyperspectral data; water resource management

---

1. Introduction

Fluvial lakes are, in general, highly dynamic shallow water bodies where different communities of primary producers coexist and are shaped by an array of physical and biological factors. They may undergo large interannual variations of residence time, depending upon water discharge, which is in turn related to rainfall patterns and water diversion for human activities. Fluvial lakes may also display fluctuating concentrations of inorganic nutrients and chlorophyll-*a* (a proxy of algal biomass), depending upon the rates of primary production and of microbial activity. Hydrological alterations (e.g., river damming, water diversion) intensify the effect of high nutrient loading from upstream: a decrease in water discharge has consequences on water renewal, thermal stratification, primary producer communities and photosynthetic rates [1], gas and nutrient dynamics [2], and benthic processes [3]. Fluvial lakes may also host dense stands of different groups of submersed, floating leaved or emergent macrophytes, each locally affecting water circulation and biogeochemical processes.

As a consequence of multiple interplaying factors, the distribution of phytoplankton in fluvial lakes is a complex issue. The capacity to predict pelagic algae distribution and abundance is, however, important in the context of increasing eutrophication, unpredictable climate, and multiple water use. The general variability of phytoplankton communities is influenced both by physical factors, such as variable wind patterns, discharge and light availability [4] and by biological processes, such as the presence of macrophyte islands and the interaction between benthic and pelagic compartments (e.g., hypoxia or anoxia events and coupled nutrient dynamics) [5,6].

According to Verhagen [7], a likely time scale for algal growth and decay is ~10 d. Therefore, having the usual advection time scales in the order of a few hours [8–11], phytoplankton can be assumed to be passively transported, except for very large shallow lakes. Time scales for phytoplankton horizontal distribution can then be assimilated to the ones of lake circulations. Thus, in shallow fluvial lakes, wind and discharge, in addition to shaping hydrodynamics, play also a major role in determining the large-scale distribution of phytoplankton and passively-advected zooplankton [8,12,13]. Microorganism arrangement is also the vector through which hydrodynamically induced patchiness is transferred to higher trophic levels [14]. Biological processes are hence relevant only over multi-daily time scales or over very small spatial scales [15].

Circulation patterns in shallow lakes can be divided into two categories [16]: "conveyor belt" type, in which prevailing circulations develop in the vertical plane, and "Livingstone-type" [17], in which water motions develop in the horizontal plane. George & Edwards [18] found that, in a lake with prevailing circulations in the vertical plane, organisms which tend to concentrate near the surface accumulate in downwelling regions downstream, while the ones which tend to aggregate near the bottom or the thermocline, hence are transported by return currents, accumulate in upwind upwelling regions. In such a situation, the horizontal distribution of plankton is linked to the vertical one [19]. According to the observations by George & Edwards [18] on a shallow reservoir with prevailing circulation in the vertical plane, cyanobacteria concentrate near the water surface because of their positive buoyancy, being mixed along the water depth by turbulence for wind speeds above 4 m/s. In this case, horizontal patchiness is observed only for low winds, as the return current homogenizes the distribution for higher intensities. For shallow lakes with Livingstone-type circulations, horizontal patchiness persists even for stronger winds, even if more mixed conditions are attained.

Located in Northern Italy, the Mantua Lakes system (Superior, Middle, and Inferior Lakes) is an interesting case study with respect to phytoplankton distribution and transport. These lakes are shallow, nutrient-enriched, host different primary producers, and undergo pronounced interannual discharge variations, coinciding with seasonal water diversion for irrigation purposes and rainfall patterns. The interaction between the continuous import of nutrients and export of particulate matter, the hydrological regime and the lake morphology leads to a mosaic of environmental conditions, allowing the simultaneous occurrence of different macrophyte typologies and high phytoplankton biomass [20]. Bolpagni *et al.* [21] found generally lower concentrations of chlorophyll-*a* (Chl-*a*) in the areas closest to the macrophyte stands compared to the outer zone and speculated allelopatic interactions or nutrient limitation. They did not include in their study water circulation, which in eutrophic fluvial lakes shapes the primary producer community [22–25].

Legendre and Demers [14] called the coupled study of biological and hydrodynamic processes "dynamic biological limnology". In this frame, biological dynamics can be used to estimate hydrodynamic patterns e.g., [26]: for example, remote-sensed images of Chl-*a* distribution can be used to validate wind-induced circulation models [7], while numerical model results can be used to assess the correlation between plankton distribution and hydrodynamic variables [8]. Thackeray *et al.* [15] pointed out that given the responsive nature of plankton distribution to wind variations, the results of field surveys taking many hours or days to complete cannot describe properly the distribution dynamics. In this sense, remote-sensed images are the most appropriate tool, as they capture instantaneously the whole water body.

The full coupling between ecological and hydrodynamic models, with a large number of equations and calibration parameters for the ecological part e.g., [27,28], is an advancing field in limnology, leading to physically-based interpretations, replacing black-box approaches [29]. The use of coupled ecological and hydrodynamic models is justifiable over long time scales (e.g., seasonal and multi-annual) and very large spatial scales (e.g., very large lakes and sea areas) [30]. Under such circumstances, in fact, there is an interplay between physical and biological processes. This is not the case for shallow fluvial lakes, where phytoplankton horizontal patchiness is an almost exclusive product of hydrodynamic factors, so that its determination can be obtained on a hydrodynamic basis only.

The main objective of this work is the analysis of the distribution of phytoplankton in the Superior Lake of Mantua, through the comparison of remotely sensed maps of Chl-*a*, derived in previous studies [21,31,32] with the results of a newly-developed numerical hydrodynamic model, in order to identify the circulation patterns responsible for algal patchiness. Four historical situations were analyzed, including a dry summer case, characterized by extremely low flow and long water residence time. The consistent relations found between the hydrodynamic variables and phytoplankton distribution also provide an indirect validation of the numerical model.

2. Materials and Methods

2.1. Study Area

The Superior Lake of Mantua (surface area 3.67 km^2, average depth 3.6 m) is the uppermost of three shallow hypertrophic fluvial lakes, located in a highly human-impacted (for agriculture and industrial purposes) sub-basin of the Po River in Northern Italy (Figure 1). The lakes system was created in the 12th century by the damming of the Mincio River. The water discharge in the Superior Lake of Mantua is regulated by the Vasarone sluice gate and by the recently-opened Vasarina gate, maintaining a constant water level of 17.5 m a.s.l. The average water residence time of the lake in the period 2000–2006 was ~8 days (mean discharge $Q = 20 \pm 6$ m^3 s^{-1}). Some small secondary tributaries along the lake left shore also contribute to the discharge, with an overall mean summer inflow of about 5 m^3 s^{-1}, mainly due to the drainage from crop fields. The Mantua Lakes system has been part of the UNESCO World Cultural Heritage since 2008 and is protected as a Natural Regional Park. Moreover, the Superior Lake is located just downstream of the "Valli del Mincio" wetlands Natural Reserve, which is also a Site of Community Importance.

Figure 1. (**a**) Geographic position of the Mantua Lakes system; (**b**) Bathymetry of the Superior Lake of Mantua (ortophotos courtesy of Lombardy Region).

The Superior Lake of Mantua receives high nutrient loads from the Mincio River, which sustain a high primary producer growth (0.1 mg/L N–NH_4^+, 2.1 mg/L N–NO_3^-, 0.1 mg/L P–TP; average data by ARPA—Regional Environmental Protection Agency—Lombardy 2003–2011). In this fluvial lake, allochtonous nutrient loads, generated upstream, largely exceed those recycled (or retained/lost) from sediments [20]. The water transparency average value is 1.1 ± 0.6 m (variations mainly due to particulate matter and phytoplankton growth), while Chl-*a* concentrations vary between a few and ~200 mg m^{-3} (data source: ARPA Lombardy 2003–2011). Daylight oxygen saturation is, on average, above the equilibrium (137% ± 59%; data source: ARPA Lombardy 2003–2011).

Eutrophic to hypertrophic conditions result in dense phytoplankton communities, such as chlorophytes, euglenophytes, chrysophytes, diatoms and cyanophytes [4], that strongly limit available light to submersed macrophytes.

The shallower areas of the Superior Lake host different groups of macrophytes (emergent e.g., *Phragmites australis,* submersed e.g., *Ceratophyllum demersum* and floating-leaved e.g., *Nelumbo nucifera*). *N. nucifera*, the lotus flower, pervasively colonises an extensive area of the lake surface forming islands, the widest one being located in the south-eastern portion of the lake (Figure 1b).

2.2. Remote Sensing Data

Remotely sensed images used in this study were acquired from both satellite and airborne platforms and include data from three hyperspectral sensors. Airborne data were acquired by MIVIS (Multispectral Infrared Visible Imaging Spectrometer) and APEX (Airborne PRISM Experiment). These sensors operate with different technologies, as MIVIS is a whisk broom scanner, while APEX is a dispersive push broom imaging spectrometer. MIVIS measures the reflected radiation in 102 bands 0.441–12.42 μm, but the ones relevant for water quality studies are the first 20 nm wide 20 bands 0.441–0.820 μm. Within this spectral range, APEX has a finer spectral resolution, providing 98 bands 0.426–0.910 μm, with a bandwidth between 3.5 and 13 nm. The satellite images were acquired from CHRIS sensors (Compact High Resolution Imaging Spectrometer) onboard of PROBA (PRoject for On-Board Autonomy). CHRIS-PROBA is a physically compact payload, as its name implies (it weighs less than 15 kg), and operates in the push broom mode. CHRIS-PROBA images have 18 bands with a bandwidth between 12 and 20 nm, useful for water related applications [33].

The airborne campaigns with MIVIS and APEX were performed on 26 July 2007 and on 21 September 2011 and provided data with a spatial resolution of 4 m and 3 m, respectively. CHRIS-PROBA images at 18 m spatial resolution were acquired in between the two airborne campaigns, on 29 June 2008 and 28 August 2011. Land features and macrophyte-covered water regions were first masked out. Then, the four images were transformed into Chl-*a* concentration distribution maps as described in [4,21,31,32], having been first corrected for atmospheric effects and then converted to Chl-*a* concentration by semi-empirical approaches. These methods, which are widely used in productive waters to assess the Chl-*a* concentration, employ band ratios between the secondary Chl-*a* absorption maximum (~675 nm) and adjacent spectral bands not affected by phytoplankton absorption, such as the near-infrared reflectance peak near 700 nm e.g., [34]. The use of the band ratios to estimate Chl-*a* from remotely sensed images was evaluated by using field data collected at the time of imagery acquisitions. The average relative error for the three sensors was 20%. Due to water transparency being much smaller than the water depth, except immediately nearshore, the remotely sensed signal is essentially not affected by bottom reflection. Details on the transformation of images into Chl-*a* concentration maps are given in Table 1.

2.3. Numerical Hydrodynamic Lake Model

A 3D numerical model of the hydrodynamics of the Superior Lake of Mantua was developed, implementing the relevant physical phenomena that shape circulations. Topography was obtained by the Kriging interpolation of a low-resolution bathymetrical survey, integrated by a 2012 airborne image by the Lombardy Region for the definition of lake shores and lotus flower island borders (Figure 1b). Because of the limited data, the upstream boundary of the model was set at the entrance to the actual lacustrine part of the lake, leaving out the channelized transitional section from the 'Valli del Mincio' wetlands.

Table 1. List of algorithms with related accuracy (R^2) used to transform remote-sensed reflectances (*Rw*) into chlorophyll-*a* (Chl-*a*) concentration maps. The rRMSE (relative Root Mean Square Error) gives the percentage error when the algorithms are applied to the images.

Sensor	Data	Algorithm	R^2	rRMSE [%]	References
APEX	21 September 2011	$54.647 \times [Rw(690,697) - Rw(670,673)] - 52.47$	0.86	27	[4,21]
CHRIS	29 June 2008	$81.257 \times [Rw(706) - Rw(669)] - 54.5$	0.74	16	[32]
MIVIS	26 July 2007	$170.81 \times [Rw(677) - Rw(710)] \times Rw(747) + 36.105$	0.77	18	[31]

The general arrangement of circulations in the Superior Lake of Mantua is more of the "Livingstone-type" rather than the 'conveyor belt' type, due to the bathymetrical inhomogeneities (Figure 1b), the complex shape and the strong through-flowing current of the Mincio River: three-dimensional effects are present only locally. This leads to noticeable phytoplankton horizontal patchiness [7]. Levasseur *et al.* [35] observed that, in estuarine environments, stronger winds are needed to homogenize the horizontal distribution because of the strong tidal current: similarity to the riverine through-flowing discharge in a fluvial lake can be derived from there.

Hourly wind data at the Mantua-Tridolino weather station for 2007 and 2008 and 5-minute wind data at the Rivalta sul Mincio weather station for 2011 were provided by ARPA Lombardy and by the CML (Lombardy Meteorological Centre), respectively. Daily discharge values were provided by AIPo (Interregional Agency for the Po River) Mantua office. In all of the simulated conditions, only the Vasarone outlet in the north-eastern portion of the lake (Figure 1b) was active.

The model adopts the Internal Boundary Layer (IBL) theory to determine the wind stress field on the water surface e.g., [36], considering also wind sheltering by vegetation at the leeward and windward shores according to [37]. Multiple embedded IBLs are implemented to solve roughness transitions between land, shore, and emergent vegetation [38]. Wind fetch is calculated over multiple direction radials and then averaged to take wind field irregularities into account, as suggested by the Shore Protection Manual [39]. No wind stress was assumed over the lotus flower island, because the leaf cover present during the summer, when all the remote-sensed data compared with the simulations results were obtained, prevents air turbulence from reaching the water surface, e.g., [40].

Hydrodynamic simulations were performed using the STAR-CCM+ v9.02 3D CFD solver [41], using a Realisable RANS k-ε model for turbulence parameterization [42]. The 3D solid of the water volume contained in the lake was discretized into a computational mesh with parallelepiped trimmed elements having 6 m horizontal and 0.3 m vertical resolution. Notwithstanding the shallowness of the lake and the mainly horizontal development of circulations, the use of a 3D model ensures the correct simulation of wind stress turbulent transmission to water, leading to better results than 2D depth-integrated models e.g., [43]: this is the case especially under low-wind conditions and when the interaction with a through-flowing current occurs, as for the Superior Lake of Mantua. Three-dimensional models are also needed when the attention focuses on the advection of particles concentrated in a restricted portion of the flow depth, as in the present work. The model adopts a fixed free surface: water level variations due to wind setup and head losses are much lower than the cell vertical resolution and can be calculated from pressure variations at the surface boundary around the null value. A constant velocity, normal to the inlet surface, was set on the fictitious upstream boundary, while a fixed discharge, equal to the inflowing one, was imposed as downstream boundary condition at the Vasarone outlet.

The lotus flower island (35.5 ha) was modelled as a porous medium, adding a Forchheimer-type source term in the momentum equation, whose coefficients have been calculated as proposed by [44]. Wind-generated waves have a negligible effect on the lake circulations by bottom friction enhancement e.g., [45], as almost the whole lake is under deep-water conditions under ordinary and storm winds as regards wave-bottom interaction: their action is therefore not included in the model. Local water renewal times were calculated according to the procedure suggested by [41], based on the use of a purely-advected passive scalar. Further details and discussion about the numerical model of the lake will be published in a separate paper.

To provide meaningful comparisons to the instantaneous remote-sensed images, steady-state simulations were performed by considering 'effective' wind conditions derived from the analysis of meteorological data up to 12 h prior to the images acquisition. This time interval was chosen because the time scale for the adaptation of the lake circulations to wind changes was found to be in the order of ~6 h. Data from the most frequent wind sector (90° wide for Mantua-Tridolino data, 22.5° wide for the Rivalta sul Mincio more frequent data) were isolated, calculating the mean wind direction and magnitude over them. Neither reversal in the wind direction, nor any intense storm, was observed in the 24 h prior to each image acquisition. With such premises, the steady-state approximation and the determined wind conditions proved to be reasonable and representative. Wind and discharge data for the four simulations are shown in Table 2.

Table 2. Discharge and wind conditions for the simulated images situations.

Parameter	MIVIS 26 July 2007	CHRIS-PROBA 29 June 2008	CHRIS-PROBA 28 August 2011	APEX 21 September 2011
Q [$m^3 s^{-1}$]	2.6	10	9.7	12.7
Wind direction [°]	74	63	270	269
W_{10} [$m s^{-1}$]	2.18	1.26	1.49	2.35

Discharge values are representative of the average summer conditions ($Q \approx 10$ $m^3 s^{-1}$), except for the 2007 drought simulation. Wind intensities are representative of the ordinary summer conditions (mean wind at 10 m height $W_{10} = 1.93$ $m s^{-1}$; data source: Rivalta sul Mincio weather station 2011–2013). The main winds on the Superior Lake of Mantua are aligned over the 80°–260° axis, parallel to the riverine flow direction. Easterly winds prevail during the summer (64% of occurrences during 2011–2013), as for the 2007 and 2008 simulations, while 2011 cases are relative to the secondary seasonal wind direction; westerly winds prevail during the winter. Wind direction inversions are relative to meso-scale air movements in the Po River Plain, while daily thermal patterns as in the subalpine lake area are not observed.

2.4. Data Analyses

A Pearson correlation analysis was performed to infer the relationships between the remote-sensed Chl-*a* distributions and the simulated hydrodynamic variables.

Mean (μ) and coefficient of variation (CV) values of the simulated local water renewal time (T_r) and of the remote-sensed Chl-*a* concentrations were calculated. The coefficient of variation gives an indication about the degree of uniformity of the variables fields over the lake surface. The Chl-*a* concentration fields used in the analyses were limited, for consistency, to the area which overlaps with the domain of the numerical model.

3. Results and Discussion

3.1. Primary Producers and Physico-Chemical Characterization

Table 3 reports the median summer values of the water quality parameters for the Superior Lake of Mantua in the years under investigation. Nitrate and soluble reactive phosphorous concentrations were lower in the 2007 dry summer compared to the other two years. Total nitrogen and phosphorous (TN, TP), mainly due to particulate forms, were instead higher in 2007 rather than 2008 and 2011.

Table 3. Water quality characteristics (median values) during the summer period for the Superior Lake of Mantua in the three years considered in the present study.

Year	pH	Conductivity [$\mu S\ cm^{-1}$]	NH4+ [$mg\ L^{-1}$]	NO3- [$mg\ L^{-1}$]	TN [$mg\ L^{-1}$]	SRP [$mg\ L^{-1}$]	TP [$mg\ L^{-1}$]	Chl-*a* [$mg\ m^{-3}$]	O_2 [% Sat]
2007	8.2	386	0.08	0.30	1.75	0.001	0.14	90	143
2008	8.6	300	0.05	0.62	0.57	0.003	0.02	38	179
2011	8.2	327	0.05	1.14	1.45	0.018	0.07	58	165

In 2007, phytoplankton communities were typical of eutrophic to hypertrophic systems rich in organic matter, including diatoms (e.g., *Synedra* spp., *Aulacoseira* spp.), cyanophytes (e.g., *Oscillatoria* spp.), and chlorophytes (e.g., *Scenedesmus* spp., *Pediastrum simplex*) [20]. Moving from 2007 to 2008, the Superior Lake registered a significant decrease in Chl-*a* concentrations due to the higher flow discharges, but always characterized by high diurnal variations [4,31]. In 2008, the mean annual biovolume in the Superior Lake of Mantua was 10 $mm^3\ L^{-1}$, mainly due to *Bacillariophyta* from March to May and to *Cryptophyta* (mainly *Cryptomonas* spp.) and *Cyanobacteria* (e.g., *Limnothrix redekei, Cuspidothrix elenkiniii* (Kisel.)) in August and September (data source: ARPA Lombardy). In 2011, phytoplankton communities were mainly represented by *Chlorophyta* (50% of the total biodiversity and 60% of the total algal biomass; main species *Pandorina morum*), *Euglenophyta* (*Phacus tortus*), and *Chrysophyta* (*Oochromonas* spp.) [4]. The cyanobacteria species with the highest density was *Planktolyngbya limnetica* [4]. Among the potentially toxic cyanobacteria species, the presence of *Cylindrospermopsis raciborskii* and *Planktothrix rubescens/agardhii* was reported [4].

The dominant macrophyte in terms of areal extension on the lake surface was the lotus flower. During the study period, the *N. nucifera* main island in the Superior Lake was wider in 2007 (~56 ha before harvesting operations) and relatively similar in 2008 and 2011 (~34 and ~38 ha, respectively).

Water quality in terms of nutrient loading, risk of hypoxia and primary productivity is surely a function of water discharge [20,25], which was lower during the 2007 dry summer and typical of summer conditions in 2008 and 2011. In June and July 2007, high water temperature and scarce water renewal, together with a high stability of meteo-climatic conditions, determined hypoxic events on the lake bottom, due to high oxygen demand by the benthic and pelagic systems. High water velocity reduces phytoplankton accumulation by flushing algae outside and by affecting negatively the primary production rates [23,46]. Conversely, Rennella and Quiros [25] found that high water flow leads to a twofold increase in algal biomass in shallow lakes with high nutrient levels, likely due to decreased grazing pressure and improved light availability. Light limitation probably inhibits algal primary production when water renewal time exceeds 8–10 days, a condition that usually occurs in the Mantua Lakes during July [22–24].

These findings are probably common to eutrophic fluvial lakes with continuous nutrient loading from upstream [47,48], but they contrast with the results reported for lentic shallow basins, in which internal load and recycling are often more important than external sources of nutrients and the uptake by primary producers reduces dissolved nutrients [49,50]. In the Superior Lake of Mantua, the longer residence time during summer 2007 resulted in higher chlorophyll-*a* and lower inorganic nutrient concentrations in the water column.

Storm wind events have a two-sided general effect on algal blooms in shallow lakes, limiting their proliferation due to turbidity increase and, on the opposite, favoring phytoplankton growth, due to nutrients resuspension caused by bottom uptake and water mixing [51]. In the Superior Lake, the former phenomenon was observed shortly after wind storms, while the second with a larger delay [4].

3.2. Chlorophyll-a Distribution Maps

In the MIVIS 26 July 2007 image, Chl-*a* concentration (Figure 2a) was very high throughout the lake, due to the drought conditions of the Mantua Lakes system during that summer. In particular, the highest Chl-*a* concentration was registered upstream, in the north-western part of the lake. Unfortunately, remote-sensed chlorophyll-*a* data for the channel south of the lotus flower island are not available due to a mirror effect.

In the CHRIS-PROBA 29 June 2008 image, Chl-*a* was also quite evenly distributed throughout the lake, with maximum concentrations observed upstream and in the water canal north of the lotus flower island (Figure 2b). The lowest concentrations were measured instead along the channel south of the macrophyte island. Chlorophyll-*a* concentration is also quite high in the proximity of the downwind shore.

Both Chl-*a* distribution maps from 2011 (CHRIS-PROBA 28 August 2011, Figure 2c; APEX 21 September 2011, Figure 2d) showed higher concentrations in the downstream part of the lake, more markedly in the APEX image. In the downstream part of the lake, a corridor with lower Chl-*a* concentration can be clearly seen along the north side of the lotus flower island, especially in the CHRIS-PROBA image. Chlorophyll-*a* accumulates nearby the northern shore, in the channel south of the island and downstream of it.

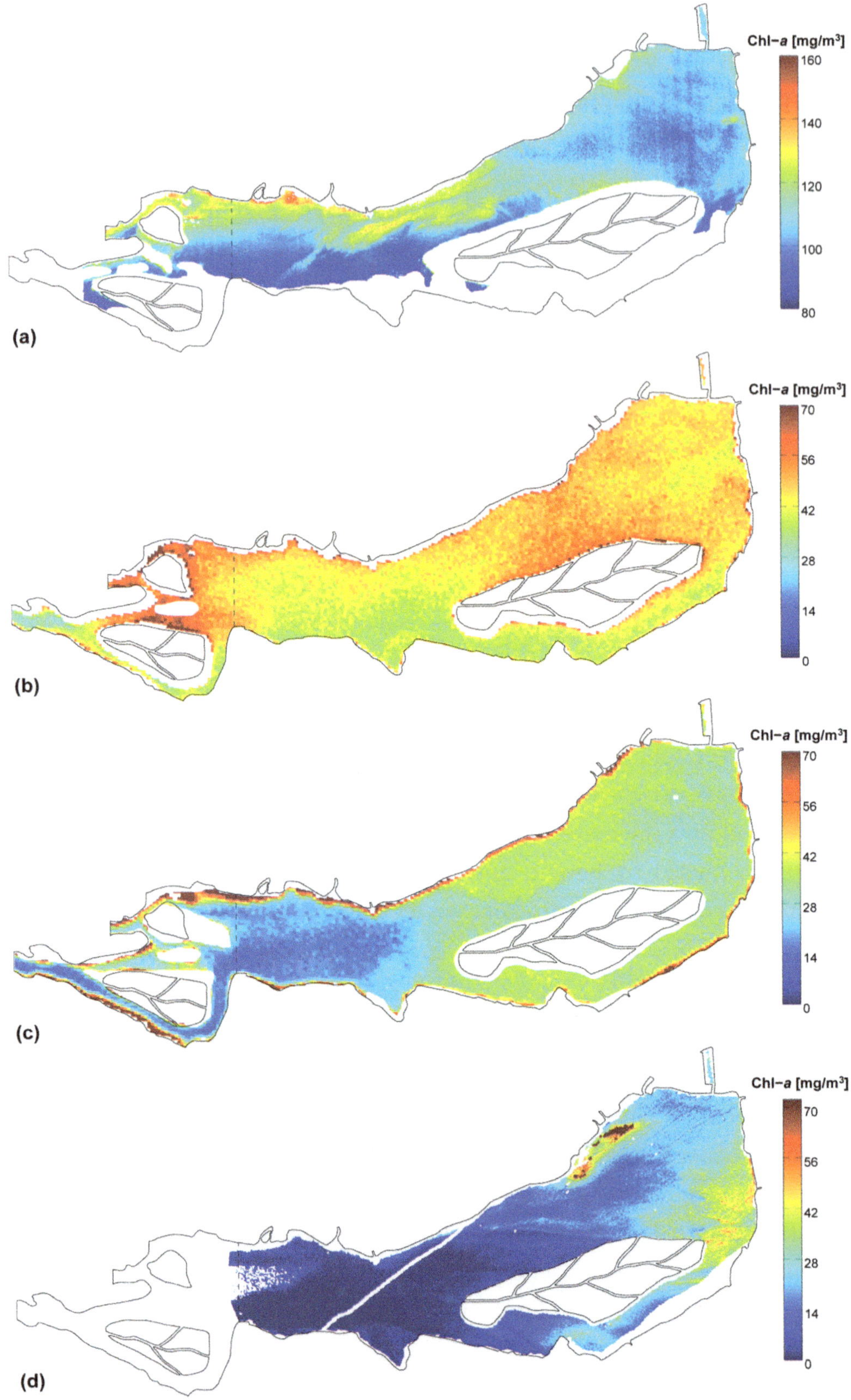

Figure 2. Chlorophyll-*a* (Chl-*a*) distribution maps for the MIVIS 26 July 2007 (**a**); CHRIS-PROBA 29 June 2008 (**b**); CHRIS-PROBA 28 August 2011 (**c**) and APEX 21 September 2011 (**d**) images.

3.3. 3D Hydrodynamic Model

The hydrodynamic variables computed with the 3D model were depth-averaged along the first meter below the free surface only, in order to compare them to the remote-sensed Chl-*a* data, which are representative of approximately one meter of water depth due to the limited light penetration. The hydrodynamic variables which were found to be related to the chlorophyll-*a* horizontal patchiness were: water level variations (Δy), wind stress (τ_S), water velocity direction and its magnitude (V), local water renewal time (T_r).

Water level variations (obtained from pressure deviations) from the control level for the Superior Lake of Mantua of 17.50 m a.s.l., fixed as top elevation for the surface cells, are representative both of wind-induced water displacement, which occurs in the wind direction and is proportional to its intensity, and of flow head losses in the downstream direction, which are proportional to flow discharge: the overlapping of these effects determines the actual surface setup. For the ordinary wind intensities of the analyzed images situations, wind-induced level differences within the model are in the order of 10^{-4} to 10^{-3} m, while head losses for the low to average discharge conditions herein examined are in the order of 10^{-6} to 10^{-5} m, so that wind action always prevails: the two 2011 simulations with westerly winds show higher water levels downstream (e.g., Figure 3). Surface setup can then be used as a general indicator for the transport direction of the water mass and its associated phytoplankton content.

Figure 3. Water level variations for the CHRIS-PROBA 28 August 2011 simulation.

Wind stress according to the adopted IBL model grows in the windward direction, with sheltered areas adjacent to the windward shores, downwind of the lotus flowers island and next to leeward vegetated shores. Simulations of the 2007 and 2008 conditions with easterly winds show higher wind stresses in the upstream part of the lake (e.g., Figure 4), while the opposite occurs in the 2011 cases with westerly winds.

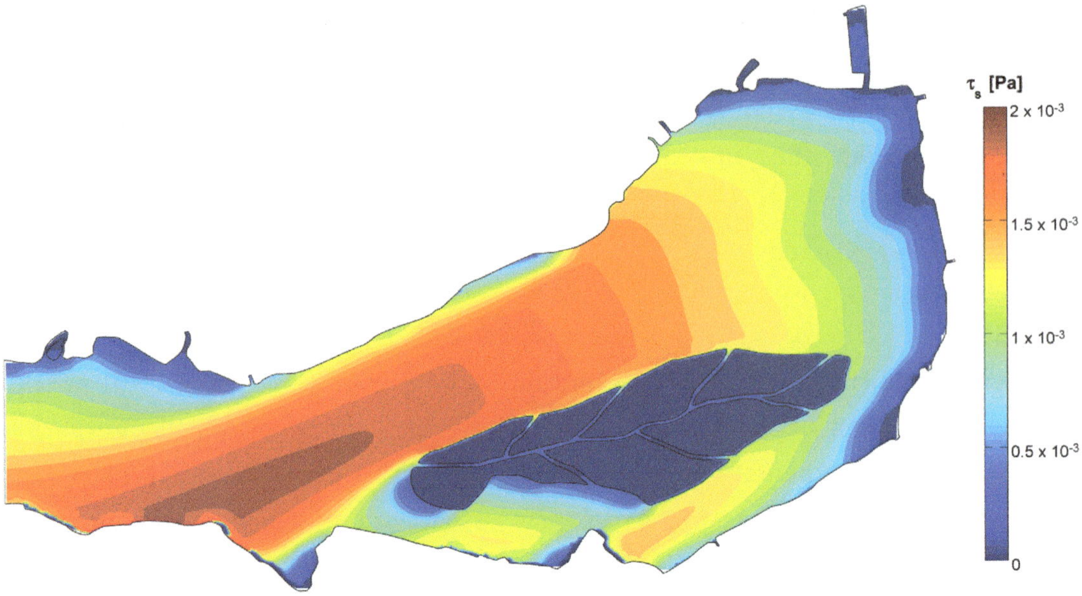

Figure 4. Wind stress field for the CHRIS-PROBA 29 June 2008 simulation.

In the MIVIS 2007 conditions with very low discharge, almost non-fluvial lake circulations are obtained (Figure 5a), as a lake-wide counter-clockwise gyre forms around the lotus flower island: an upstream backflow, parallel to the wind direction, is present in the wind-exposed channel north of the island, while a downstream flow forms in the sheltered southern channel. In the upstream part of the lake, a clockwise gyre develops, which limits the development of the main gyre. The exact layout of this upstream gyre may be possibly influenced by the fictitious upstream boundary of the numerical model, but its presence is indeed realistic, because of the differential wind stress between the sheltered northern and the wind-exposed southern shores. Inside the lotus flower island, water flows opposite to the wind, a character common to all simulations because of sheltering. A stagnation area also forms in the downstream part of the lake, north of the large gyre. Local water renewal times are higher inside recirculation areas, especially in the main one, but an almost uniform distribution is attained elsewhere because of low flow and wind-induced circulations (Figure 6a).

In the CHRIS-PROBA 2008 conditions, instead, wind and through-flowing discharge contrast one another in determining the circulations layout (Figure 5b). The riverine current, which is opposite to the wind direction, flows in wind-sheltered areas (Figure 4), first channeling along the northern shore in the upstream part of the lake, then shifting into the channel south of the island and finally flowing towards the Vasarone sluice gate along the downstream shore. Multiple recirculation structures form nearby the northern shore and downstream of the lotus flower island. The local simulated T_r distribution echoes the circulations layout: lower values are attained in the stronger flow areas, higher values in the stagnation regions (Figure 6b).

The 2011 simulations with westerly winds, concordant to the riverine flow direction, give similar results (Figure 5c,d): the flow fields have the same structure, although higher velocity magnitudes are attained in the APEX case because of the higher wind intensity. A river-like behavior is evident in the upstream part of the lake, with a flow parallel to the shores; circulations start to develop where the lake cross section widens. Downstream, the main flow concentrates in the most wind-exposed central region of the lake, north of the lotus flower island. Multiple large recirculation zones develop close to the northern shore, which is sheltered by wind and vegetation. A relevant stagnation area is present about halfway along that shore, caused by a deeper region in the lake bathymetry (Figure 1b). The flow in the channel south of the lotus flower island is much less intense than north of it, especially in its sheltered downstream half. Another stagnation area is present downstream of the island, due to wind sheltering and to wake effect. The local water renewal times are consistent with the described flow field: higher values are obtained inside gyres and in low velocity zones (Figures 6c,d).

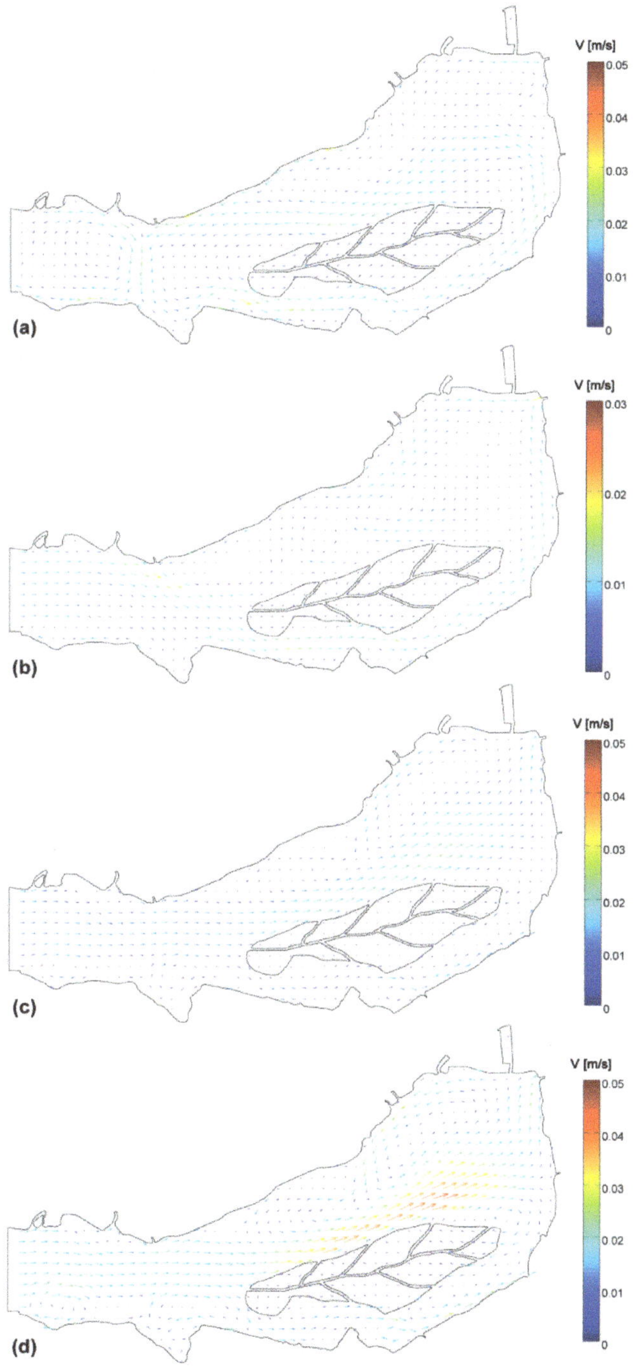

Figure 5. Surface 1 m depth-integrated flow fields for the MIVIS 26 July 2007 (**a**); CHRIS-PROBA 29 June 2008 (**b**); CHRIS-PROBA 28 August 2011 (**c**) and APEX 21 September 2011 (**d**) simulations.

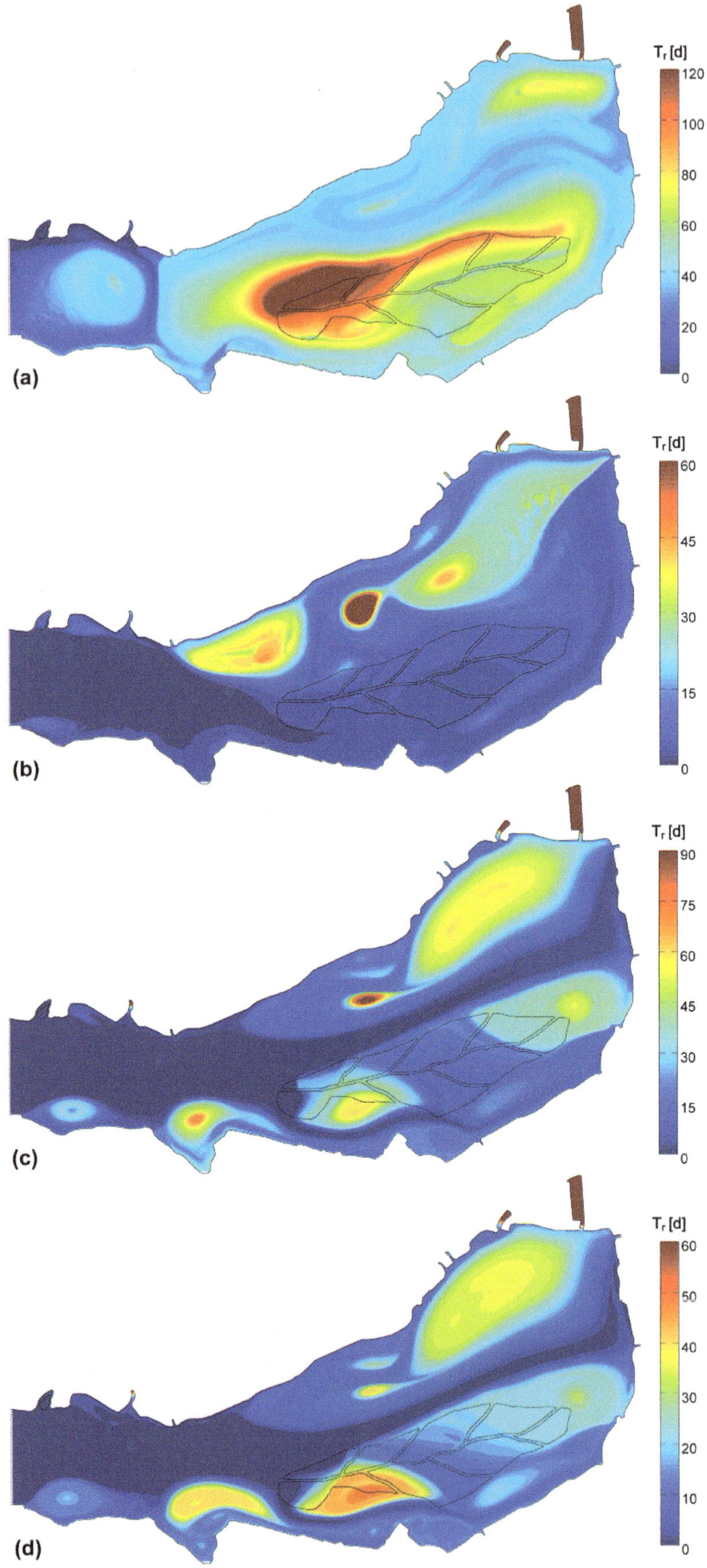

Figure 6. Surface 1 m local water renewal times for the MIVIS 26 July 2007 (**a**); CHRIS-PROBA 29 June 2008 (**b**); CHRIS-PROBA 28 August 2011 (**c**) and APEX 21 September 2011 (**d**) simulations.

3.4. Data and Simulation Analyses

Table 4 reports the μ and CV values of the simulated local T_r and of the remote-sensed Chl-*a* concentrations. The mean values of T_r can be compared to the theoretical water renewal times (also shown in Table 4), obtained as the ratio between the volume of the model (9.68 Mm^3) and the discharge. Low values of renewal time CV imply better mixing conditions; the same can be said for the CV of the Chl-*a* concentration, which also hints at phytoplankton distribution inhomogeneity between the upstream and the downstream part of the lake, owing to advection effects e.g., [10].

Table 4. Theoretical T_r estimates and μ and CV values of simulated T_r and remote-sensed Chl-*a* concentration for the simulated images conditions.

Parameter		MIVIS 26 July 2007	CHRIS-PROBA 29 June 2008	CHRIS-PROBA 28 August 2011	APEX 21 September 2011
Theoretical T_r [d]		43.6	11.2	11.6	8.8
Simulated T_r	μ [d]	45.9	10.0	20.0	16.1
	CV	59%	150%	129%	118%
Remote-sensed Chl-*a*	μ [mg/m^3]	107.8	45.9	32.3	15.1
	CV	9%	12%	30%	86%

A qualitative comparison of the simulated hydrodynamic fields to the remote-sensed Chl-*a* distributions shows that the general characters of the phytoplankton horizontal patchiness can be deduced from the numerical model results.

In the MIVIS 2007 case, the simulated local water renewal time distribution (Figure 6a) is approximately constant, matching with the generally uniform trophic state of the Superior Lake (Figure 2a, see also low CV in Table 4). The higher Chl-*a* concentration in the upstream part of the lake can be due to wind transport from downstream, in particular along the sheltered northern shore, where currents from upstream and downstream merge (Figure 5a), advecting phytoplankton from both directions. In this area, the small inflow from two drainage channels (not reproduced in the model) can contribute to the formation of local phytoplankton patches due to additional nutrients discharge [9,19]. Phytoplankton accumulation seems to follow the path of the significant backflow induced by the main gyre. The results in the channel south of the lotus flower island are most likely affected by the fact that during summer 2007 the lotus flower island was 35% larger than that reproduced by the model, so that a flow weaker than the simulated one may have occurred in that area (Chl-*a* data are not available in the image). In fact, a macrophytes control plan was started by local authorities in late 2007, aiming at the reduction of the macrophyte islands extension.

In the CHRIS-PROBA 2008 conditions, the Chl-*a* concentration field (Figure 2b) results from the contrasting actions of the riverine current, conveying phytoplankton downstream, and of the wind, pushing it upstream. In particular, the mean estimated T_r is even smaller than the theoretical one (Table 4), highlighting how easterly winds produce a better mixing, as opposed to the CHRIS-PROBA 2011 case, which is characterized by similar discharge and wind intensity, but by opposite wind direction (Tables 2 and 4). Apart from the maximum in the upstream part of the lake, most likely due to wind advection, the highest Chl-*a* concentrations were observed in correspondence of the deepest part of the lake, where the numerical simulation highlights a central recirculation structure, receiving water from both upstream and downstream (Figure 5b); in the same area, the highest local water renewal times are found (Figure 6b), hence supporting phytoplankton accumulation. The quite high concentration adjacent to the downwind shore may be due to the well-defined flow towards the Vasarone outlet (Figure 5b). The lowest Chl-*a* concentrations are instead reported, as shown by the simulation, in the sheltered regions where the riverine flow channels.

In the 2011 CHRIS-PROBA and APEX conditions, Chl-*a* concentrations (Figure 2c,d) were higher in the downstream part of the lake due to the wind action and to the discharge from the Mincio River, with lower values in the upstream riverine part, where the simulations show a channeled flow (Figure 5c,d). Phytoplankton was concentrated further downstream in the APEX image, because of the higher wind intensity (Table 2), resulting in the highest *CV* value for Chl-*a* concentration in Table 4. Phytoplankton packed where the strongest backflow develops in the simulation (*i.e.*, nearby the northern shore), as well as in the wind-sheltered downstream half of the channel south of the lotus island and downstream of it, where recirculation structures, weak currents and high local water renewal times (Figure 6c,d) are simulated. Table 4 also shows that the formation of recirculation structures in the downstream part of the lake, where the phytoplankton concentration is the highest, determines the mean simulated local renewal times at almost double the theoretical ones.

The consistency in the identified relations between the hydrodynamic variables and the remote-sensed Chl-*a* fields suggest that the developed model is indeed robust and validated. However, some limits of the numerical simulations, which are deemed responsible for the deviations between the remote-sensed chlorophyll-*a* and the hydrodynamic fields, must be acknowledged. First of all, phytoplankton patchiness is the product of transient past processes, of which the steady-state hydrodynamic conditions here assumed are an approximation e.g., [7]. Secondly, a high-resolution bathymetry of the Superior Lake of Mantua, scanned with echosounder techniques, is not available at the present moment, so that the geometry of the numerical model includes some imperfections. The fictitious upstream boundary of the model, which was chosen because of the lack of topographic data for the western part of the lake, also limits the correct development of circulations in the upstream reach of the model. The intrinsic approximations of the schemes used to model the lake physics and the uncertainties in the choice of some of their parameters must be taken into account as well. Moreover, phytoplankton patchiness is remotely conditioned by biochemical factors which, even if acting over long time and very small spatial scales, are not included in the model. Finally, defects in the Chl-*a* concentration detection algorithms should also be taken into consideration (see Table 1).

The performed analysis suggests that hydrodynamic phenomena determine phytoplankton horizontal patchiness in the Superior Lake of Mantua over three levels:

1. The generalized transport in the wind direction is responsible for major phytoplankton arrangement, as observed, for instance, by George *et al.* [18]. In the Superior Lake of Mantua, the main wind axis is aligned to the riverine flow, so that, according to the wind direction, the two forces may sum up or oppose one another, resulting either in packing on one of the ends of the lake or in a more uniform distribution.
2. The stagnation in correspondence of gyres traps water and its phytoplankton content, favoring species that prefer motionless waters, such as cyanobacteria [52], to proliferate under stable

hydrodynamic conditions. Opposed to this phenomenon are the lower concentrations observed where strong currents develop, because of turbulence disturbance and faster water renewal.

3. The small-scale transport by local currents is responsible for the formation of phytoplankton assemblies on top or outside of the generalized wind advection.

Such hypotheses, deduced from qualitative evaluations, are confirmed by the analysis of the Pearson correlation coefficients between the remote-sensed Chl-*a* distributions and the simulated hydrodynamic fields, shown in Table 5.

Table 5. Correlation coefficients between the remote sensed Chl-*a* distributions and the simulated hydrodynamic variables for the simulated images conditions (correlations with $p > 0.001$ are typed in italics).

Variable	MIVIS 26 July 2007	CHRIS-PROBA 29 June 2008	CHRIS-PROBA 28 August 2011	APEX 21 September 2011
Δy	0.071	−0.368	0.455	0.792
τ_s	−0.055	−0.087	−0.472	−0.193
T_r	−0.175	0.276	0.215	0.276
V	0.170	−0.159	−0.183	*0.008*

The calculated values of the correlation coefficients are generally low, due to the aforementioned limits of the numerical model and the non-linearity of the relations between hydrodynamics and phytoplankton patchiness occurring in the real world. However, their sign and the trend of the values over the four cases are meaningful to understand the underlying mechanisms relating the two processes. Elaborate statistical models e.g., [8,15] would better describe the complex, composite and, to a large extent, site-specific relations of the Chl-*a* concentrations to the hydrodynamic variables. Yet, such an analysis would overcome the purpose of the present research article, which is to demonstrate the feasibility of the use of a hydrodynamic model to forecast the extension and location of the areas of most probable phytoplankton accumulation and the consequent environmental risks in shallow fluvial lakes.

Given the mentioned remarks, quite high positive values are found for the correlation between Chl-*a* concentration and free surface level variations for the 2011 simulations, in which the downstream water level is higher (e.g., Figure 3) and wind and riverine current advections support one another, proving the major influence of these phenomena. The largest correlation is found for the APEX 2011 case, which is characterized by higher wind and discharge. A weaker positive correlation is obtained for the MIVIS 2007 simulation, as the very low discharge causes more uniform mixing conditions to occur inside the lake because of wind-induced currents (Figure 6a and Table 4). A negative correlation is instead found for the CHRIS-PROBA 2008 simulation: this can be due to the fact that the section in which riverine and wind-related currents clash, advecting phytoplankton from both ends of the lake, is placed in the downstream part of the lake.

Generally weaker negative correlations are found in all cases for the wind stress, showing that wind-sheltered areas (*i.e.*, where wind stress is minimum) are more subject to phytoplankton packing. It must be anyway underlined that in a fluvial lake, such as the one studied here, the riverine through-flowing current tends to channel in wind-sheltered regions, provoking phytoplankton flushing, which may explain the less robust link.

Positive correlations are found in three out of four cases for the simulated local water renewal time, showing the definite effect of water stagnation on phytoplankton packing and proliferation. Small negative correlations are finally found in two cases for velocity magnitude, as phytoplankton tends to accumulate in weak flow areas. In the APEX 2011 simulation, a non-significant correlation is found for this variable. Due to the complexity of the drought condition, it is not immediately evident to explain the opposite correlation of the Chl-*a* concentration to water velocity and renewal time found for the MIVIS 2007 situation compared to the others. It seems that water stagnation caused phytoplankton decay and that where minimum water flow persisted algae were favored. In this case, the low discharges and higher water renewal times, causing a longer phytoplankton residence inside the lake, may allegedly result in biochemical factors such as photo-inhibition, CO_2 or nutrients limitation and prey-predator dynamics to play a relevant role in determining phytoplankton distribution, reducing the physical factors dominance. Under such conditions, therefore, the use of a coupled ecologic-hydrodynamic model may be justifiable.

4. Conclusions

Shallow fluvial lakes are rapidly evolving and menaced ecosystems due to eutrophic conditions coupled to reduced water flows, rapid infilling and occurrence of algal blooms, including potentially harmful cyanobacteria. They are extremely patchy, with a mosaic of very different hydrodynamic, physico-chemical and biological conditions, and for these reasons difficult to analyze comprehensively. This study demonstrates the capabilities of hyperspectral remote-sensing products to validate numerical hydrodynamic lake models, in addition to detect most probable algal blooms location and extension. Remote sensing techniques, despite the still existing spatial resolution limits, may contribute to the monitoring of fluvial lakes, offsetting the limits of occasional, conventional samplings performed on a single location. On the other hand, the costs of airborne campaigns (e.g., MIVIS, APEX) or calibration issues may prevent the mapping of chlorophyll-*a* in inland waters. Numerical models of lake hydrodynamics, simulating the relevant physical phenomena responsible for the circulations layout, can be used to effectively estimate the environmental risk connected to local stagnation and phytoplankton accumulation (e.g., anoxia events) under given hydrological and meteorological conditions. Relations of the hydrodynamic variables to the Chl-*a* distribution can be meaningfully assessed by comparing the simulations results to remote-sensed images, providing also an indirect validation of the numerical model, in which the implementation of a variable wind stress model (IBL) has been proved essential to obtain realistic flow fields.

The monitoring and protection of inland water quality may be therefore guaranteed by the integration of ecological, hydraulic, and remote-sensing techniques, with important improvements in management actions by local authorities and policy makers.

The simulation of several flow conditions in the Superior Lake of Mantua, a peculiar fluvial lake case, shows that the main hydrodynamic effects which influence Chl-*a* distribution are related to the combined effect of advection due to wind force and riverine current, as well as to the presence of large gyres which induce recirculation and stagnation regions, favoring phytoplankton accumulation.

Acknowledgments

The research leading to these results received funding from the European Community's Seventh Framework Programme [FP7/2007–2013] under grant agreement No. 606865, INFORM project. Thanks to European Facility for Airborne Research (EUFAR) project and VITO for APEX data. MIVIS data were acquired by CGR-CISIG Parma (Italy). CHRIS data were made available through the ESA projects MELINOS (AO-553). We acknowledge ARPA Lombardy and AIPo Mantua office for providing historical wind, biological, physico-chemical data and water discharge data, respectively. We also thank MeteoMincio and the CML for providing the other wind data. The contributions of three anonymous reviewers to the improvement of the original manuscript are likewise acknowledged.

Author Contributions

Monica Pinardi performed the biochemical evaluations, wrote the relative part of the paper and joined the different contributions, writing also the introduction and the conclusions. Andrea Fenocchi developed the hydrodynamic model, conceived the simulations, wrote the connected portion of the manuscript and prepared the figures. Claudia Giardino contributed to the remote-sensing analysis and to the writing of the relative section of the paper. Stefano Sibilla supervised the hydrodynamic modelling as advisor for A. Fenocchi's Ph.D. and revised the paper. Marco Bartoli participated to the ecological analysis and revised the paper. Mariano Bresciani produced the Chl-*a* distribution maps and wrote the relative part of the paper.

Conflicts of Interest

The authors declare no conflicts of interest.

References

1. Teodoru, C.; Wehrli, B. Retention of sediments and nutrients in the Iron Gate I Reservoir on the Danube River. *Biogeochemistry* **2005**, *76*, 539–565.
2. Humborg, C.; Ittekkot, V.; Cociasu, A.; Bodungen, B.V. Effect of Danube River dam on Black Sea biogeochemistry and ecosystem structure. *Nature* **1997**, *386*, 385–388.
3. McCartney, M. Living with dams: Managing the environmental impacts. *Water Policy* **2009**, *11*, 121–139.
4. Bresciani, M.; Rossini, M.; Morabito, G.; Matta, E.; Pinardi, M.; Cogliati, S.; Julitta, T.; Colombo, R.; Braga, F.; Giardino, C. Analysis of within- and between-day chlorophyll-a dynamics in Mantua Superior Lake, with a continuous spectroradiometric measurement. *Mar. Freshwater Res.* **2013**, *64*, 303–316.
5. Harris, G.P. Phytoplankton productivity and growth measurements: Past, present and future. *J. Plankton Res.* **1984**, *6*, 219–237.
6. Reynolds, C.S. Scales of disturbance and their role in plankton ecology. *Hydrobiologia* **1993**, *249*, 157–171.
7. Verhagen, J.H.G. Modeling phytoplankton patchiness under the influence of wind-driven currents in lakes. *Limnol. Oceanogr.* **1994**, *39*, 1551–1565.

8. De Souza Cardoso, L.; da Motta Marques, D. Hydrodynamics-driven plankton community in a shallow fluvial lake. *Aquat. Ecol.* **2009**, *43*, 73–84.
9. George, D.G.; Heaney, S.I. Factors Influencing the Spatial Distribution of Phytoplankton in a Small Productive Lake. *J. Ecol.* **1978**, *66*, 133–155.
10. Jones, R.I.; Fulcher, A.S.; Jayakody, J.K.U.; Laybourn-Parry, J.; Shine, A.J.; Walton, M.C.; Young, J.M. The horizontal distribution of plankton in a deep, oligotrophic lake—Loch Ness, Scotland. *Freshwater Biol.* **1995**, *33*, 161–170.
11. Ciraolo, G.; Lipari, G.; Napoli, E.; Józsa, J.; Krámer, T. Three-dimensional numerical analysis of turbulent wind-induced flows in the lake Balaton (Hungary). In *Shallow Flows—Research Presented at the International Symposium on Shallow Flows*, Delft, Netherlands, 16–18 June 2003; Uijttewaal, W.S.J.; Jirka, G.H., Eds.; CRC Press/Balkema: Leiden, The Netherlands, 2004.
12. Wiens, J.A. Spatial Scaling in Ecology. *Funct. Ecol.* **1989**, *3*, 385–397.
13. Pinel-Alloul, B. Spatial heterogeneity as a multiscale characteristics of zooplankton community. *Hydrobiologia* **1995**, *104*, 17–42.
14. Legendre, L.; Demers, S. Towards Dynamic Biological Oceanography and Limnology. *Can. J. Fish. Aquat. Sci.* **1984**, *41*, 2–19.
15. Thackeray, S.J.; George, D.G.; Jones, R.I.; Winfield, I.J. Quantitative analysis of the importance of wind-induced circulation for the spatial structuring of planktonic populations. *Freshwater Biol.* **2004**, *49*, 1091–1102.
16. George, D.G. Wind-induced water movements in the South Basin of Windermere. *Freshwater Biol.* **1981**, *11*, 37–60.
17. Livingstone, D.A. On the orientation of lake basins. *Am. J. Sci.* **1954**, *252*, 547–554.
18. George, D.G.; Edwards, R.W. The Effect of Wind on the Distribution of Chlorophyll *A* and Crustacean Plankton in a Shallow Eutrophic Reservoir. *J. Appl. Ecol.* **1976**, *13*, 667–690.
19. George, D.G.; Winfield, I.J. Factors influencing the spatial distribution of zooplankton and fish in Loch Ness, UK. *Freshwater Biol.* **2000**, *43*, doi:10.1046/j.1365-2427.2000.00539.x.
20. Pinardi, M.; Bartoli, M.; Longhi, D.; Viaroli, P. Net autotrophy in a fluvial lake: the relative role of phytoplankton and floating-leaved macrophytes. *Aquat. Sci.* **2011**, *73*, 389–403.
21. Bolpagni, R.; Bresciani, M.; Laini, A.; Pinardi, M.; Matta, E.; Ampe, E.M.; Giardino, C.; Viaroli, P.; Bartoli, M. Remote sensing of phytoplankton-macrophyte coexistence in shallow hypereutrophic fluvial lakes. *Hydrobiologia* **2014**, *737*, 67–76.
22. Søballe, D.M.; Bachmann, R.W. Influence of reservoir transit on riverine algal transport and abundance. *Can. J. Fish. Aquat. Sci.* **1984**, *41*, 1803–1813.
23. Walz, N.; Welker, M. Plankton development in a rapidly flushed lake in the River Spree system (Neuendorfer See, Northeast Germany). *J. Plankton Res.* **1998**, *20*, 2071–2087.
24. Welker, M.; Walz, N. Plankton dynamics in a river-lake system—On continuity and discontinuity. *Hydrobiologia* **1999**, *408*, 233–239.
25. Rennella, A.M.; Quiros, R. The effects of hydrology on plankton biomass in shallow lakes of the Pampa Plain. *Hydrobiologia* **2006**, *556*, 181–191.
26. Gower, J.F.R.; Denman, K.L.; Holyer, R.J. Phytoplankton patchiness indicates the fluctuation spectrum of mesoscale oceanic structure. *Nature* **1980**, *288*, 157–159.

27. Blauw, A.N.; Los, H.F.J.; Bokhorst, M.; Erftemeijer, P.L.A. GEM: A generic ecological model for estuaries and coastal waters. *Hydrobiologia* **2009**, *618*, 175–198.
28. Salacinska, K.; El Serafy, G.Y.; Los, H.F.J.; Blauw, A. Sensitivity analysis of the two dimensional application of the Generic Ecological Model (GEM) to algal bloom prediction in the North Sea. *Ecol. Model.* **2010**, *221*, 178–190.
29. Mooij, W.M.; Trolle, D.; Jeppesen, E.; Arhonditsis, G.; Belolipetsky, P.V.; Chitamwebwa, D.B.R.; Degermendzhy, A.G.; DeAngelis, D.L.; de Senerpont Domis, L.N.; Downing, A.S.; *et al.* Challenges and opportunities for integrating lake ecosystem modelling approaches. *Aquat. Ecol.* **2010**, *44*, 633–667.
30. Los, H.F.J.; Villars, M.T.; Van der Tol, M.W.M. A 3-dimensional primary production model (BLOOM/GEM) and its applications to the (southern) North Sea (coupled physical-chemical-ecological model). *J. Mar. Syst.* **2008**, *74*, 259–294.
31. Bresciani, M.; Giardino, C.; Longhi, D.; Pinardi, M.; Bartoli, M.; Vascellari, M. Imaging spectrometry of productive inland waters. Application to the lakes of Mantua. *Ital. J. Remote Sens.* **2009**, *41*, 147–156.
32. Bresciani, M.; Giardino, C.; Bartoli, M.; Longhi, D.; Pinardi, M. Assessment of chlorophyll-*a* and algal blooms in inland waters from hyperspectral data. In Proceedings of the Hyperspectral 2010 Workshop, ESA SP-683; Frascati, Italy, 17–19 March 2010.
33. Cutter, M.A. A low cost hyperspectral mission. *Acta Astron.* **2004**, *55*, 631–636
34. Gitelson, A.A.; Schalles, J.F.; Hladik, C.M. Remote chlorophyll-*a* retrieval in turbid, productive estuaries: Chesapeake Bay case study. *Remote Sens. Environ.* **2007**, *109*, 464–472.
35. Levasseur, M.; Therriault, J.-C.; Legendre, L. Tidal currents, winds and the morphology of phytoplankton spatial structures. *J. Mar. Res.* **1983**, *41*, 655–672.
36. Józsa, J. On the internal boundary layer related wind stress curl and its role in generating shallow lake circulations. *J. Hydrol. Hydromech.* **2014**, *62*, 16–23.
37. Ottesen Hansen, N.-E. Effects of Boundary layers on mixing in small Lakes. *Dev. Water Sci.* **1979**, *11*, 341–356.
38. Krámer, T. Solution-Adaptive 2D Modelling of Wind-Induced Lake Circulation. Ph.D. Thesis, Budapest University of Technology and Economics, Budapest, Hungary, 2006.
39. Coastal Engineering Research Center (CERC). *Shore Protection Manual*, 4th ed.; Waterways Experiment Station: Vicksburg, MS, USA, 1984.
40. Józsa, J. *Shallow Lake Hydrodynamics—Theory, Measurement and Numerical Model Applications*, Mundus-Euroaquae lecture notes; Budapest University of Technology and Economics: Budapest, Hungary, 2006.
41. CD-adapco. *STAR-CCM+ 9.02 User Guide*; online manual; CD-adapco: Melville, NY, USA, 2014.
42. Shih, T.H.; Liou, W.W.; Shabbir, A.; Yang, Z.; Zhu, J. *A New* k-ε *Eddy Viscosity Model for High Reynolds Number: Model Development and Validation*; NASA Technical Memorandum 106721, NASA Lewis Research Center: Cleveland, OH, USA, 1994.
43. Teeter, A.M.; Johnson, B.H.; Berger, C.; Stelling, G.; Scheffner, N.W.; Garcia, M.H.; Parchure, T.M. Hydrodynamic and sediment transport modeling with emphasis on shallow-water, vegetated areas (lakes, reservoirs, estuaries and lagoons). *Hydrobiologia* **2001**, *444*, 1–23.

44. Zinke, P. Application of a porous media approach for vegetation flow resistance. In *River Flow 2012*, San José, Costa Rica, 5–7 September 2012; Murillo Muñoz, R., Ed.; CRC Press/Balkema: Leiden, The Netherlands, 2012.
45. Grant, W.D.; Madsen, O.S. Combined wave and current interaction with a rough bottom. *J. Geophys. Res.* **1979**, *84*, 1797–1808.
46. Dickman, M. Some effects of lake renewal on phytoplankton productivity and species composition. *Limnol. Oceanogr* **1969**, *14*, 660–666.
47. Krivtsov, V.; Sigee, C. Importance of biological and abiotic factors for geochemical cycling in a freshwater eutrophic lake. *Biogeochemistry* **2005**, *74*, 205–230.
48. Yamamuro, M.; Hiratsuka, J.; Ishitobi, Y.; Hosokawa, S.; Nakamura, Y. Ecosystem shift resulting from loss of eelgrass and other submerged aquatic vegetation in two estuarine lagoons, Lake Nakaumi and Lake Shinji, Japan. *J. Oceanogr.* **2006**, *62*, 551–558.
49. Ekholm, P.; Malve, O.; Kirkkala, T. Internal and external loading as regulators of nutrient concentrations in the agriculturally loaded Lake Pyhajarvi (southwest Finland). *Hydrobiologia* **1997**, *345*, 3–14.
50. Burger, D.F.; Hamilton, D.P.; Pilditch, C.A.; Gibbs, M.M. Benthic nutrient fluxes in a eutrophic, polymictic lake. *Hydrobiologia* **2007**, *584*, 13–25.
51. Luettich, R.A.; Harleman, D.R.F.; Somlyody, L. Dynamic Behavior of Suspended Sediment Concentrations in a Shallow Lake Perturbed by Episodic Wind Events. *Limnol. Oceanogr.* **1990**, *35*, 1050–1067.
52. Becker, V.; de Souza Cardoso, L.; da Motta Marques, D. Development of *Anabaena* Bory ex Bornet & Flahault (Cyanobacteria) blooms in a shallow, subtropical lake in southern Brazil. *Acta Limnol. Bras.* **2004**, *16*, 306–317.

Drought Mitigation Ability Index and Application Based on Balance between Water Supply and Demand

Shaohua Liu [1], Denghua Yan [1,*], Jianhua Wang [1], Baisha Weng [1], Gang Wang [2] and Meijian Yang [1]

[1] State Key Laboratory of Simulation and Regulation of Water Cycle in River Basin, China Institute of Water Resources and Hydropower Research (IWHR), Beijing 100038, China; E-Mails: shliuiwhr@hotmail.com (S.L.); wjh@iwhr.com (J.W.); baishaweng@126.com (B.W.); oymj2082@163.com (M.Y.)

[2] Beijing Municipal Research Institute of Environmental Protection, Beijing 100037, China; E-Mail: gangwnan@163.com

* Author to whom correspondence should be addressed; E-Mail: yandh@iwhr.com;

Academic Editor: Athanasios Loukas

Abstract: Drought impacts not only nature, but also the socioeconomic system and results from the imbalance of water supply and demand. It is necessary to estimate the ability of drought mitigation on the basis of water allocation. In this research, the drought mitigation ability index (DMAI) was constructed by the ratio of theoretical water supply and demand obtained from the optimal water allocation. Then, the DMAI was applied to Daqinghe watershed for temporal-spatial validation compared with the agricultural affected areas and the distribution of groundwater overexploitation. In addition, the historical (1961–2010) and future (2020) DMAI in Daqinghe watershed were analyzed. The results show that DMAI can overall represent the temporal-spatial characteristics of the drought mitigation ability. Furthermore, the DMAI of Daqinghe watershed is significantly decreased in the last 50 years, and groundwater resources play a prominent role in drought mitigation. Although the water from the South-to-North Water Transfer Project (SNWTP) can improve DMAI of Daqinghe watershed, the drought risk continues to be serious in the future.

Keywords: drought mitigation ability index (DMAI); theoretical water supply and demand; optimal water allocation; temporal-spatial validation; Daqinghe watershed

1. Introduction

Droughts are recurrent and world-wide natural hazards, impacting the development of ecosystems, the environment and socio-economic systems [1]. Drought occurrences have been more frequent, and their effects have been aggravated in recent years [2–4]. On the one hand, climate changes have already altered, and will continue to alter, the hydrological cycle, and the hydrological uncertainty has increased significantly [5]. On the other hand, growing urban population, increasing agriculture production and socioeconomic development have contributed to the rise of water demand, especially in China [6–8]. As a result, droughts have affected socioeconomic development and become one of the threats against sustainable development in many countries [9–12].

As a complex and dangerous phenomenon, droughts have attracted the attention of various researchers over the last few decades [7]. A lot of research has been done on drought monitoring [13,14] and drought assessment [15,16]. Among this research, various drought indices have been developed to quantify the drought, including multiple drought indices, like the standardized precipitation index (SPI) [17,18], the Palmer drought severity index (PDSI) [19], the surface water supply index (SWSI) [20] and the soil moisture drought index (SMDI) [21]. Meanwhile, due to the complex impacts of drought [22] and regional diversity in the water supply system, there is no index on drought mitigation, which is required and more important to cope with drought disasters. Drought mitigation measures have been classified into three categories: supply-oriented measures, demand-oriented measures and impact-minimization measures [23]. Hence, the drought mitigation ability can be improved by increasing water supply, reducing water demand and optimal water allocation based on the water balance of the supply and demand system. Some studies have demonstrated the effectiveness of the optimal water allocation for drought mitigation [12,24,25]. However, the drought mitigation ability is significantly different among various regions or basins and depends on the local water resource condition, water supply system and socioeconomic status [26–28]. Consequently, drought mitigation is a local and watershed issue, and the drought mitigation ability index (DMAI) should be constructed based on the local condition of water demand and supply [29,30], which is necessary for and meaningful to drought risk management.

In this paper, the Daqinghe watershed, a typical semi-arid basin under great drought stress, is selected as the study area (Section 2.1). Additionally, the materials and the calculation of the DMAI, constructed based on the ratio of theoretical water demand and water supply, are demonstrated in Sections 2.2 and 2.3, respectively. Then, Section 3 is a case study including the application and validation of the DMAI at temporal-spatial scales in the Daqinghe watershed. Finally, conclusions are given in Section 4.

2. Materials and Methods

2.1. Study Area

The Daqinghe watershed is located in the middle of the Haihe River Basin with the most serious water stress in China. The Daqinghe River originates in Taihang Mountain, and the watershed with a sector-shape is surrounded by Taihang Mountain in the west, Bohai Sea in the east, Yongdinghe River in the north and Ziyahe River in the south. The river basin area is over 40,000 km^2, of which mountain and hillside area (MHA) account for 43% and plain areas for 57%, and is divided into two partitions by Baiyangdian Lake (the eastern and western Baiyangdian plain (EBP and WBP, respectively)). Based on the joint distribution of the natural river system and the administrative division (Baoding city, Shijiazhuang city, Cangzhou city, Langfang city, Hengshui city, Zhangjiakou city, Tianjin city, Beijing city and Shanxi province), the Daqinghe watershed was divided into 13 basic units (Figure 1). According to the requirement of water allocation, the water supply network was constructed (Figure 2) by abstracting and conceptualizing the hydraulic relation between basic units, river channel system and reservoirs.

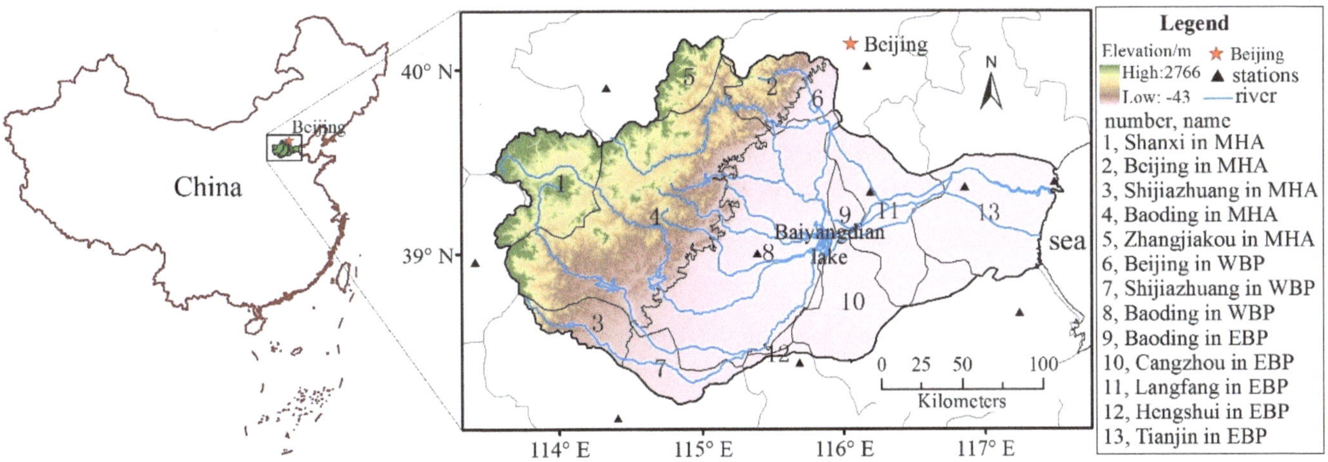

Figure 1. Location of the basic units in the Daqinghe watershed. MHA, mountain and hillside area; WBP, western Baiyangdian plain; EBP, eastern Baiyangdian plain.

2.2. Materials

Two datasets of water resource quantity and water demand on a monthly scale during 1961–2010 in the Daqinghe watershed were used in the research. According to the principles of equality and sustainability, the water resources were optimally allocated with the goal of minimizing the total water shortage, and the monthly balance between water supply and demand were calculated by the general algebraic modeling system (GAMS), a mathematical programming and optimization model system. The dataset of water resource quantity was acquired from the National Comprehensive Planning of Water Resources (1961–2000) and the Water Resources Bulletin in Haihe River Basin (2000–2010). The domestic, industrial and ecological water demand data in the dataset of water demand were acquired from the Water Resources Bulletin of the base year (2010); the agricultural water demand was calculated by deviation between theoretical agricultural water demand and effective precipitation during the crop growth period. Additionally, daily meteorological data, such as precipitation, evaporation, humidity, wind speed, sunshine hours and vapor pressure, of 10 meteorological stations (Figure 1 and Table 1) within and around the watershed were gained from China Meteorological Data Sharing Service System during 1961–2010 [31].

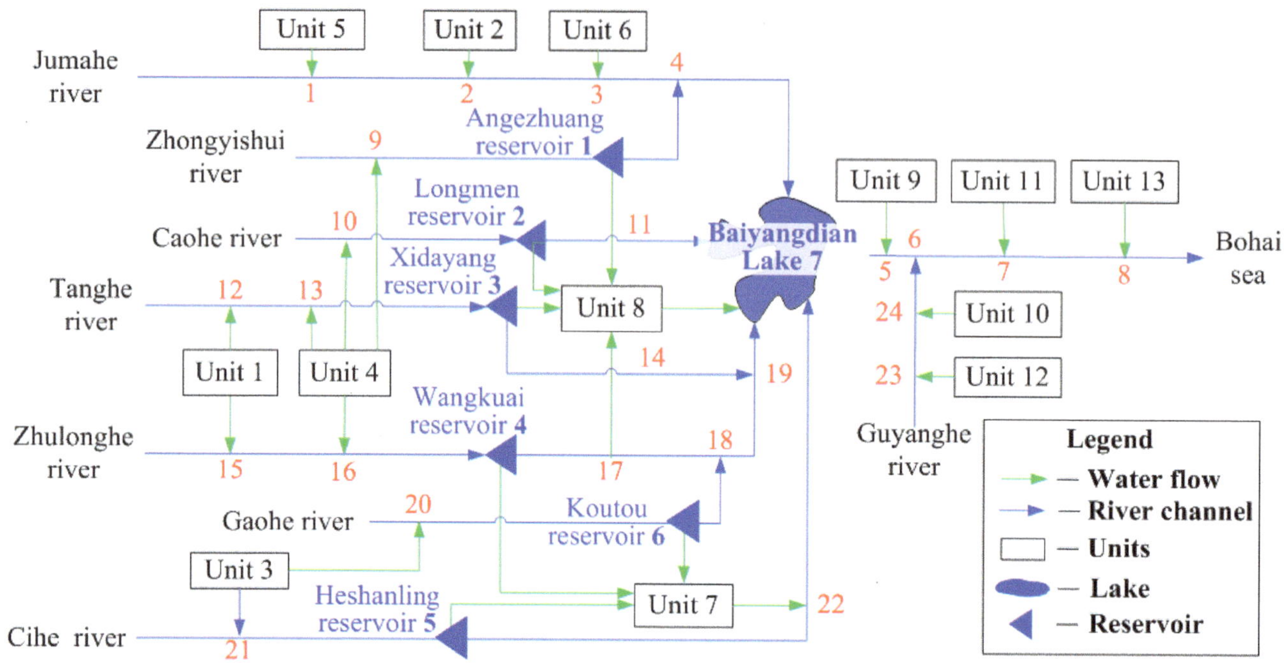

Figure 2. Water supply network in the Daqinghe watershed. The red and blue digits are the number of river channels and reservoirs, respectively. With the impact of water flows, the river channel has been divided into sections, and these can be used as catchments in the water allocation model. Furthermore, all of the units have local ground water supply in the Daqinghe watershed, which means the number of ground water supply can be equal to the number of units. Thus, the number of ground water supply is not depicted to simplify the network.

Table 1. Information of the meteorological stations within and around the watershed.

Stations	Name	Number	Longitude/°	Latitude/°	Elevation/m	Period
1	Beijing	54511	116.5	39.8	31.3	1951–2010
2	Wutaishan	53588	113.5	39.0	2208.3	1951–2010
3	Shijiazhuang	53698	114.4	38.0	81.0	1955–2010
4	Yuxian	53593	114.6	39.8	909.5	1954–2010
5	Bazhou	54518	116.4	39.1	9.0	1957–2010
6	Tianjin	54527	117.1	39.1	2.5	1954–2010
7	Baoding	54602	115.5	38.9	17.2	1955–2010
8	Raoyang	54606	115.7	38.2	19.0	1957–2010
9	Tanggu	54623	117.7	39.1	4.8	1954–2010
10	Huanghua	54624	117.4	38.4	6.6	1960–2010

2.3. Methods

Drought mitigation ability means to prohibit drought occurrence or to reduce drought impacts through reducing water demand, increasing water resources or optimal water allocation in regions or watersheds. However, total water demand consists of domestic demand, productive demand (industry and agriculture) and ecological demand; it is always difficult and infeasible to reduce water demand. On the other hand, the water resources mainly depend on precipitation, which is uncertain and uncontrollable. Therefore, the optimal water allocation is usually the effective method for drought mitigation. Consequently, the DMAI, as an integrated index to estimate the regional drought mitigation ability, is defined by the theoretical ratio of water supply and demand obtained from the optimal water allocation, and the formula of DMAI can be constructed as follows:

$$DMAI_t = \frac{W_{supply,t}}{W_{demand,t}} \tag{1}$$

where $DMAI_t$ is the drought mitigation ability in the period t; $W_{supply,t}$ is the theoretical water supply obtained from optimal water allocation in the period t; $W_{demand,t}$ is the total theoretical water demand in the period t, including domestic water demand, productive water demand and ecological water demand, and it is more than the actual water demand. According to the formula, the $DMAI_t$, ranging from 0 to 1, means maximizing the water supply by optimal water allocation, which is closely related to water supply engineering, the water resource situation and integrated water resource management. Accordingly, the regional DMAI is not a constant and varies under the different water resource situations (flood, normal, or drought year) and the water supply system.

2.3.1. Calculation of Water Demand

Regional water demand consists of four main types: domestic water demand, industry water demand, theoretical agricultural water demand and ecological water demand. In the base year of 2010, the water shortages were mainly caused by agricultural water demand, which accounts for about 80% of the total water demand. Consequently, the domestic water demand, industry water demand and ecological water demand only account for 20% and were acquired from the base year in this study. Accordingly, the agricultural water demand changes significantly under different water resource situations (flood, normal, or drought year), and the inner-annual water supply-demand processes are highly influenced by runoff processes. Therefore, the agricultural water demand was calculated as the deviation between the theoretical agricultural water demand and effective precipitation during the crop growth period, as:

$$I_{s,m} = \sum_{n=1}^{N} ET_{c,n} - \sum_{n=1}^{N} P_{e,n} \tag{2}$$

where N is the number of crop growth days in the m-th month; $I_{s,m}$ is the total supplementary water in irrigation in the m-th month, mm/month; $ET_{c,n}$ is the theoretical crop water demand in the n-th day, mm/day. The method recommended by the Soil Conservation Board of the United States Department of Agriculture is adopted in this article to calculate the effective precipitation, as Formula (3) [32,33], where P is the measured daily precipitation, mm/d.

$$\begin{cases} P_e = P(4.17-0.2P)/4.17 & P<8.3 \\ P_e = 4.17+0.1P & P\geq 8.3 \end{cases} \tag{3}$$

Additionally, crop water demand is calculated by the FAO Penman–Monteith equation, as follows:

$$ET_c = K_c \cdot ET_0 \tag{4}$$

$$ET_0 = \frac{0.408\Delta(R_n - \mathrm{G}) + \gamma\frac{900}{T+273}U_2(\mathrm{e_a}-\mathrm{e_d})}{\Delta+\gamma(1+0.34U_2)} \tag{5}$$

where ET_c is the theoretical water demand of crop, mm/d; K_c is the crop coefficient; ET_0 is the reference vegetation evapotranspiration, mm/d. Assume the vegetation height as 0.12 m, the canopy resistance as 70 s/m and the ground surface reflectivity as 0.23 when calculating ET_0; R_n is the net radiation at the crop surface, MJ/m^2·d; G is the soil heat flux density, MJ/m^2·d; T is the mean daily air temperature at 2 m height, °C; U_2 is wind speed at 2 m height, m/s; $\mathrm{e_d}$ is the saturation vapor pressure, kPa; $\mathrm{e_a}$ is the actual measured vapor pressure, kPa; Δ is the slope vapor pressure curve, kPa/°C; γ is the psychrometric constant.

Finally, the regional theoretical agricultural water demand (TAWD) can be acquired from the multiplication of regional crop water demand and farmland area, as:

$$I_{w,m} = I_{s,m} \times A_{agr,m} \tag{6}$$

where $I_{w,m}$ is the TAWD in the m-th month, 1000 m^3/month; $I_{s,m}$ is the total supplementary water in irrigation in the m-th month, mm/month; $A_{agr,m}$ is the regional effective irrigation area, km^2. According to the calculation method of TAWD, it can be seen that the TAWD is absolutely more than actual agricultural water demand, since the ET_c is relatively more than actual vegetation evapotranspiration. Nevertheless, the ET_c can also manifest the potential water demand of crop under the ideal soil moisture condition. In addition, the TAWD can represent the potential agricultural water demand with the adequate water supply scenario. Then, with the above material input and calculations of water demand, the monthly water demand within the Daqinghe watershed from 1961 to 2010 was obtained. Moreover, the average annual ecological, domestic, industry and agricultural water demands during 1961–2010 are shown in Table 2.

Table 2. Average annual ecological, domestic, industry and agricultural water demand within the Daqinghe watershed. (Unit: 10,000 m^3).

Units	Ecological Water Demand	Domestic Water Demand	Industry Water Demand	Agricultural Water Demand
Unit 1	34.12	622.72	358.28	4197.40
Unit 2	29.76	543.05	312.44	9503.21
Unit 3	238.28	4348.63	2501.95	32,117.37
Unit 4	27.19	496.28	285.53	1596.28
Unit 5	402.14	2380.93	1506.43	10,750.44

Table 2. *Cont.*

Units	Ecological Water Demand	Domestic Water Demand	Industry Water Demand	Agricultural Water Demand
Unit 6	778.66	4610.16	2916.88	22,614.60
Unit 7	5119.20	30,308.91	19,176.68	160,680.06
Unit 8	70.65	1289.32	741.80	8120.21
Unit 9	775.44	3673.83	3247.97	15,162.34
Unit 10	3259.42	15,442.15	13,652.15	64,608.18
Unit 11	3422.81	16,216.27	14,336.53	67,564.54
Unit 12	181.60	860.37	760.64	3369.58
Unit 13	4560.73	21,607.38	19,102.72	57,188.71

2.3.2. Calculation of Water Supply

Regional theoretical water supply is the water supply ability to satisfy the water demand of every unit through optimal water allocation model. Based on the water supply network, the linear programming model of GAMS was applied to obtain the theoretical water supply in the Daqinghe watershed. The constraints consist of the water balance of units (Equation (7)), water balance of reservoirs (Equations (8) and (9)) and water balance of river channels (Equations (10) and (11)), and the objective function is Equation (12).

$$Demand_{i,t} - \alpha_i Supply_{i,t}^{g} - \beta_i Supply_{i,t}^{s} - \eta_i Supply_{i,t}^{R_k} - \lambda_i Supply_{i,t}^{r_j} = Shortage_{i,t} \tag{7}$$

$$S_{j,t-1} + Qin_{j,t} - \sum_{i=1}^{13} Supply_{i,t}^{r_j} - Qout_{j,t} - L_{j,t}^{S} = S_{j,t} \tag{8}$$

$$S_{j,min} \le S_{j,t} \le S_{j,max} \tag{9}$$

$$R_{k,t-1} + \sum_{i=1}^{13} surface_{i,t}^{k} - \sum_{i=1}^{13} Supply_{i,t}^{s} + \sum_{j=1}^{7} Rin_{k,t} - \sum_{i=1}^{13} Supply_{i,t}^{R_k} - \sum_{j=1}^{7} Rout_{k,t} - L_{k,t}^{R} = R_{k,t} \tag{10}$$

$$R_{k,min} \le R_{k,t} \le R_{k,max} \tag{11}$$

$$obj = \min\left(\sum_{t=1}^{n} \sum_{i=1}^{13} \omega_i Shortage_{i,t}\right) \tag{12}$$

where t is the monthly time step; n is the termination time; i is the number of units; j is the number of reservoirs (Baiyangdian Lake was considered as a reservoir); k is the number of river channels; $Demand_{i,t}$ is the total water demand (consisting of domestic water demand, agricultural water demand, industry water demand and ecological water demand); $Supply_{i,t}^{g}$ is the local ground water resources to supply the i-th unit; $Supply_{i,t}^{s}$ is local surface water resources derived from the local precipitation to supply the i-th unit; $Supply_{i,t}^{R_k}$ is the remaining water resources of the upstream river channel, derived from the precipitation of upstream units, to supply the i-th unit after the water supply of upstream units; $Supply_{i,t}^{r_j}$ is the r_i-th reservoir water supply to the i-th unit; $Shortage_{i,t}$ is the i-th unit water shortage; α_i, β_i, η_i and λ_i are the water use efficiency of ground water, local surface water, river channel water and reservoir water, accordingly; $S_{j,t}$ is the effective water storage of the j-th reservoir; $S_{j,min}$ and $S_{j,max}$

are the minimum and maximum reservoir storage, respectively; $Qin_{j,t}$ is the inflow into the j-th reservoir; $Qout_{j,t}$ is the water release to the downstream river channel from the reservoir; $L^S_{j,t}$ is the water loss in the reservoir, including evaporation and infiltration losses. $R_{k,t}$ is the effective water storage of the river channel; $R_{k,min}$ and $R_{k,max}$ are the minimum and maximum water storage in the river channel for ecological water demand and flood protection, respectively; $Rin_{k,t}$ and $Rout_{k,t}$ are the inflow and outflow of the k-th river channel, accordingly; $L^R_{k,t}$ is the water loss in the channel; ω_i, determining the priority of units water supply, is the weight of units of water shortage, severally, the higher weight meaning a prior water supply opportunity. Additionally, all of the weights are equivalent to 1, which implies the justification of the water supply between units in this study, and all the variables in the equations above are non-negative.

3. Results and Discussion

3.1. Index Validation

3.1.1. Temporal Scale

Baoding is located in the central part of the Daqinghe watershed and accounts for more than half of the watershed. Research [34,35] shows that there is an exponential relationship between agricultural drought-affected areas (ADAA) and the magnitude of drought. Based on the ADAA over the years from the Economic Statistical Yearbook of Baoding (1995–2011), the ADAA was processed by logarithmic transformation (inverse to exponential transformation) to acquire the logarithmic indexes of agriculture drought-affected areas (LIADAA), which was compared with the agricultural DMAI (aDMAI) of Baoding (Figure 3). The result shows that the LIADAA basically agrees with the reverse trends of aDMAI over the years in Baoding, with a correlation coefficient of 0.67. Additionally, due to the differences of annual hydrological processes, planting structure, and so forth, the LIADAA are significantly inconsistent with aDMAI in the years of 1995 and 1996. Nevertheless, the aDMAI can overall reveal the regional response to agricultural drought and represent the ability to mitigate the drought impact effectively.

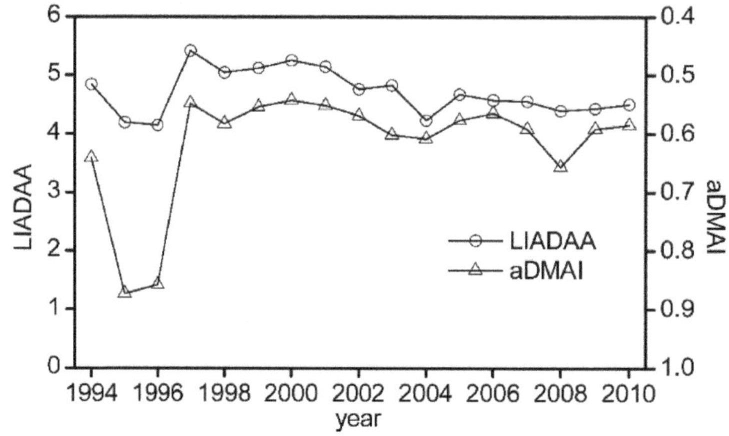

Figure 3. Comparison of logarithmic indexes of agriculture drought-affected areas (LIADAA) and agricultural drought mitigation ability index (aDMAI) in Baoding City.

3.1.2. Spatial Scale

Groundwater resources are the primary sources for water resource exploitation and utilization under serious water resource stress in the Daqinghe watershed. The drought mitigation mainly depends on pumping groundwater particularly in the place with low drought mitigation ability. Therefore, according to the average annual groundwater overexploitation data of every city from 2001 to 2010, the rationality of the spatial distribution of the drought mitigation ability was validated by the comparison of the average annual DMAI distribution (Figure 4a) and the groundwater overexploitation distribution. The result manifests that the distribution of drought mitigation ability is to some extent consistent with the distribution of groundwater overexploitation (Figure 4b), and the units with low drought mitigation ability in plain areas are also the severest groundwater overexploitation areas.

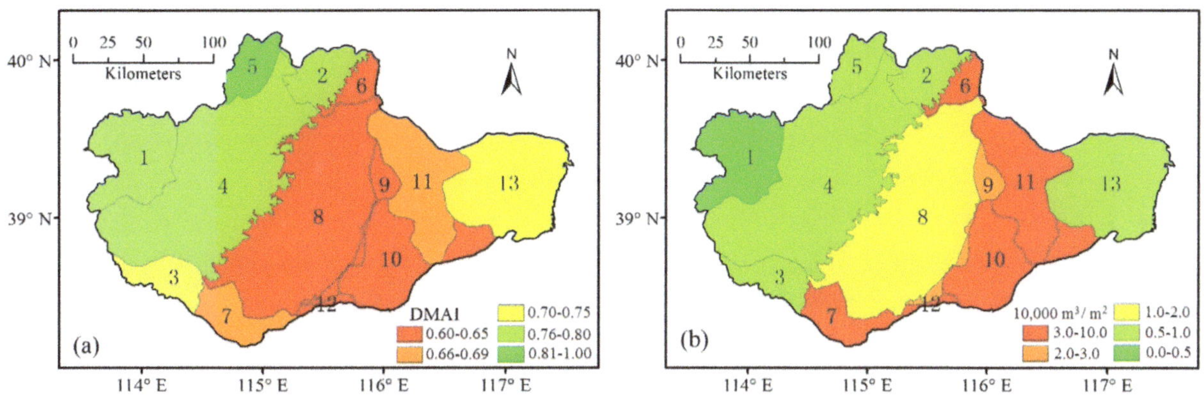

Figure 4. Distribution of the (**a**) drought mitigation ability index (DMAI) and (**b**) groundwater overexploitation.

The temporal-spatial validation of the drought index is a hot and tough issue, due to the fuzzy onset and end characteristics [36] and complex impacts of drought [22]. Therefore, spatial validation is infrequent and unquantifiable in the current research. According to the above result, although DMAI had not been perfectly validated, it can overall represent the regional ability for drought mitigation in the temporal-spatial scales within the Daqinghe watershed, which means that the DMAI is obviously beneficial to the drought risk management at the regional/basin scale.

3.2. Evolution of the Drought Mitigation Ability Index (DMAI) Characteristics in the Daqinghe Watershed

3.2.1. Annual Evolution Characteristic of the Drought Mitigation Ability Index (DMAI)

The trend of the DMAI for the overall Daqinghe watershed was calculated during 1961–2010 (Figure 5a). The result reveals that there is a decreasing trend of the DMAI over the last 50 years. Then, the contribution rate of groundwater and surface water to the DMAI were computed from the ratios of groundwater and surface water supply to the total water supply in DMAI (Figure 5b,c). As a result, the sum of the contribution rates of groundwater and surface water equal one. The contribution rate of groundwater to the DMAI shows an increasing trend, while it is opposite for the surface water. The groundwater contribution rate was significantly high, even more than 90%, particularly in typical drought years (such as 1965, 1968, 1972, *etc*.). This indicates that the decrease of the DMAI will aggravate the drought risk of the Daqinghe watershed in the future, and groundwater overexploitation, which has caused a series of environmental-ecological problems [37], will be more serious.

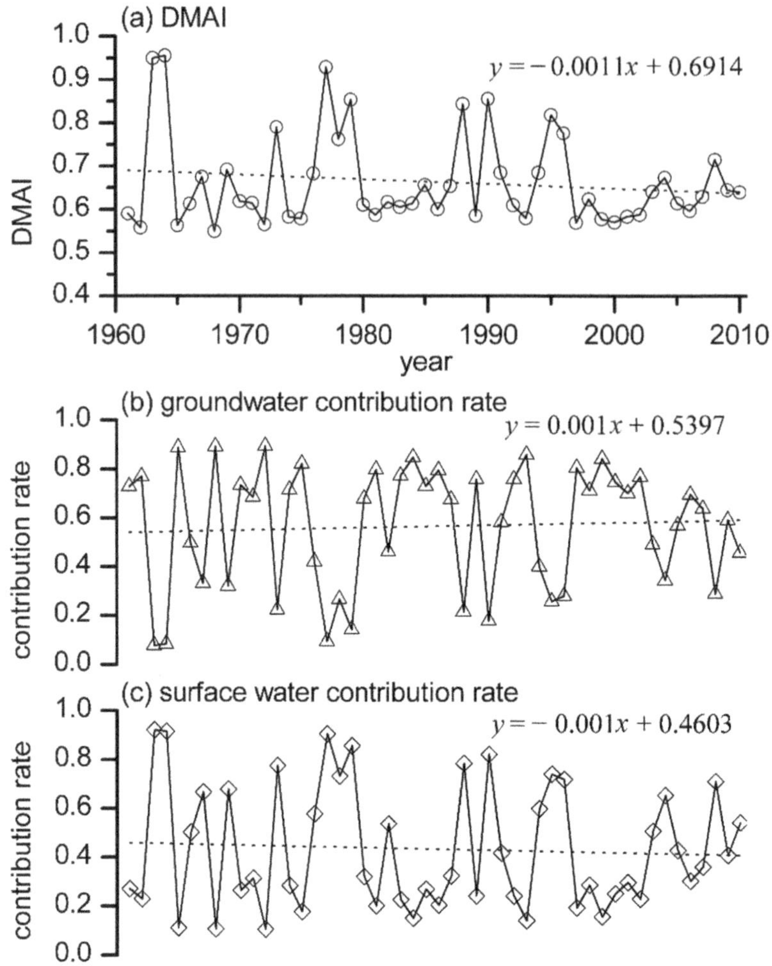

Figure 5. Trends of annual (**a**) drought mitigation ability index (DMAI); (**b**) groundwater and (**c**) surface water contribution rate in the Daqinghe watershed.

3.2.2. Evolution Characteristics of the Drought Mitigation Ability Index (DMAI) in Typical Drought Years

In order to better understand the DMAI characteristics in typical drought years, the exceedance probability of X exceeding or equaling a specific value x(m) is defined as:

$$P(X \geq x(\mathrm{m})) = \frac{m}{n+1} \tag{13}$$

where X is the annual water resources in the Daqinghe watershed, a random variable during 1961–2010; x(m) is the m-th largest observation value of X; and n is the total number of observations; a higher exceedance probability means a more serious drought year. Additionally, the exceedance probabilities of annual water resources were obtained based on historical data from 1961 to 2010. According to the exceedance probability of annual water resources, annual DMAI of each unit had been analyzed in typical drought years (1986, 1997 and 1965 are the 75%, 90% and 95% exceedance probabilities of annual water resource quantity, respectively) (Figure 6a). The result shows that the DMAI of

Units 6–12 in plain areas is somewhat lower than the DMAI of Units 1–5 in MHA, except for Unit 13, which is located in Tianjin with higher precipitation and next to the Bohai Sea. In addition, the DMAI deviation between 75%, 90% and 95% exceedance probabilities and average annual water resources were acquired (Figure 6b). The result displays that the DMAI of Units 1–8 in WBP and MHA decrease more dramatically in drought years, especially in Unit 7, with the most water demand (Table 2). The reason is that the units in EBP can use the water from the upstream to improve their DMAI. However, the DMAI of units in MHA and WBP mostly rely on the local water resources. Consequently, the DMAI changes are accompanied with the changes of water resources in the MHA and WBP areas.

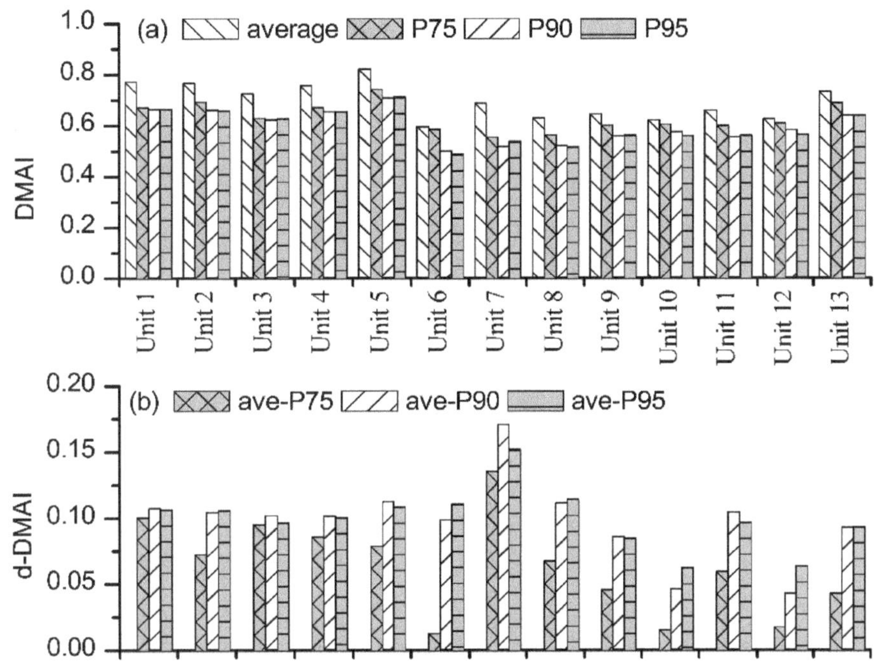

Figure 6. DMAI of units in typical drought years: (**a**) "*average*" represents the average annual drought mitigation ability index (DMAI); P75, P90 and P95 represent the 75%, 90% and 95% exceedance probabilities of annual water resources, respectively; (**b**) ave-P75, ave-P90 and ave-P95 represent the DMAI deviation between 75%, 90%, 95% exceedance probabilities and average annual value.

3.2.3. Inner-Annual Evolution Characteristics of the Drought Mitigation Ability Index (DMAI)

Average monthly DMAI over the years of 1961–2010 is illustrated in Figure 7. It shows that the months with low DMAI are from March to June, August and September accordingly. Unfortunately, the primary crop types in the Daqinghe watershed are winter wheat and summer maize. Their growth periods are from March to June and from August to October, separately, which almost coincides with the low DMAI months. On the one hand, the high agricultural water demand is the reason for the low DMAI in the growing season. On the other hand, the agricultural drought will still be a big challenge for drought mitigation in the Daqinghe watershed, which is consistent with the severe agricultural drought problems in reality.

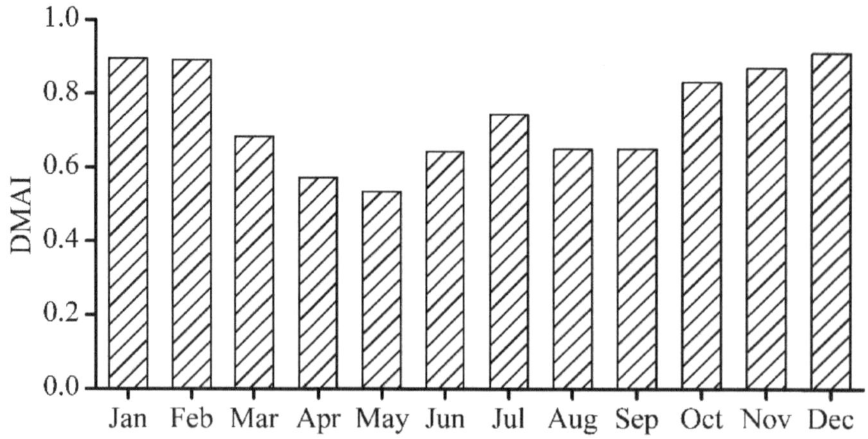

Figure 7. Distribution of monthly drought mitigation ability index (DMAI) within the Daqinghe watershed.

3.3. Drought Mitigation Ability Index (DMAI) Changes in the 2020

3.3.1. Water Resources Decrease Impact on Drought Mitigation Ability Index (DMAI)

According to the trends of the water resources and demand in the last 50 years (Figure 8), water demand shows a slight decreasing trend, and water resources (surface water and groundwater) have a significant decreasing trend attributed to climate change and local human activities, which is the prominent reason for the DMAI decrease. Climate change has led to not only the decrease in precipitation, but also the increase of precipitation intensity, which are not conducive to the transformation from precipitation to water resources [38]. Meanwhile, water demand variation mainly depends on the change of agricultural water demand. Thus, the socioeconomic water demand variation is not taken into consideration in this paper, since the socioeconomic water demand is more complicated and uncertain. On the one hand, the socioeconomic development, the improvement in living standards and urbanization can increase the water demand. On the other hand, the development of water saving technology and enhancement of water use efficiency will decrease the water demand contrarily. Therefore, water demand in 2020, which is very unpredictable, was assumed to equal water demand in 2010. Then, with the decrease of water resources (-5.73×10^7 m^3 per year), the water resources can be calculated in 2020 on the basis of data in 2010. Furthermore, the DMAI were obtained from the optimal water allocation model (Figure 9). The result shows that the reduction of water resources results in the DMAI decrease for all if the units, especially the DMAI of Units 6–13 in the downstream plain areas, and this points out that the plain areas is more sensitive to drought.

3.3.2. Interbasin Water Transfer Project Impact on the Drought Mitigation Ability Index (DMAI)

To combat against water shortage in north China, China launched the South-to-North Water Transfer Project (SNWTP) in 2002, which planned three routes (eastern, middle and western) to transfer 45 billion m^3 of water per year to north China. Among them, the middle route of SNWTP was mainly designed to solve the water shortage and drought issues in the North China Plain, especially in Beijing and Tianjin. After more than 12 years of engineering construction, the first stage of the middle route, with the target to transfer 9.5 billion m^3 of water per normal year, was initially completed on

12 December 2014. Additionally, the first stage of the middle route of SNWTP was planned to divert about one billion m^3 of water per normal year into the Daqinghe watershed, mostly depending on Wangkuai reservoir and Xidayang reservoir. Hence, it will result in a significant DMAI increase of the Daqinghe watershed in 2020. Therefore, with the effectiveness of the middle route of SNWTP, the DMAI was obtained by the water allocation model in 2020 (Figure 9). The result shows that the DMAI increases overall for the watershed, especially in the units with low DMAI in plain areas. The reason is that the rising water level in reservoirs cannot only improve the water supply of Units 7 and 8 directly, but also can be beneficial to downstream units' water supply indirectly by optimal water allocation.

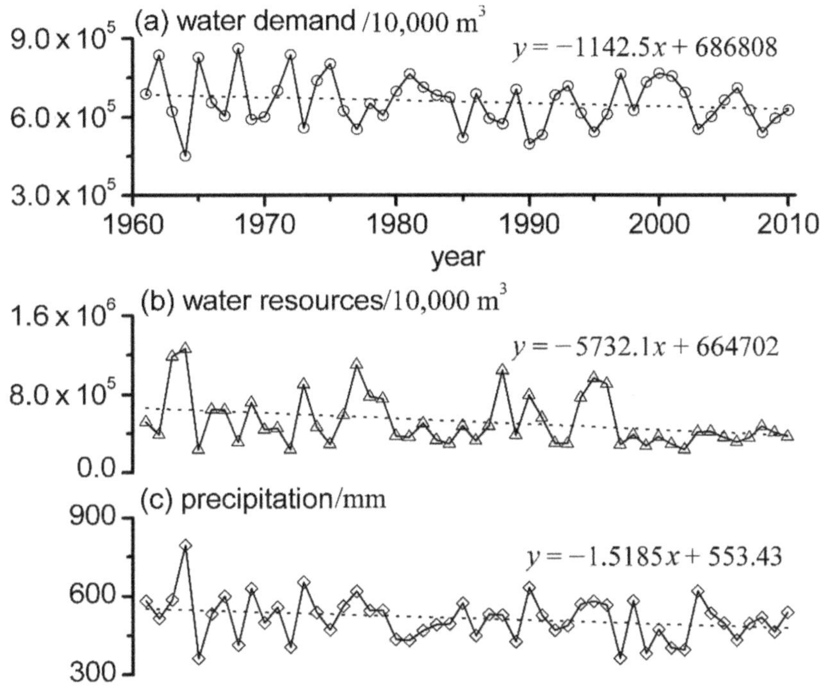

Figure 8. Trends of annual (**a**) water demand and (**b**) and resources annual (**c**) precipitation within the Daqinghe watershed.

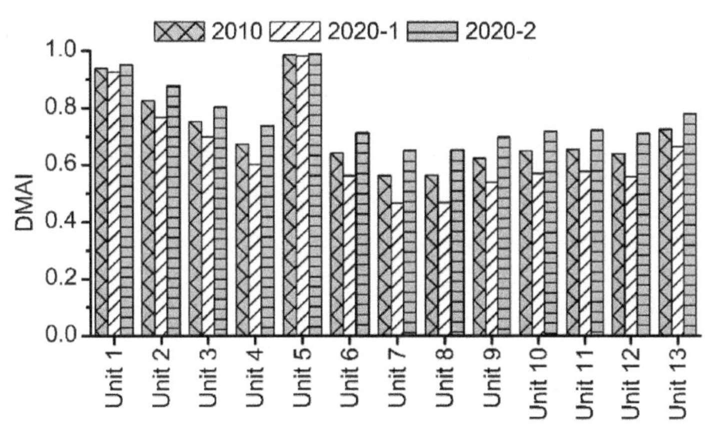

Figure 9. DMAI in different scenarios: 2010 represents the DMAI in 2010; 2020-1 represents the DMAI with the impact of the water resource decrease in 2020; 2020-2 represents the DMAI with the function of South-to-North Water Transfer Project (SNWTP) on the basis of the water resource decrease in 2020.

4. Conclusions

Droughts, as an imbalance of the performance of water resource systems, have been a worldwide obstacle to the development of socioeconomic systems, especially in developing countries. Additionally, it is inevitable and urgent to quantify the drought mitigation ability for drought risk management. Therefore, the DMAI was constructed to represent the regional ability of drought mitigation. In addition, the DMAI was applied in the Daqinghe watershed. First, the results show that the DMAI can represent the temporal-spatial characteristics of drought mitigation in the Daqinghe watershed by the comparison with the agricultural affected areas and the distribution of groundwater overexploitation. Second, the DMAI of the Daqinghe watershed is significantly decreased in the last 50 years, and the groundwater plays a prominent role in the response to drought mitigation, particularly in drought years. Additionally, the plain areas are more sensitive to drought impact than mountain areas, and the agricultural drought will still be a challenge to the Daqinghe watershed. Furthermore, the DMAI in 2020 was calculated in the scenarios of a water resource decrease and SNWTP. The result indicates that without SNWTP, there will be a high drought risk in 2020 due to the water resource decrease. Meanwhile, although the water from SNWTP can slightly improve the DMAI of the Daqinghe watershed, the drought risk continues to be serious. Generally speaking, although there is still some uncertainty in DMAI validation and application due to the limitation of data material and the complexity of drought mitigation and the water supply system, the DMAI is an effective index to quantify the ability of drought mitigation, which is indispensable for drought risk management.

Acknowledgments

This work was supported by the General Program of the National Natural Science Foundation of China under Grant 51279207 and Grant 51409266; and the National Science and Technology Support Program Project under Grant 2012BAC19B03 and Grant 2013BAC10B01.

Author Contributions

Denghua Yan had the original idea and supervised the research. Jianhua Wang and Baisha Weng performed the data collection. Shaohua Liu performed the statistical analysis and prepared the manuscript. Gang Wang performed partial statistical analysis, and Meijian Yang reviewed the manuscript.

Conflicts of Interest

The authors declare no conflict of interest.

References

1. Heim, R.R., Jr. A review of twentieth-century drought indices used in the United States. *Bull. Am. Meteorol. Soc.* **2002**, *83*, 1149–1165.
2. Mishra, A.K.; Singh, V.P. Drought modelling—A review. *J. Hydrol.* **2011**, *403*, 157–175.
3. Shiau, J.T.; Hsiao, Y.Y. Water-deficit-based drought risk assessments in Taiwan. *Nat. Hazards* **2012**, *64*, 237–257.

4. Yan, D.; Weng, B.; Wang, G.; Wang, H.; Yin, J.; Bao, S. Theoretical framework of generalized watershed drought risk evaluation and adaptive strategy based on water resources system. *Nat. Hazards* **2014**, *73*, 259–276.
5. Hughes, D.A.; Kingston, D.G.; Todd, M.C. Uncertainty in water resources availability in the Okavango River basin as a result of climate change. *Hydrol. Earth Syst. Sci.* **2011**, *15*, 931–941.
6. Alcamo, J.; Flörke, M.; Märker, M. Future long-term changes in global water resources driven by socio-economic and climatic changes. *Hydrol. Sci. J.* **2007**, *52*, 247–275.
7. Mishra, A.K.; Singh, V.P. A review of drought concepts. *J. Hydrol.* **2010**, *391*, 202–216.
8. Flörke, M.; Kynast, E.; Bärlund, I.; Eisner, S.; Wimmer, F.; Alcamo, J. Domestic and industrial water uses of the past 60 years as a mirror of socio-economic development: A global simulation study. *Glob. Environ. Change* **2013**, *23*, 144–156.
9. Liu, J.; Savenije, H.H. Food consumption patterns and their effect on water requirement in China. *Hydrol. Earth Syst. Sci. Discuss.* **2008**, *12*, 887–898.
10. Pereira, L.S.; Cordery, I.; Iacovides, I. *Coping with Water Scarcity: Addressing the Challenges*; Springer Science and Business Media: Berlin, Germany, 2009.
11. Steduto, P.; Faurès, J.M.; Hoogeveen, J.; Winpenny, J.; Burke, J. *Coping with Water Scarcity: An Action Framework for Agriculture and Food Security*; Food and Agriculture Organization of the United Nations: Rome, Italy, 2012.
12. Chang, F.J.; Wang, K.W. A systematical water allocation scheme for drought mitigation. *J. Hydrol.* **2013**, *507*,124–133.
13. Wilhite, D.A.; Svoboda, M.D.; Hayes, M.J. Monitoring drought in the United States: Status and trends. In *Monitoring and Predicting Agricultural Drought: A global Study*; Oxford University Press: New York, NY, USA, 2005; pp. 121–131.
14. Luo, L.; Wood, E.F. Monitoring and predicting the 2007 US drought. *Geophys. Res. Lett.* **2007**, *34*, doi:10.1029/2007GL031673.
15. Tsakiris, G.; Pangalou, D.; Vangelis, H. Regional drought assessment based on the Reconnaissance Drought Index (RDI). *Water Resour. Manag.* **2007**, *21*, 821–833.
16. Weng, B.S.; Yan, D.H.; Wang, H.; Qin, T.L.; Yin, J. Generalized drought assessment in Dongliao river basin based on water resources system. *Nat. Hazards Earth Syst. Sci. Discuss.* **2014**, *2*, 6703–6746.
17. McKee, T.B.; Doesken, N.J.; Kleist, J. The relationship of drought frequency and duration to time scales. In Proceedings of the 8th Conference on Applied Climatology, Anaheim, CA, USA, 17–22 Januray 1993.
18. McKee, T.B.; Doesken, N.J.; Kleist, J. Drought monitoring with multiple time scales. In *Ninth Conference on Applied Climatology*; American Meteorological Society: Boston, MA, USA, 1995.
19. Palmer, W.C. *Meteorological Drought*; US Department of Commerce, Weather Bureau: Washington, DC, USA, 1965.
20. Shafer, B.A.; Dezman, L.E. Development of a surface water supply index (SWSI) to assess the severity of drought conditions in snowpack runoff areas. In Proceedings of the Western Snow Conference, Reno, NV, USA, 19–23 April 1982.
21. Hollinger, S.E.; Isard, S.A.; Welford, M.R. A new soil moisture drought index for predicting crop yields. In Proceedings of the 8th Conference on Applied Climatology, San Francisco, CA, USA, 17–22 January 1993.

22. Wilhite, D.A.; Svoboda, M.D.; Hayes, M.J. Understanding the complex impacts of drought: A key to enhancing drought mitigation and preparedness. *Water Resour. Manag.* **2007**, *21*, 763–774.
23. Yevjevich, V.; Hall, J.D.; Salas, W. *Drought Research Needs*; Water Resources Publications: Littleton, CO, USA, 1978.
24. Huang, W.C.; Chou, C.C. Drought early warning system in reservoir operation: Theory and practice. *Water Resour. Res.* **2005**, *41*, doi:10.1029/2004WR003830.
25. Sechi, G.M.; Sulis, A. Drought mitigation using operative indicators in complex water systems. *Phys. Chem. Earth* **2010**, *35*, 195–203.
26. Rossi, G.; Castiglione, L.; Bonaccorso, B. Guidelines for planning and implementing drought mitigation measures. In *Methods and Tools for Drought Analysis and Management*; Springer Netherlands: Dordrecht, The Netherlands, 2007; pp. 325–347.
27. Vicente-Serrano, S.M.; Beguería, S.; Gimeno, L.; Eklundh, L.; Giuliani, G.; Weston, D.; Pegram, G.G. Challenges for drought mitigation in Africa: The potential use of geospatial data and drought information systems. *Appl. Geogr.* **2012**, *34*, 471–486.
28. Vogt, J.V.; Somma, F. *Drought and Drought Mitigation in Europe*; Kluwer Academic Publishers: Norwell, MA, USA, 2013.
29. Shih, J.S.; ReVelle, C. Water-supply operations during drought: Continuous hedging rule. *J. Water Resour. Plan. Manag.* **1994**, *120*, 613–629.
30. Wilhite, D.A. *Drought and Water Crises: Science, Technology, and Management Issues*; CRC Press: Boca Raton, FL, USA, 2014.
31. China Meteorolgical Data Sharing System. Available online: http://cdc.nmic.cn/home.do (accessed on 7 April 2015).
32. Smith, M. *CROPWAT: A Computer Program for Irrigation Planning and Management*; Food and Agriculture Organization of the U.N. Rome: Rome, Italy, 1992.
33. Döll, P.; Siebert, S. Global modelling of irrigation water requirements. *Water Resour. Res.* **2002**, *38*, doi:10.1029/2001WR000355.
34. Qin, Y.; Yang, D.; Lei, H.; Xu, K.; Xu, X. Comparative analysis of drought based on precipitation and soil moisture indices in Haihe basin of North China during the period of 1960–2010. *J. Hydrol.* **2014**, in press.
35. Sun, K.K.; Chen, J.; Jin, J.; Li, J.Q.; Xu, J.J.; Fei, Z.Y. Calculation method of agricultural drought loss risk curve under the actual drought resistance ability condition in Southern China. *J. Hydraul. Eng.* **2014**, *7*, 809–815. (In Chinese)
36. Mishra, A.K.; Singh, V.P.; Desai, V.R. Drought characterization: A probabilistic approach. *Stoch. Environ. Res. Risk Assess.* **2009**, *23*, 41–55.
37. Wang, Z.G.; Luo, Y.Z.; Zhang, M.H.; Xia, J. Quantitative evaluation of sustainable development and eco-environmental carrying capacity in water-deficient regions: A case study in the Haihe River Basin, China. *J. Integr. Agric.* **2014**, *13*, 195–206.
38. Zhang, X.; Cong, Z. Trends of precipitation intensity and frequency in hydrological regions of China from 1956 to 2005. *Glob. Planet. Change* **2014**, *117*, 40–51.

5

A New Way for Incorporating GCM Information into Water Shortage Projections

Seung Beom Seo [1,*], Young-Oh Kim [2] and Cho-Rong Kim [3]

1 Department of Civil, Construction, and Environmental Engineering, North Carolina State University, Raleigh, NC 27606, USA

2 Department of Civil and Environmental Engineering, Seoul National University, 1 Gwanak-ro, Gwanak-gu, Seoul 151-741, Korea; E-Mail: yokim05@snu.ac.kr

3 Korea Institute of Civil Engineering and Building Technology, 283 Goyangdae-ro, Ilsanseo-gu, Goyang-si, Gyeonggi-do 411-712, Korea; E-Mail: crkim13@kict.re.kr

* Author to whom correspondence should be addressed; E-Mail: sseo@ncsu.edu;

Academic Editor: Athanasios Loukas

Abstract: Climate change information is essential for water resources management planning, and the majority of research available uses the global circulation model (GCM) data to project future water balance. Despite the fact that the results of various GCMs are still heterogeneous, it is common to utilize GCM values directly in climate change impact assessment models. To mitigate these limitations, this study provides an alternative methodology, which uses GCM-based data to assign weights on historical scenarios rather than to directly input their values into the assessment models, thereby reducing the uncertainty involved in the direct use of GCMs. Therefore, the real innovation of this study is placed on the use of a new probability weighting scheme with multiple GCMs rather than on the direct input of GCM-driven data. Applied to make future projections of the water shortage in the Han River basin of Korea, the proposed methodology produced conservative but realistic projection results (15% increase) compared to the existing methodologies, which projected a dramatic increase (144%) in water shortage over 10 years. As a result, it was anticipated that the amount of water shortages in the Han River basin would gradually increase in the next 90 years, including a 57% increase in the 2080s.

Keywords: water resources management; water balance; Global Circulation Model (GCM); climate change; water shortage

1. Introduction

An accurate water balance projection is essential to a successful water resources planning for an uncertain future, especially in a nonstationary world [1]. Water balance projections simply calculate the difference between the total water demand and the total supply capability for a basin. When the demand exceeds the supply, water shortage occurs. To be systematic, short- and medium-range water resource planning alternatives, which can reduce potential water shortages, should be established under a predefined long-range water resource plan encompassing at least 20 years. For such a long-range time scale, the impacts of climate change cannot be ignored.

This study began with a broad review of studies that focused on water balance projection methodologies reflecting climate change impacts. Most studies utilized multiple general circulation model (GCM) climate change scenarios, and the average number of GCM scenarios used in these water balance projection studies ranges from four to five. For instance, some studies relied on only a single GCM scenario [2–4], whereas Islam *et al.* [5] applied the most number of GCM scenarios, which is 12 in total. In terms of water shortage calculation, water balance analysis models have been commonly used with GCM-driven streamflow scenarios. Thus, most studies that project future water balance have been using GCM-driven runoff series as water supply scenarios. They also used multiple GCM scenarios to overcome the heterogeneity that arises from GCMs. Moreover, various water balance analysis models have been utilized, which include CalSim II [6], WEAP [7,8], CALVIN [3], WRAP [9], Mospa [10], Sibuc [11], IQQM [12], Riverware [13], and WSM [14]. In Korea, the K-WEAP and the K-MODSIM models have been popularly employed for water balance analysis, and there have been several efforts to incorporate GCM scenarios into water balance analysis to reflect future climate change impact [15–17]. Thus, most studies made use of different models for their own purposes.

The assessment of future water resources using GCM-driven streamflow, which is the general analysis framework of most previous studies, would provide an evaluation of possible climate change impacts. However, there are two possible problems to be considered when GCM-driven streamflow scenarios are used for water balance analysis. First, Korea, for example, has established its own national water resource plan based on historical data. Although it is imperative that the impacts of climate change on long-range water resources be considered, an inconsistency in methodologies may occur if historical runoff data are suddenly substituted with GCM-driven runoff scenarios. Furthermore, it would be risky to establish water resource plans based on the results of GCM-driven scenarios when they are directly inputted into the water balance analysis models because of the heterogeneity and uncertainty that arise from GCMs.

This paper describes a new alternative approach that uses GCM climate information by assigning weights to historical runoff scenarios rather than directly inputting GCM-driven runoff into water balance models. By maintaining most of the existing water balance analysis framework, the proposed methodology is able to mitigate the risk created by the uncertainty of GCM-driven values.

The remainder of this paper is organized in the following manner. The first section demonstrates the previous methodology for water shortage projection. The next section describes the new water shortage projection methodology, which incorporates climate change impacts and compares it to the previous methodology. The application of this approach is described in the following section. The application results are analyzed by comparing them with the results of the previous methodology, and future water shortage projection results in each future period are presented. This paper concludes with a summary of the results and a discussion of possible avenues for further studies.

2. Materials and Methods

2.1. Previous Methodology

2.1.1. Water Vision 2020: The National Water Resources Planning Report in Korea

In general, most countries have established their own long-term water resources plans. In Korea, the long-term water resources plan is established every 20 years and updated every 5 years. The current plan entitled "*Water Vision 2020*" was established in 2001 and updated in 2011. Water balance analysis predicts possible water shortages by calculating the balance between water supplies—Such as surface runoff, reservoir storage, and groundwater—And multiple water demands, including municipal, agricultural, and industrial demands. In this regard, water balance analysis is one of the core parts of the long-term water resources plan. A rainfall-runoff model calculates the conversion of rainfall into runoff using the observed rainfall data. Afterward, a water balance model estimates the amount of water shortages using this calculated runoff data as fundamental water supply.

In the case of Korea, the water shortage for each watershed is calculated every five days because agricultural water demands temporally vary during the farming season. When it comes to a model used in Water Vision 2020, the TANK model, a conceptual rainfall-runoff model that is composed of multiple tanks laid vertically, and the K-WEAP model (briefly introduced in Section 2.4) have been officially applied as the rainfall-runoff model and the water balance model, respectively. Water Vision 2020 [18] declared that the water shortage of the Korean Peninsula will amount to 862 million m^3 in 2020 (Table 1).

Table 1. Water shortage projection of previous studies in Korea.

Report	Water Vision 2020 [a]	The Strategic Report [b]
Target year	2020	2040s (2031 ~ 2060)
Maximum water shortage (10^6 m^3)	862	3300

Notes: The same water demand scenario was applied to both reports. [a] "*Water Vision 2020 (2006–2020)*", which is the national water resources plan in Korea, used the observation-driven flow in 1967–2003 as water supply scenarios [19]; [b] The Strategic Report, "*Adaptation to Climate Change in Future Water Resources Management*", used a fine resolution RCM-driven runoff series in 2031–2060 as water supply scenarios.

2.1.2. An Example of Water Balance Analysis Using Climate Change Information

Studies on climate change impacts have sporadically been conducted in Korea since the mid-1990s. However, "*The Strategic Report,*" which is the first national report on water resources planning that considered climate change, was published based on the research [15]. The Strategic Report estimated the future water balance of the Korean Peninsula, taking into consideration climate change information, as shown in Table 1, but there remained two issues to be resolved.

First, the continuity of the existing water balance analysis methodology, that is, the methodology of the current national plan, was not considered. The Strategic Report utilized rainfall data from a single RCM, while the typical national water resources plans, Water Vision 2020, used the observed rainfall data as an input variable as mentioned above. Moreover, The Strategic Report suddenly changed the rainfall-runoff model from the TANK (used in Water Vision 2020) to the Semi-Distributed Land Use-Based Runoff Processes (SLURP) which has made the discrepancies in methodology larger. The TANK is a conceptual rainfall-runoff model, whereas the SLURP is a semi-distributed model that makes use of graphical data input and a graphical display of output. Both models simulate vertical water balance for runoff calculation, but the SLURP routes runoff from each element to the outlet. In this regard, the TANK is computationally less intensive, whereas the SLURP can produce spatial maps of runoff. Although the SLURP model has some advantages for spatial representation, this inconsistency in the use of a hydrologic model can mask a fair and effective comparison between different methodologies.

The other issue was the lack of reflection of the uncertainty that arises from climate change scenarios. The Strategic Report used rainfall data from a single RCM called RegCM3 under an A2 emission scenario. However, the projected monthly rainfall series from various GCMs provided by the Intergovernmental Panel on Climate Change (IPCC) lead to highly heterogeneous results [19], as shown in Figure 1. Given that these uncertainties are rather noticeable, a national water resources plan based on a single GCM scenario seems unrealistic. Moreover, the direct use of GCM datasets through hydrologic and water balance models could be risky in practice because GCMs are still highly uncertain despite the fact that many climate change studies rely on their direct use.

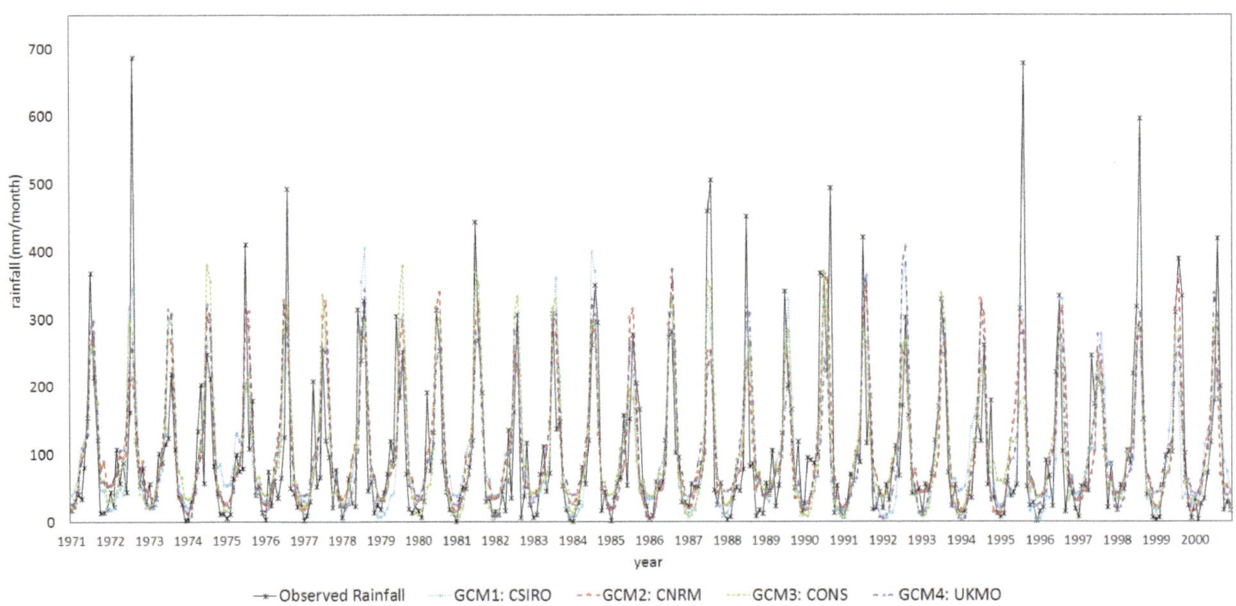

Figure 1. Time series plots of the observed and simulated monthly rainfall of four GCMs under the A2 scenario for the Korean Peninsula.

2.2. New Approach

2.2.1. Overview

This study has proposed a new alternative approach to solve these two problems, which are the discontinuity of the existing water balance analysis methodology for the current national water resources management plan and the direct use of GCM projection values, which involves a huge uncertainty. To ensure the continuity of the existing methodology, the TANK model which has been used for "*Water Vision*," the national water resources management plan, was used as a rainfall-runoff model. Second, by considering the runoff series driven by observed rainfall time series (rather than GCM) as an input climate variable for the water balance model, the continuity of the methodology could be maintained. In addition, four GCM datasets under the AR4 A2 emission scenario were used to consider the uncertainty that arises from GCMs. Then, GCM information was incorporated to estimate the likelihood of observation-driven runoff scenarios to minimize the risks when GCM values are directly utilized. The proposed methodology assigns more probability to an observation-driven scenario, which is closer to a GCM scenario.

2.2.2. Weighting Scheme

In this study, a nonparametric *K*-nearest neighbor (*K*-nn) algorithm [20–22] was used as a weighting scheme for clustering/sampling objects based on the closest training samples. This algorithm has been widely applied because of its simple and useful scheme to weight the contribution of the neighbors so that the closer neighbors can contribute more to the expectation than the more distant ones [20].

2.2.3. Procedure

The procedure is described as follows. Suppose that there are m years (e.g., 30 years of available GCM data in a target period) of the runoff series simulated by inputting GCM data (hereinafter GCM-driven flow) where the runoff series simulated by inputting historical observed rainfall data (hereinafter observation-driven flow) can be obtained. In the case of the ith year of the target period, this study selects K number of nearest neighbors to a given GCM-driven flow of the ith year among a whole annual series of observation-driven series, and inputs these K series into a water balance model to estimate the water shortage value of the ith target year. Here, the metric distance to be compared is the total sum of monthly flows for nine months (from October to the following June), that is, the non-flood season in Korea. Scenario selecting and weighting schemes are based on the *K*-nn algorithm. By calculating the weighted average value across the selected K series, the water shortage value (S_i) of the ith year of the target period can be obtained. The weight of the jth nearest series (w_j) can be estimated using the following equation (1):

$$w_j = \frac{1/j}{\left(\sum_{j=1}^{K} 1/j\right)} \tag{1}$$

where $j = 1$ represents the nearest neighbor while $j = k$ represents the kth nearest neighbor.

The aforementioned steps are repeated from the first year to the last year of the target period, and the whole procedure is carried out using all GCMs.

2.3. Study Basin

The proposed methodology was applied to the Han River basin located at the center of the Korean Peninsula (Figure 2). The area of the Han River is approximately 26,200 km^2, which is 23% of the national territory, and the total length of the basin is approximately 481 km. Prior to the construction of a number of major dams, the river was known for its huge coefficient of river regime, with a ratio between the maximum and the minimum amount of flow of 1:390. The annual mean precipitation and runoff based on a 30-year period from 1978 to 2007 are 1260 mm and 17.4 billion m^3, respectively. Because of the monsoon climate, 62% of the annual precipitation occurs during the three-month flood season from late June to late September. The Han River basin is occupied by several types of land covers, such as urban, paddy fields, farms, forests, grassland, wetlands, barren, and water, but more than 70% of the basin is covered with forests. The major soil type in the basin (as of 2001) is lithosol, which typically has a shallow overburden and lacks horizon development because of the steep slopes.

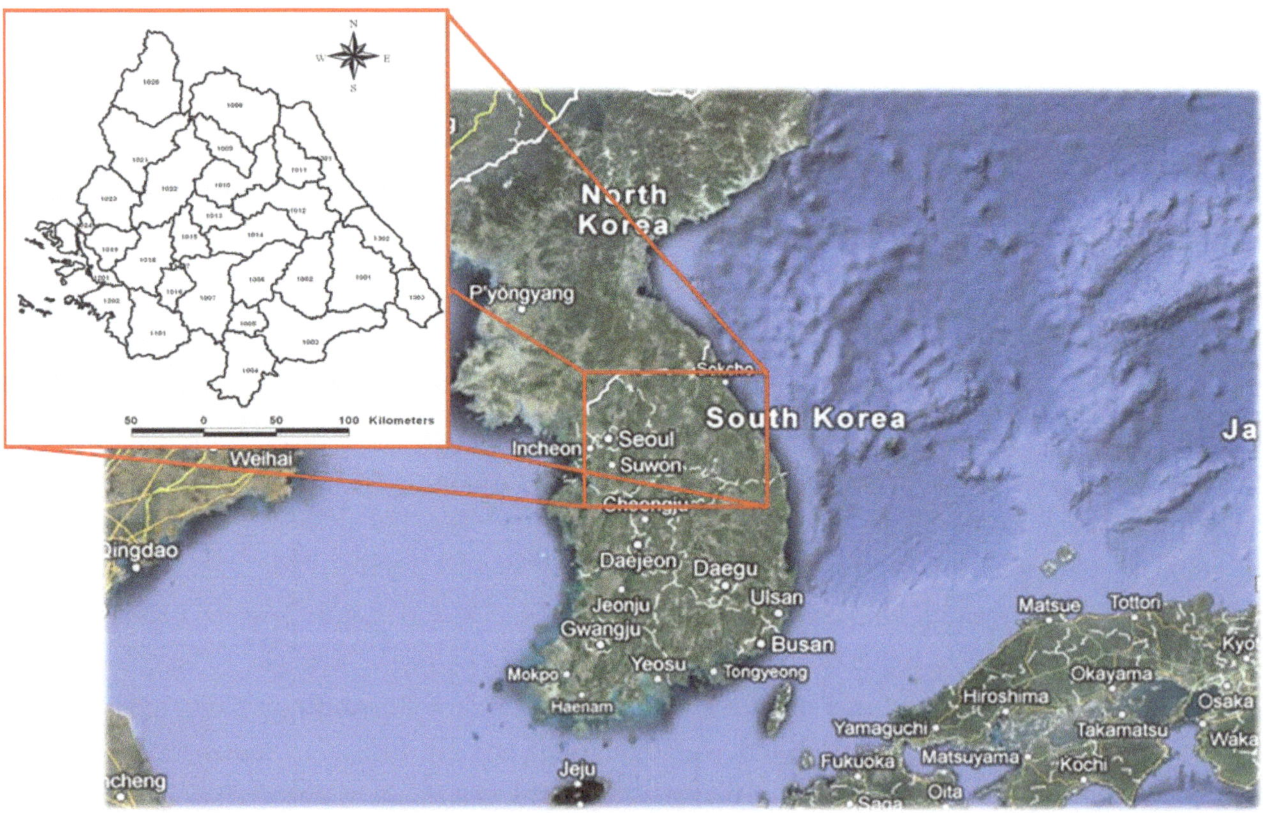

Figure 2. Han River basin, Korea.

2.4. Water Balance Model

The K-WEAP water resources management model was used to project potential water shortages for each sub-watershed of the Han River basin. The K-WEAP model was developed through a cooperation between the Korea Institute of Construction Technology (KICT) and the Stockholm Environment

Institute—Boston Center (SEI-B). Before becoming officially adopted in Water Vision 2020, the K-WEAP was tested by several studies to prove its applicability and efficiency [23]. A linear program, built in the model, is applied to solve water allocation problems at each time step by maximizing the satisfaction of water demands subject to supply priorities, demand site preferences, and other constraints (see [24] for details). Figure 3 presents a snapshot of the Han River basin's water allocation network realized in the K-WEAP model.

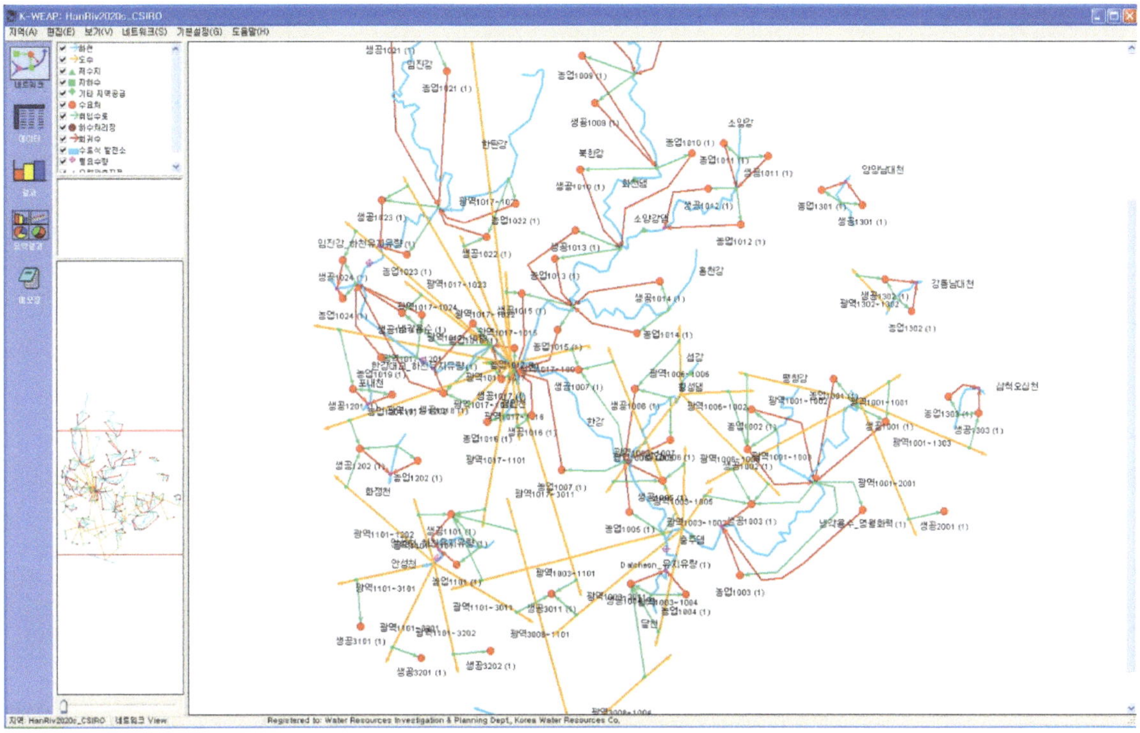

Figure 3. Schematic network of the Han River basin, the K-WEAP model.

2.5. Climate Change Scenarios

To take into account GCM uncertainty, this study utilized four GCM datasets under an A2 emission scenario from IPCC AR4: CSIRO, CNRM, CONS, and UKMO (Table 2). Although recent studies on climate change projection use RCP (representative concentration pathways) scenarios from IPCC AR5, which is up-to-date climate change information, we preferred to use the former dataset—A2 emission scenario from IPCC AR4—for an exhaustive comparison with the current national water resources management plan of Korea. The A2 emission scenario is used here under the assumption that there will be radical climate change due to the continuously increasing population and a very heterogeneous world. These four GCM datasets were selected based on the research that addresses a selection of principal GCM scenarios [25]. Lee [25] provided principal scenarios that reflects overall uncertainty across multiple GCM datasets. Thus, the selected GCM datasets can preserve most of the uncertainty even though all of the GCM datasets available are not used. According to Kim and Lee [20], those GCMs were known to more suitably represent the Korean rainfall patterns in the low-flow (non-flood) period (October to June), along with better performance on temperature. The GCM data on monthly temperature and precipitation were downloaded from the IPCC data distribution center [26]. Figure 4 illustrates the monthly precipitation and its standard deviation of both the observed data and the CSIRO: MK3.0 GCM hindcast precipitation data.

As shown in this figure, the GCM mean series follow well those of the observed data, while the GCM standard deviation series tend to underestimate those of the observed data.

Table 2. GCMs used in this study.

No.	Model (Agency: Version)	Abbreviation	Country	Resolution (km)	
				Atmosphere	Ocean
1	CSIRO: MK3.0	CSR	Australia	192 × 96	192 × 189
2	CNRM: CM3	GNR	France	128 × 64	182 × 152
3	CONS: ECHO-G	MIU	Germany/Korea	96 × 48	128 × 117
4	UKMO: HadCM3	UKC	United Kingdom	192 × 144	360 × 216

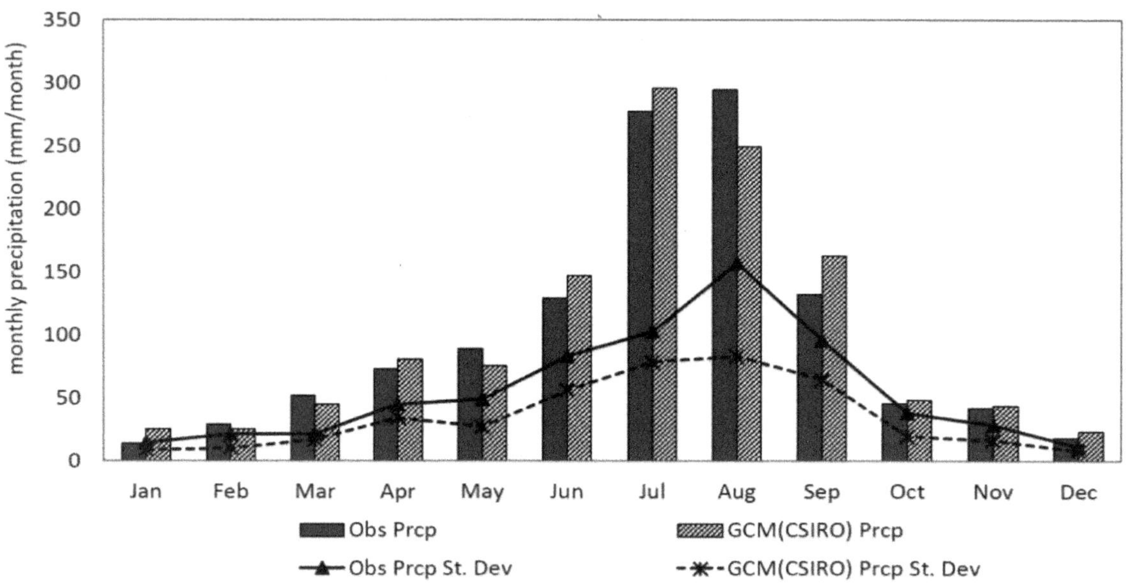

Figure 4. Mean monthly precipitation and standard deviation of observed and GCM-simulated series (1971–2000).

However, these GCM data should be spatially downscaled to local scale and temporally disaggregated to daily series to be used as an input forcing data of the TANK model. Hence, these GCM data are spatially downscaled and temporally disaggregated by Bae *et al.* [27]. They spatially downscaled the GCM data using the cyclostationary empirical orthogonal function (CSEOF) and multi-linear regression analysis, and temporally disaggregated from monthly to daily series using a weather generator called WXGEN [27].

2.6. Bias Correction

The GCM-driven flows usually contain a systematic bias [28], which should be corrected before being applied to a water balance model. This study employed a quantile mapping (QM) technique, which has been frequently applied to correct the bias of flow series [10,29,30], and is based on a bias correction technique for the downscaled GCM output described by Wood *et al.* [31]. Unlike a simple bias correction method such as the delta method, which applies a constant correction factor for all events within a single duration (e.g., a month), the QM determines correction factors by mapping hindcast/historical

GCM-driven flows to observation-driven flows through the empirical distribution of variables, and same correction factors are applied to the projected future GCM-driven flows of the target period. In other words, the non-exceedance probability of the future GCM value is mapped in the empirical distribution of the GCM hindcast dataset with a duration that is similar to the observed dataset [28]. Then, future GCM values are corrected through the Equation (2) as follows. More details about the QM technique are described on Wiley [28].

$$Z_i = F_{oi}^{-1}\left(F_{si}\left(\hat{Y}_i\right)\right) \tag{2}$$

where $\hat{Y}_i$ is the projected future GCM value, Z_i is the bias-corrected value, $F_{si}(\cdot)$ is the empirical CDF of the GCM hindcast data set, and $F_{oi}(\cdot)$ is the empirical CDF of the observed dataset.

3. Results

3.1. Flow Projections

The daily precipitation and potential evaporation projections in the basin scale were inputted into the TANK rainfall-runoff model, which was officially utilized in the development of Water Vision 2020 in Korea. The GCM-driven daily flow series for 130 years (1970–2099) were simulated by the TANK model using the downscaled precipitation and potential evaporation, and they were bias-corrected. A 100-year flow series (2000–2099) was corrected based on a 30-year control period (1970–1999). Figure 5 presents the annual flow series of observation-driven flow, the GCM-driven flow without bias correction (CSIRO MK3.0 GCM model), and the GCM-driven flow with bias correction. As shown in Figure 5, without bias correction, the annual mean value and the variation of the GCM-driven flow were somewhat underestimated than the observation-driven flow in the historical period. However, after the QM-based bias correction, both the annual mean value and the variation of the GCM-driven flow were reliably realized as to those of the observation-driven flow.

Figure 6 shows the monthly pattern of the bias-corrected GCM-driven flow, averaged over the four GCMs, in the Han River basin in four time periods: the basis (1967–2003), 2020s (2010–2039), 2050s (2040–2069), and 2080s (2070–2099). In the 2020s, as shown in Figure 6, whereas a dramatic increase of flow is anticipated for the flood season, a moderate decrease is anticipated for the fall and winter seasons (October—January), which is in accordance with Figure 5b, which shows a decreasing trend for 2020s. Moreover, this decreasing trend in the low-flow season lasts until the 2080s. Given that water shortage is mainly affected by the low-flow season, we only focused on this non-flood annual series of GCM-driven flow in this study.

3.2. Water Shortage Projections

A set of time series of observation-driven flows for 37 years from 1967 to 2003 was inputted into the K-WEAP water balance analysis model. Note that the GCM-driven flow projections were not directly inputted but rather utilized to assign a weight to each yearly scenario of the 37-year observation-driven flows. In other words, the closest seven observation-driven flow scenarios to the value of the given year of each GCM-driven flow are selected through the *K*-nn technique. Then, using Equation (1), the corresponding weights of the selected seven scenarios were estimated, and these weights were used to calculate the weighted average of the seven water shortages values driven by the seven selected flows. Consequently, zero weight was assigned to each of the remaining 30 scenarios, that is, not used in the analysis of the given year. This procedure was repeated for the four GCMs, and the final water shortage projection in the projection year was calculated as the average of the four GCM results. In this study, it is assumed that the water demand for the year 2020, estimated by Water Vision 2020, remains constant throughout the entire projection period. Table 3 summarizes the average water shortage (AWS) in the Han River basin in three future periods, such as the 2020s, 2050s, and 2080s. Moreover, Figure 7 presents a trend of water shortage projections, which is the AWS toward the end of the 21st century. It was anticipated that the amount of water shortages in the Han River basin would gradually increase over the next 90 years, and water shortages would increase up to 56% on average (ranging from 33% to 78%) in the 2080s compared to the current water shortage amounts projected by Water Vision 2020.

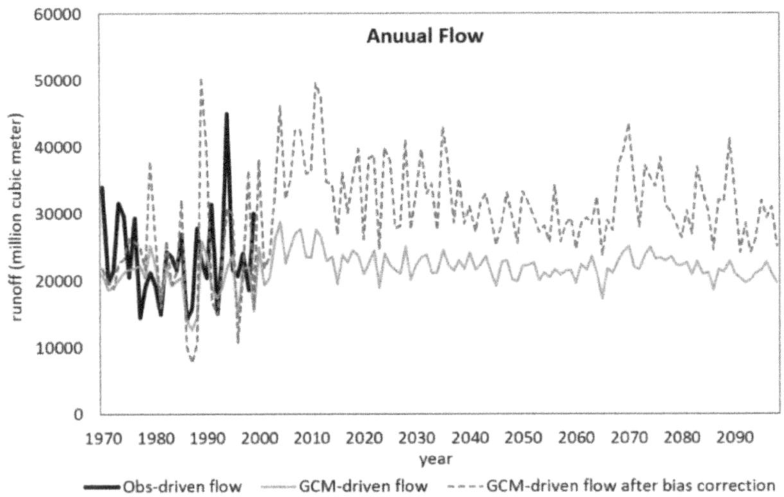

(a)

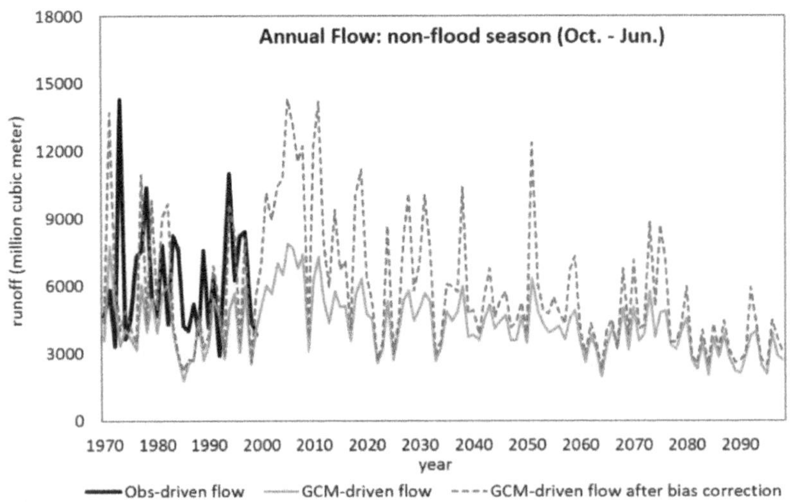

(b)

Figure 5. Annual flow comparison of observation-driven flow, GCM-driven flow (without bias correction), and GCM-driven flow after bias correction: (**a**) total annual flow; (**b**) non-flood season flow.

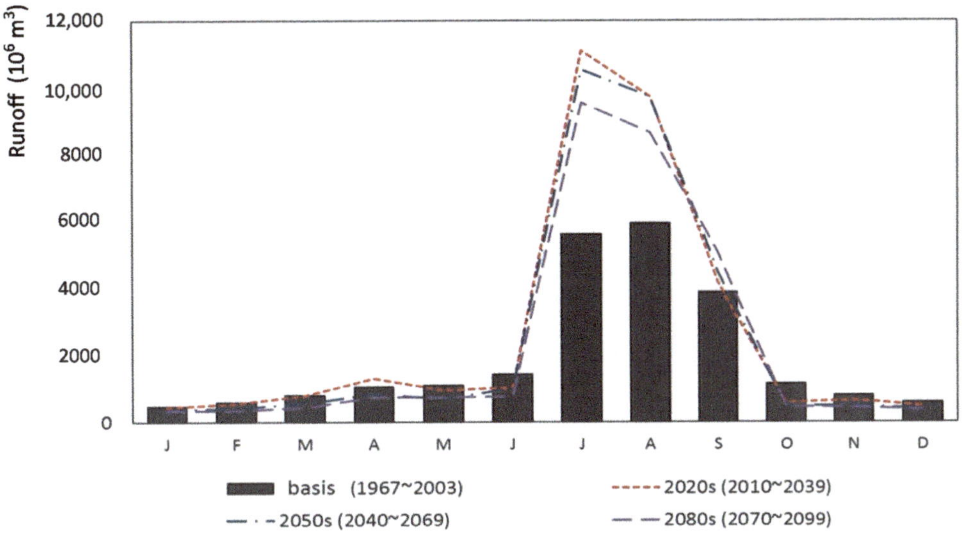

Figure 6. Monthly flow variations in the projections for the Han River basin, Korea.

Table 3. The average water shortage projections in the Han River basin (unit: 10^6 m^3).

Water Vision 2020 [a] **(Target Year: 2020)**	**Future Projection**			
	GCM Models	**2020s**	**2050s**	**2080s**
42.0	CSIRO: MK3.0	42.3	49.6	55.8
	CNRM: CM3	47.1	55.7	72.8
	UKMO: HadCM3	52.9	70.9	74.6
	CONS: ECHO-G	51.6	59.8	60.6
Average **(increase ratio from the current)**		48.5 **(15% ↑)**	59.0 **(40% ↑)**	65.4 **(56% ↑)**

Note: [a] Water Vision 2020 used the observation-driven flow in 1967 ~ 2003 as water supply scenarios [19].

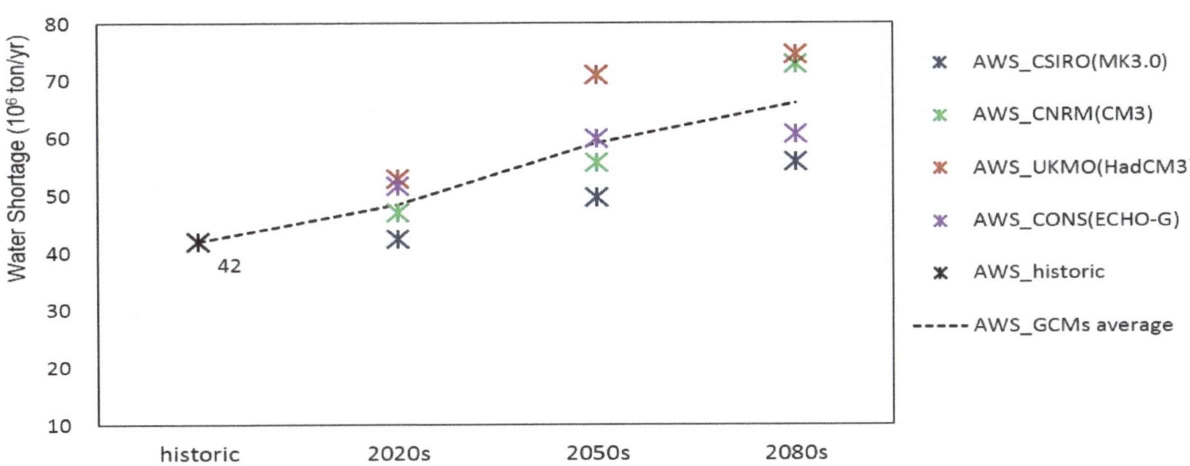

Figure 7. Trend of water shortage projections in the Han River basin: 2020s (2010–2039), 2050s (2040–2069), and 2080s (2070–2099).

Figure 8 illustrates schematically the amount of water shortages in each sub-watershed, warning that the downstream areas where the capital city of Korea, Seoul, is located would be vulnerable to future water shortages.

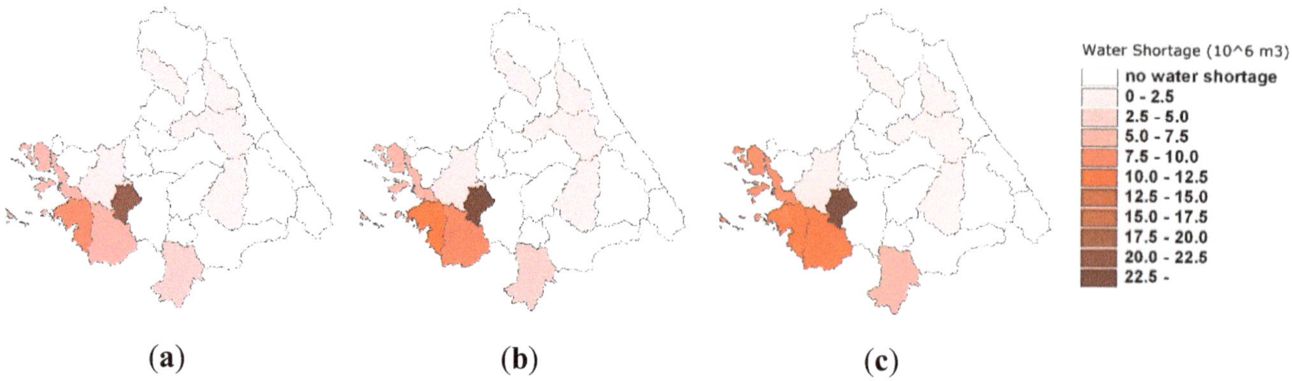

Figure 8. Average water shortage in each sub-watershed of the Han River basin (unit: 10^6 m^3/year) (**a**) 2020s; (**b**) 2050s; (**c**) 2080s.

3.3. Comparison with the Previous Methodology

This study compared the proposed methodology with the previous methodology, the GCM value input method (described in Section 2.2), which directly inputs the values of GCM-driven flows into the K-WEAP model. For fair comparison, both methodologies began with the same GCM selections for consistency. Therefore, the only difference between the two methodologies is whether the GCM-driven data were directly inputted into the K-WEAP model while the other settings remained constant. Note that the Strategic Report, an example of the GCM value input method, used only a single RCM scenario. Thus, it is different from the GCM value input method described in this section.

Table 4 reports a significant difference between the results from both methodologies. The GCM value input method anticipated an unrealistically huge increase (144%) in water shortage, while the proposed methodology produced a rather conservative result, that is, a 15% increase with an overall range of 1% to 26%. Whether or not the projection results of the proposed methodology are significantly different from those of the previous methodology was validated with a statistical hypothesis test: the two-tailed *t*-test was performed with the null hypothesis that there is no difference between the two population mean values. As a result, it turned out that the both methodologies have different mean values with all the four GCM scenarios (*i.e.*, all the null hypotheses were rejected with significance level of 0.05). Furthermore, in Table 4, the results produced by the GCM value input method show a huge heterogeneity between the GCMs. Although it is impossible to prove which methodology produces a better prediction because the future is uncertain, a 144% increase in 10 years seems extremely unlikely.

4. Conclusions

This study proposed a new methodology for water shortage projection under climate change impacts, which is essential to national water resources planning, to overcome the limitations of the GCM value input method that inputs GCM-driven flow directly. The new methodology uses GCM-driven flow scenarios to assign different weights to each observation-driven flow scenario rather than directly inputting GCM-driven flow scenarios into a water balance analysis model.

Table 4. The average water shortage projections for 2020s in the Han River basin (unit: 10^6 m^3).

Water Vision 2020 [a] (Target Year: 2020)	Future Projection (Target Period: 2020s)		
	GCM Models	**Proposed Methodology**	**Previous Methodology [b]**
42.0	CSIRO: MK3.0	42.3	63.5
	CNRM: CM3	47.1	17.4
	UKMO: HadCM3	52.9	137.6
	CONS: ECHO-G	51.6	192.2
Average **(increase ratio from the current)**		48.5 **(15% ↑)**	102.7 **(144% ↑)**

Notes: [a] Water Vision 2020 used the observation-driven flows in 1967 ~ 2003 as water supply scenarios [18]; [b] The previous methodology, the GCM value input method, inputted the GCM-driven flow directly into the K-WEAP model.

The proposed methodology was applied to project future water shortages in the Han River basin in Korea. The *K*-nn scheme selected seven observation-driven flows, which were most similar to the GCM-driven flows in the low-flow season. By calculating the weighted average values of the simulation results of the selected seven observation-driven flows, the water shortages of each target period were projected.

Compared to most previous studies, which assess climate change impacts on water resources by directly using GCM-driven flows, we have focused on how to provide reliable projection results for decision-makers in the water resources management field. As mentioned in the introduction, a radical change of methodology may confuse people with uncertain projection results. The comparison between the proposed method and the GCM value input method also showed a considerable difference in the projections. The results of the GCM value input method predicted a dramatic increase of 144% in the 2020s, which seems unrealistic. On the other hand, the proposed methodology used GCM-driven flows to assign different weights to each historical scenario rather than directly inputting them into a model so that we could reflect a more conservative climate change impact in water shortage projections, a 15% increase in the 2020s. Thus, this study successfully utilized observation-driven flow series as water supply scenarios and also reflected climate change impacts by assigning different weights to those observation-driven flows in accordance with the given GCM. The proposed methodology projected that the water shortage in the next 90 years in the Han River basin would gradually increase by 15%, 41%, and 57% in the 2020s, 2050s, and 2080s, respectively.

It is difficult to argue that the implementation of this conservative approach is the best way to assess climate change impacts. However, the blind use of GCM-driven flows should be reevaluated. Because the proposed methodology successfully reflected climate change impacts on water resources using GCM-driven flows, it may become one of the appropriate alternatives to be considered when addressing the impact of climate change. Nonetheless, as we proposed a rather conservative approach given that each annual water shortage value is projected within the limits of historical data, efforts to project extreme values beyond the observed record will be pursued in future studies.

In the near future, the current results of this study should be updated with the AR5 RCP scenarios that have been presented to the public in 2014 and are, thus, consequently being tested in Korea. The primary reason that we used the A2 emission scenario from AR4 was because of a consistent comparison, that is, because both Water Vision 2020 and the Strategic Report utilized the A2 emission scenario. Moreover, the QM technique was used to correct systematic bias on GCM datasets in this study. Other up-to-date bias correction schemes [32,33], however, can also be considered for future studies.

The assumption of a constant water demand until 2100 made in this study because of a consistency with Water Vision 2020. Note that the choice of the demand scenario used in Water Vision 2020 was a very valuable product that results from a long negotiation process between the Korean government and non-governmental organizations. However, the suitability of this future demand scenario should be re-studied and updated to be more realistic as a future study. For instance, Koutroulis *et al.* [34] developed a simple future water demand storyline based on future irrigation extension plans, population trends, *etc*. Above all, however, it needs to be reiterated that the main purpose of this study is to alert decision makers of the risk of the direct use of GCM-driven data without thorough analysis.

The proposed methodology is likely to be more useful for regions where the rainfall regime is highly variable and thus hard to find a consensus between GCMs. Korea, affected by the monsoon climate, belongs to this category. The proposed methodology may perform differently for other rainfall regimes such as semiarid regions.

To conclude, this approach could provide decision makers a reliable and realistic alternative for future water resources management plans. In addition, various approaches that take into consideration multiple variables, such as weighting factors, and project extreme water shortage values should be pursued in future studies.

Acknowledgments

This study has been supported by "the Construction Technology Innovation Program (09-Tech-Inovation-C01) through the Climate Change Assessment and Projection for Hydrology in Korea (CCAPH-K)" at the Korea Institute of Construction and Transportation Technology Evaluation and Planning (KICTEP) of the Ministry of Land, Transport and Maritime Affairs (MLTM), Korea.

This study is also supported by an on-going project, "the Water Management Research and Development (R&D) through the Climate Change Adaptation for Water Resources (Project ID: 14AWMP-B082564-01)", at the Korea Agency for Infrastructure Technology Advancement (KAIA) of the Ministry of Land, Infrastructure and Transport (MOLIT), Korea.

Author Contributions

Seung Beom Seo mainly contributed to the design and development of this manuscript under the supervision of Young-Oh Kim. Cho-Rong Kim contributed to the data collection and analysis.

Conflicts of Interest

The authors declare no conflict of interest.

References

1. Milly, P.C.D.; Betancourt, J.; Falkenmark, M.; Hirsch, R.M.; Kundzewicz, Z.W.; Lettenmaier, D.P.; Stouffer, R.J. Stationarity is dead: Whither water management? *Science* **2008**, *319*, 573–574.

2. Christensen, N.S.; Wood, A.W.; Voison, N.; Lettenmaier, D.P.; Palmer, R.N. The effect of climate change on the hydrology and water resources of the Colorado River Basin. *Clim. Chang.* **2004**, *62*, 337–363.
3. Medellin-Azuara, J.; Harou, J.J.; Olivares, M.A.; Madani, K.; Lund, J.R.; Howitt, R.E.; Tanaka, S.K.; Jenkins, M.W.; Zhu, T. Adaptability and adaptations of California's water supply system to dry climate warming. *Clim. Chang.* **2008**, *87*, 75–90.
4. Minville, M.; Brissette, F.; Leconte, R. Impacts and uncertainty of climate change on water resource management of the Peribonka River System (Canada). *J. Water Resour. Plan. Manag.* **2010**, *136*, 376–385.
5. Islam, M.S.; Aramaki, T.; Hanaki, K. Development and application of an integrated water balance model to study the sensitivity of the Tokyo Metropolitan area water availability scenario to climate changes. *Water Resour. Manag.* **2005**, *19*, 423–445.
6. Vicuna, S.; Maurer, E.P.; Joyce, B.; Dracup, J.A.; Purkey, D. The sensitivity of California water resources to climate change scenarios. *J. Am. Water Resour. Assoc.* **2007**, *43*, 482–498.
7. Hoff, H.; Swartz, C.; Yates, D.; Tielborger, K. Water management under extreme water scarcity: Scenario analyses for the Jordan River Basin, using WEAP21. *Water Sav. Mediterr. Agric. Future Res. Needs* **2007**, *2*, 321–331.
8. Purkey, D.R.; Joyce, B.; Vicuna, S.; Hanemann, M.W.; Dale, L.L.; Yates, D.; Dracup, J.A. Robust analysis of future climate change impacts on water for agriculture and other sectors: A case study in the Sacramento Valley. *Clim. Chang.* **2008**, *87*, 109–122.
9. Wurbs, R.A.; Muttiah, R.S.; Felden, F. Incorporation of climate change in water availability modelling. *J. Hydrol. Eng.* **2005**, *10*, 375–385.
10. Fowler, H.J.; Kilsby, C.G.; Stunell, J. Modeling the impacts of projected future climate change on water resources in North-West England. *Hydrol. Earth Syst. Sci.* **2007**, *11*, 1115–1126.
11. Fujihara, Y.; Tanaka, K.; Watanabe, T.; Nagano, T.; Kojiri, T. Assessing the impacts of climate change on the water resources of the Seyhan River Basin in Turkey: Use of dynamically downscaled data for hydrologic simulations. *J. Hydrol.* **2008**, *353*, 33–48.
12. Kirono, D.G.C.; Podger, G.; Franklin, W.; Siebert, R. Climate change impact on rous water supply. *J. Aust. Water Assoc.* **2007**, *34*, 68–72.
13. Vano, J.A.; Scott, M.; Voisin, N.; Stockle, C.O.; Hamlet, A.F.; Mickelson, K.E.B.; Elsner, M.M.; Lettenmaier, D.P. Climate change impacts on water management and irrigated agriculture in the Yakima River Basin, Washington, USA. *Clim. Chang.* **2010**, *102*, 287–217.
14. Zhu, T.; Ringler, C. Climate change impacts on water availability and use in the Limpopo River Basin. *Water* **2012**, *4*, 63–84.
15. Kim, S.J.; Kim, B.S.; Jun, H.D.; Kim, H.S. The evaluation of climate change impacts on the water scarcity of the Han River Basin in South Korea using high resolution RCM data. *J. Korea Water Resour. Assoc.* **2010**, *42*, 295–308. (In Korean)
16. Kang, S.U.; Rieu, S.Y.; Cheong, T.S.; Ko, I.H. Evaluation water resources in Geum River Basin using generalized water budget analysis model. In Proceedings of the Korea Society of Civil Engineers Conference, Gwangju, Korea, 11 October 2006; Volume 10, pp. 1653–1656. (In Korean)
17. Choi, S.J.; Kim, J.H.; Lee, D.Y. Decision of the water shortage mitigation policy using multi-criteria decision analysis. *KSCE J. Civ. Eng.* **2012**, *16*, 247–253.

18. Ministry of Construction and Traffic. *National Water Resources Plan (2006–2020)*; Ministry of Construction and Transportation: Gwacheon, Korea, 2006. (In Korean)
19. Kim, Y.O.; Lee, J.K. Addressing heterogeneities in climate change studies for water resources in Korea. *Curr. Sci.* **2010**, *98*, 1077–1083.
20. Groves, D.G.; Yates, D.; Tebaldi, C. Developing and applying uncertain global climate change projections for regional water management planning. *Water Resour. Res.* **2008**, *44*, W12413.
21. Sharif, M.; Burn, D.H. Simulating climate change scenarios using an improved K-nearest neighbor model. *J. Hydrol.* **2006**, *325*, 179–196.
22. Yates, D. A Technique for generating regional climate scenarios using a nearest-neighbor algorithm. *Water Resour. Res.* **2003**, *39*, W1199.
23. Ministry of Construction and Traffic. *National Water Resources Security against Climate Change*; Ministry of Land, Transport and Maritime Affairs: Sejong Special Autonomous City, Korea, 2009. (In Korean)
24. Korea Institute of Construction Technology. *K-WEAP User Guide*; Korea Institute of Construction Technology: Goyang, Korea, 2005. (In Korean)
25. Lee, J.K. Scenario selection and uncertainty quantification for climate change impact assessments in water resources. In *Doctoral Dissertation*; Seoul National University: Seoul, Korea, 2013. (In Korean)
26. IPCC Data Distribution Centre: Climate Model Output—AR4 (2007): SRES Scenarios. Available online: http://www.ipcc-data.org/sim/gcm_monthly/SRES_AR4/index.html (accessed on 15 May 2015).
27. Bae, D.H.; Jung, I.W.; Lee, B.J. Outlook on variation of water resources in Korea under SRES A2 scenario. *J. Korean Water Resour. Assoc.* **2007**, *40*, 921–930. (In Korean)
28. Wiley, M.W. Analysis Techniques to Incorporate Climate Change Information into Seattle's Long Range Water Supply Planning. M.Sc. Thesis, University of Washington, Seattle, WA, USA, 2004.
29. Hamlet, A.F.; Lettenmaier, D.P. Effects of climate change on hydrology and water resources objectives in the Columbia River Basin. *J. Am. Water Resour. Assoc.* **1999**, *35*, 1597–1624.
30. Hashino, T.; Bradley, A.A.; Schwartz, S.S. Evaluation of bias-correction methods for ensemble streamflow volume forecasts. *Hydrol. Earth Syst. Sci.* **2007**, *11*, 939–950.
31. Wood, A.W.; Maurer, E.P.; Kumar, A.; Lettenmaier, D.P. Long-range experimental hydrologic forecasting for the Eastern United States. *J. Geophys. Res. Atmos.* **2002**, *107*, 4429–4443.
32. Ines, A.V.M.; Hansen, J.W. Bias correction of daily GCM rainfall for crop simulation studies. *Agric. For. Meteorol.* **2006**, *138*, 44–53.
33. Fabre, J.; Ruelland, D.; Dezetter, A.; Grouillet, B. Simulating past changes in the balance between water demand and availability and assessing their main drivers at the river basin scale. *Hydrol. Earth Syst. Sci.* **2015**, *19*, 1263–1285.
34. Koutroulis, A.G.; Tsanis, I.K.; Daliakopoulos, I.N.; Jacob, D. Impact of climate change on water resources status: A case study for Crete Island, Greece. *J. Hydrol.* **2013**, *479*, 146–158.

A Heuristic Dynamically Dimensioned Search with Sensitivity Information (HDDS-S) and Application to River Basin Management

Jinggang Chu *, Yong Peng, Wei Ding and Yu Li

School of Hydraulic Engineering, Dalian University of Technology, Dalian 116024, China; E-Mails: pengyong@dlut.edu.cn (Y.P.); weiding@mail.dlut.edu.cn (W.D.); landyte.ly.dut@gmail.com (Y.L.)

* Author to whom correspondence should be addressed; E-Mail: jgchu@mail.dlut.edu.cn;

Academic Editor: Athanasios Loukas

Abstract: River basin simulation and multi-reservoir optimal operation have been critical for river basin management. Due to the intense interaction between human activities and river basin systems, the river basin model and multi-reservoir operation model are complicated with a large number of parameters. Therefore, fast and stable optimization algorithms are required for river basin management under the changing conditions of climate and current human activities. This study presents a new global optimization algorithm, named as heuristic dynamically dimensioned search with sensitivity information (HDDS-S), to effectively perform river basin simulation and multi-reservoir optimal operation during river basin management. The HDDS-S algorithm is built on the dynamically dimensioned search (DDS) algorithm; and has an improved computational efficiency while maintaining its search capacity compared to the original DDS algorithm. This is mainly due to the non-uniform probability assigned to each decision variable on the basis of its changing sensitivity to the optimization objectives during the adaptive change from global to local search with dimensionality reduced. This study evaluates the new algorithm by comparing its performance with the DDS algorithm on a river basin model calibration problem and a multi-reservoir optimal operation problem. The results obtained indicate that the HDDS-S algorithm outperforms the DDS algorithm in terms of search ability and computational efficiency in the two specific problems. In addition; similar to the DDS algorithm; the HDDS-S algorithm is easy to use as it does not require any parameter tuning and automatically adjusts its search to find good solutions given an available computational budget.

Keywords: dynamically dimensioned search; non-uniform probability; dimensionality reduction; sensitivity analysis; Soil and Water Assessment Tool (SWAT); multi-reservoir operation

---

1. Introduction

It has been increasingly recognized that climate change and human activities have a profound impact on river basin management [1–3]. The law of hydrologic cycle evolution, which could be explored by river basin simulation effectively, has a profound impact on river basin management in terms of efficiency of system optimization [4–8]. Because the intense interactions between climate change and human activities lead to great changes in river quantity and quality regimes, river basin simulation becomes more difficult under the influence of many parameters. On the other hand, due to the uneven distribution of water resources in terms of time and space, a large number of reservoirs have been constructed in China as well as in many other countries. Although the joint operation of multi-reservoir systems could assist water resource optimal distribution in river basins, the complex topological structure of multiple reservoir systems in river basins makes the joint operation more difficult [9–14].

Because a large number of decision variables and constraints exist in the river basin model and the multi-reservoir operation model, developing high-performance optimization algorithms is critical for river basin model calibration problems and multi-reservoir optimal operation problems in river basin management. Use of conventional optimization algorithms to solve these optimization problems might result in slow convergence and low accuracy issues. Heuristic algorithms, which mimic human thinking, have become popular in recent years, as they are able to provide near-optimal solutions and can improve the speed of search [15,16].

There are many heuristic algorithms that are developed in recent years, such as Genetic Algorithms [17–23], Particle Swarm Optimization [24,25], Shuffled Complex Evolution (SCE) [26,27] and Ant Colony Optimization [28–30]. The SCE algorithm is one of the popular optimization algorithms in the river basin model calibration over the past 10 years given that more than 300 different publications referenced the original SCE publications [26,27,31]. However, various kinds of optimization algorithms need to utilize an extraordinary number of function (or model) evaluations to approximate the global optimum more accurately and it would considerable time to solve the complex optimization problems.

Generally, a wide variety of such problems could be solved in a simulation-optimization framework. The time required for one single simulation can vary from seconds to hours; the budget should be spent as efficiently as possible as it usually is limited. A global optimization algorithm called dynamically dimensioned search (DDS), which was proposed by Tolson and Shoemaker [32], is able to automatically refine and update optimal search space, and thus reduce the number of iterations required to reach satisfying solutions. Tolson and Shoemaker [32] compared the DDS and SCE algorithms from two aspects—The number of decision variables and the number of iterations, and found that the DDS algorithm is able to converge faster and get better solutions within a specified number of iterations, which proves its effectiveness for solving complex optimization problems. Due to the relatively good performance, simplicity and parsimony of the DDS algorithm, *i.e.*, its ability to automatically scale the search space for good solutions within a specified computational time limit without requiring algorithm parameter adjustment, the DDS algorithm is ideally suited for solving computationally expensive optimization problems.

Most optimization algorithms treat each decision variable with an equal and constant probability to search and do not consider the variation of the probability in the search process, e.g., the SCE algorithm searches in the global space without dimension reduced. However, the DDS algorithm searches globally at the start of the search and conducts a more localized search as the number of iterations approaches the maximum allowable number of function evaluations, which is achieved by dynamically and probabilistically reducing the number of dimensions in the neighborhood and the reducing probability is distributed to each decision variable equally. Meanwhile, Hansen and Ostermeier [33] developed an adaptive evolutionary strategy to improve search efficiency, where decision variable co-variant matrix could be constructed and their automatic adjustment could be implemented in the search process.

This study presents a further developed DDS algorithm to solve high-dimensional optimization problems, *i.e.*, river basin model calibration problem and multi-reservoir optimal operation problem in river basin management, with many parameters more efficiently. The newly developed global optimization algorithm, named as heuristic dynamically dimensioned search with sensitivity information (HDDS-S), is built on the DDS algorithm and could also search from global to local space with a reduced number of dimensions. However, during the adaptive change from global to local search with dimensionality reduced, instead of the reducing probability being distributed to each decision variable equally in the DDS algorithm, the non-uniform probability is assigned to each decision variable on the basis of its changing sensitivity to the optimization objectives in the HDDS-S algorithm. To evaluate the efficiency in solving high-dimensional and computationally expensive optimization problems in river basin management, the HDDS-S algorithm is tested extensively against the DDS algorithm on the river basin model calibration problem for the Tang-Wang River Basin in Jilin province, China, and the optimal operation problem for the DHF-GYG-SW (Dahuofang reservoir (DHF), Guanyinge reservoir (GYG), and Shenwo reservoir (SW)) multi-reservoir in the northeast of China.

2. Methods

2.1. Benchmark Optimization Algorithm

The optimization problem can be represented in the following mathematical way:

$$\begin{aligned} y &= minf(x) = minf(x_1, \cdots, x_p) \\ s.t.\, g_i(x_1, \cdots, x_p) &\le 0, i = 1,2, \cdots, m \\ h_j(x_1, \cdots, x_p) &\ge 0, j = 1,2, \dots, n \end{aligned} \tag{1}$$

where y is the objective function of one optimization problem, $x = (x_1, \cdots, x_p)$ is the parameter set, $g_i\,(x_1, \cdots, x_p)\ (i = 1,2, \cdots, m)$ and $h_j\,(x_1, \cdots, x_p)\ (i = 1,2, \cdots, n)$ are constraint functions.

This study presents the HDDS-S algorithm developed on the basis of the DDS algorithm to effectively perform river basin simulation and multi-reservoir optimal operation during river basin management. The original DDS algorithm is chosen as the benchmark optimization algorithm, that is, the HDDS-S algorithm is tested extensively against the DDS algorithm in solving high-dimensional and computationally expensive optimization problems in river basin management. The DDS algorithm was developed for the purpose of finding good global solutions, which are opposed to globally optimal solutions, within a specified number of maximum function evaluations. The key feature of the DDS algorithm was motivated by past experience from manual calibration of river basin models and reservoir operation models. A number of model parameters should be modified simultaneously early in the optimization exercise due to the fact that only relatively poor solutions could be obtained more often at this moment. However, as the optimization results improve, it becomes necessary to only modify a few parameters simultaneously so that the current gain in optimization results would not be lost. Additionally, through the observation of multi-dimensional decision variable optimization process, Tolson and Shoemaker [32] found that the number of decision variables that need to be optimized for improving the quality of solutions is decreasing as the values of decision variables approach their best values gradually. Therefore, unlike other heuristic algorithms that optimize each decision variable within a fixed space all the time, the DDS algorithm automatically sets the searching region size according to the current iteration count and user-specified maximum number of function (or model) evaluations to reflect the phenomenon that the number of decision variables to be optimized is decreasing, which is achieved by dynamically and probabilistically reducing the number of dimensions in the neighborhood and the reduced probability is distributed to each decision variable equally. The probability of choosing each decision variable to search based on a function of iteration count in the DDS algorithm could be calculated as:

$$P(i,j) = 1 - ln(j)/ln(N) \tag{2}$$

where $P(i,j)$ is the probability of choosing decision variable x_i to search on iteration j, N is the total number of iterations.

On the other hand, in the original DDS algorithm, candidate solutions are created by perturbing the current solution values in the randomly selected dimensions only. These perturbation magnitudes are randomly sampled from a normal distribution with a mean of zero to keep the DDS algorithm as simple and parsimonious as possible. The DDS algorithm is a greedy type of algorithm since the current best solution is never updated with a solution that has an inferior value of the objective function. See Tolson and Shoemaker [32] for a detailed description of the DDS algorithm.

2.2. Heuristic Dynamically Dimensioned Search with Sensitivity Information

The regular used algorithms adopt equal probability search approach without dimension reduction, e.g., the SCE algorithm. The DDS algorithm could select dimensions to search randomly with a function of iteration count, and the number of selected dimensions is decreasing gradually with the increase of iteration count. Similar to the DDS algorithm, covariance matrix adaptation (CMA) and a multi-algorithm, genetically adaptive multi-objective (AMALGAM) algorithms could create the number of offspring points self-adaptively based on experiences, and all the points are searched with the same probabilities [33–36].

Heuristic dynamically dimensional search with sensitivity information (HDDS-S) improves the equal probability random search with dimension reduction in the DDS, CMA and AMALGAM algorithms, realizing the non-uniform probability search with the changing sensitivity of different dimensions to the objectives. The search approaches of different algorithms are shown in Figure 1.

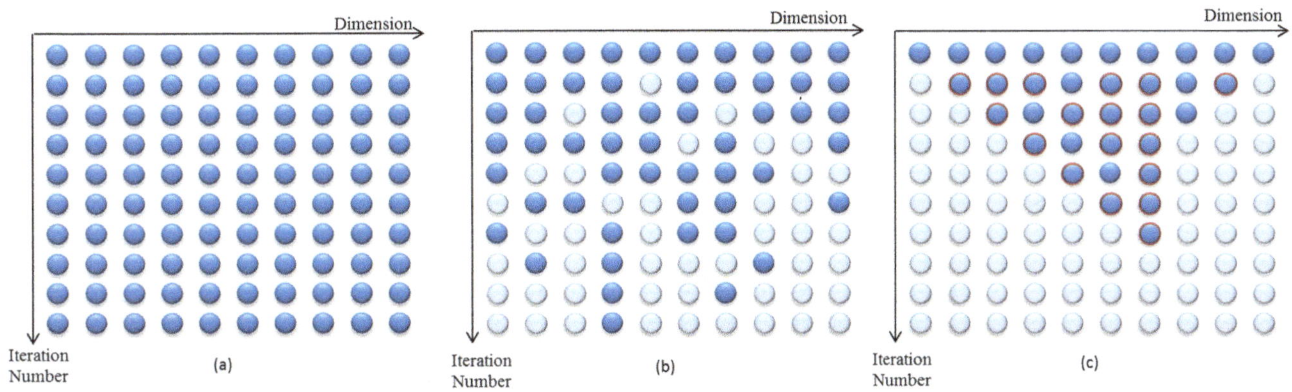

Figure 1. The search approaches of different algorithms. (**a**) Equal probability search approach without dimension reduction; (**b**) Equal probability search approach with dimension reduction; (**c**) Non-uniform probability search with dimension reduction.

Because the number of decision variables that need to be optimized for improving the quality of solutions is decreasing as the values of decision variables approach their best values gradually, the changing process of decision variable sensitivity works on two principles: (1) the sensitivity of decision variable reduces with the increase of decision variables selected for searching on each iteration; (2) the sensitivity of decision variables is continuously cumulative with the increase of iteration number.

In the HDDS-S algorithm, to reflect the two principles, firstly, the sensitivity of each decision variable could be calculated as:

$$\begin{cases} s_{i,j} = 1/n_j \ x_i \in X_j \\ s_{i,j} = 0 \ x_i \notin X_j \end{cases} \tag{3}$$

where $s_{i,j}$ is the sensitivity of decision variable $x_i(i = 1, \cdots, D)$ on iteration $j(j = 1, \cdots, N)$, n_j is the number of decision variables selected for searching on iteration j, X_j is the collection of decision variables selected for searching on iteration j, D and N are the total number of decision variables and iterations respectively.

Secondly, the cumulative sensitivity of decision variable could be calculated as:

$$sc_{i,j} = \sum_{l=1}^{j} (N - j + l)/m \times s_{i,l} \tag{4}$$

where $sc_{i,j}$ is the cumulative sensitivity of decision variable x_i from 1 to iteration j.

Similar to the DDS algorithm, the HDDS-S algorithm could search from global to local space with a reduced number of dimensions. However, instead of the reducing probability being distributed to each decision variable equally in the DDS algorithm, the non-uniform probability is assigned to each decision variable on the basis of its changing sensitivity to the optimization objectives in the HDDS-S algorithm. The probability of choosing a decision variable to search based on its cumulative sensitivity could be calculated as:

$$P(i,j+1) = (1 - ln(j)/ln(N)) \times ((sc_{i,j} - sc_j^{min})/(sc_j^{max} - sc_j^{min})) \quad (5)$$

where $P(i,j+1)$ is the probability of choosing decision variable x_i to search on iteration $j+1$, sc_j^{max} and sc_j^{min} are the maximum and minimum cumulative sensitivity values among all decision variables from 0 to the iteration j. The calculation of $P(i,j+1)$ reflects the dynamic adjustment of decision variable search strategy, that is, the non-uniform probability search with the changing sensitivity of different dimensions to the objectives.

Therefore, the HDDS-S algorithm has the following characteristics: firstly, changing of the sensitivity matrix of multi-dimension parameters (decision variables) and cumulative sensitivity of each parameter are updated dynamically during the process of optimization according to the relationship between dimension selection and optimizing efficiency; secondly, the probability of choosing each parameter to optimize is calculated according to its cumulative sensitivity; finally, the optimization strategy of choosing parameters with non-uniform probability during the process of parameter optimization is realized to improve convergence rate and optimization efficiency. Additionally, similar to the DDS algorithm, the only algorithm parameter to set in the HDDS-S algorithm is the scalar neighbourhood size perturbation that defines the random perturbation size standard deviation as a fraction of the decision variable range. Tolson and Shoemaker [32] recommended the value of r parameter as 0.2 (and used in this study) because this yields a sampling range that practically spans the normalized decision variable range for a current decision variable value halfway between the minimum and maximum.

The complete HDDS-S algorithm pseudo-code for minimization can be seen in the Appendix Pseudo-code of HDDS-S.

3. Model Description and Study Area

Almost all river basin models contain effective physical and/or conceptual model parameters that are either difficult or impossible to directly measure. Optimization algorithms could be used to automate the calibration, *i.e.*, automatic calibration is defined as an optimization algorithm-based search for a set of river basin model parameter values that minimize the model prediction errors relative to available measured data for the system being modeled process. Tolson and Shoemaker [32] compared the performance of the DDS algorithm to that of the SCE algorithm, which is currently one of the most commonly applied algorithms for automatic calibration of river basin models, with multiple optimization test functions as well as real and synthetic SWAT2000 model automatic calibration formulations. In that study, numerical results demonstrate that the DDS algorithm is a more computationally efficient and robust optimization algorithm than the SCE algorithm in the context of river basin model automatic calibration. Therefore, the HDDS-S algorithm developed on the basis of DDS is evaluated by comparing its performance with the DDS algorithm on a river basin model calibration problem in this paper.

Meanwhile, river basin simulation is closely related to multi-reservoir optimal operation. Firstly, the future runoff series under the changing conditions could be generated on the basis of the law of hydrologic cycle evolution, which could be explored with river basin model effectively, e.g., SWAT model. Secondly, the future runoff series need to be used as the inputs of the multi-reservoir operation model to explore the impacts of changing conditions on multi-reservoir operation, determine the operation rules, and assist in making multi-reservoir optimal operation decisions under changing conditions. As with river basin simulation, multi-reservoir optimal operation has been critical for river basin management; the multi-reservoir operation model contains large number of decision variables to be calibrated, *i.e.*, large number of the curves and the number of time periods, which increases the complexity and problem, and poses a significant challenge for most optimization tools currently available [9–14]. To evaluate the universality of the HDDS-S algorithm in river basin management, the HDDS-S algorithm is also evaluated by comparing its performance with the DDS algorithm on a multi-reservoir optimal operation problem in this paper.

Two cases are used to extensively test our new HDDS-S algorithm against the DDS algorithm in this paper: (1) the river basin model calibration for Tang-Wang River Basin; and (2) the optimal operation for the DHF-GYG-SW multi-reservoir. The main difference between the multi-reservoir operation model and river basin model is that the decision variables in multi-reservoir operation models are reservoir operation rule curves, which consist of a series of storage volumes or levels at different periods, and the aim of multi-reservoir optimal operation is to establish reservoir operation rule curves at the planning/design stage and provide long-term operation guidelines for reservoir managers to meet expected water demands [37–39]. Therefore, the decision variables, which will be optimized in the two optimization problems, are parameters of the river basin model and multi-reservoir optimal operation model, and the dimension being varied is the number of model parameter values being changed to generate a new search neighbourhood.

3.1. Model Description

3.1.1. River Basin Model

Overview of the SWAT Model

The SWAT model is a catchment-scale semi-distributed river basin model developed by the Agricultural Research Service of the United States Department of Agriculture; its detailed description can be seen in many references, such as [40,41]. The SWAT model can be run at a daily time step when the Soil Conservation Service (SCS) curve number method is used to calculate surface runoff, and it is not designed to accurately simulate detailed, single-event flood routing. In this paper, the river basin model calibration problem is formulated based on the set of the SWAT2005 flow-related parameters, which impact snowmelt, surface runoff, groundwater, and lateral flow simulation.

Because SWAT is a semi-distributed river basin model and Hydrological Responding Units (HRUs) divided with geographic information are its basic units, some parameters are not regarded as spatial variables but rather constant values across all spatial units and some other parameters are spatially varied among different HRUs. For example, those parameters related to snowmelt have constant values in all spatial units, and could be calibrated one by one. However, some parameters, such as SCS curve numbers, could be assigned different values for different spatial units, which significantly increase the complexity and computational model requirement of parameter calibration. In this paper, a single factor is used to represent spatial variation, by increasing or decreasing spatially varying parameter values from their base or default values. For spatially varied parameter, this approach maintains the relative differences in the base or default parameter values assigned to different spatial units.

The flow-related parameters of the SWAT model and their ranges are listed in Table 1, where the parameter ranges are based mainly on the default ranges in the SWAT2005 model documentation.

Table 1. Parameter list.

No.	Name	Brief Description (units)	Minimum	Maximum
1	SFTMP	snow fall temperature (°C)	−5	5
2	SMTMP	snowmelt temperature threshold (°C)	−5	5
3	SMFMX	melt factor for snow on June 21 (mm/°C)	1.5	8
4	TIMP	snowpack temperature lag factor	0.01	1
5	ESCO	soil evaporation compensation factor	0.001	1
6	SURLAG	surface runoff lag coefficient	1	24
7	GW_DELAY	groundwater delay time (days)	0.001	500
8	ALPHA_BF	base flow alpha factor	0.001	1
9	GWQMN	threshold groundwater depth for return flow (mm)	0.001	500
10	LAT_TTIME	lateral flow traveltime (days)	0.001	180
11	CN2 [a]	SCS runoff curve number multiplicative factor for moisture condition II	0.75	1.25

Note: [a] Parameters are multiplicative factors used to adjust the spatial variation across all model units.

Evaluation Criterion and Objective Function

The mean relative error (MRE), the coefficient of determination (R^2), and the Nash–Sutcliffe efficiency (NSE) are used to evaluate the simulated streamflows with the observed streamflows.

MRE is computed according to Equation (6):

$$MRE = \left(\overline{Q_p} - \overline{Q_t}\right)/\overline{Q_t} \times 100\% \tag{6}$$

where $\overline{Q_p}$ and $\overline{Q_t}$ are the means of the simulated streamflows and the observed streamflows, respectively. "MRE values of 0 indicate a perfect fit. Positive values indicate model overestimation bias, and negative values indicate model underestimation bias [42]."

R^2 is computed according to Equation (7):

$$R^2 = \sum_{i=1}^{n}(Q_{pi} - \overline{Q_p})(Q_{ti} - \overline{Q_t}) \Big/ \sqrt{\sum_{i=1}^{n}\left(Q_{pi} - \overline{Q_p}\right)^2 \sum_{i=1}^{n}(Q_{ti} - \overline{Q_t})^2} \tag{7}$$

where Q_{pi} is the ith simulated value for the streamflows; Q_{ti} is the ith observation for the streamflows; and n is the total number of observations. "R^2 describes the portion of the variance in measured data explained by the model. R^2 ranges from 0 to 1, with lower values indicating more error variance, and typically $R^2 = 1$ is considered the optimal value [43]."

NSE is computed according to Equation (8):

$$NSE = 1 - \sum_{i=1}^{n}\left(Q_{pi} - Q_{ti}\right)^2 \Big/ \sum_{i=1}^{n}(Q_{ti} - \overline{Q_t})^2 \tag{8}$$

The NSE ranges between $-\infty$ and 1.0, with $NSE = 1.0$ as the optimal value. According to Luo *et al.* [44], values between 0.0 and 1.0 are generally viewed as acceptable levels of performance, whereas a value less than 0.0 means that the mean observed value is a better predictor than the simulated value, which indicates unacceptable performance.

To evaluate the HDDS-S algorithm by comparing its performance with that of the DDS algorithm on river basin model calibration, the sum of squared differences between the simulated streamflows and the observed streamflows is used as the objective function, which is shown in Equation (9):

$$SSE = \sum_{i=1}^{n} \left(Q_{pi} - Q_{ti}\right)^2 \tag{9}$$

where SSE is the sum of squared differences between the simulated streamflows and the observed stream flows.

3.1.2. Multi-Reservoir Optimal Operation Model

Multi-Reservoir Operation Rule

Multi-reservoir operation rule consists of operation rule curves and rationing factors for each water demand. Details of the operation rule curves and their corresponding water supply rationing factors are illustrated in Figure 2 and Table 2. The forms of reservoir operation rule curves based on 36 10-day periods are shown schematically in Figure 2. The operation rule curves are defined according to reservoir storage, that is, the dynamic water storage of the reservoir is taken as the single, influential factor for reservoir operation.

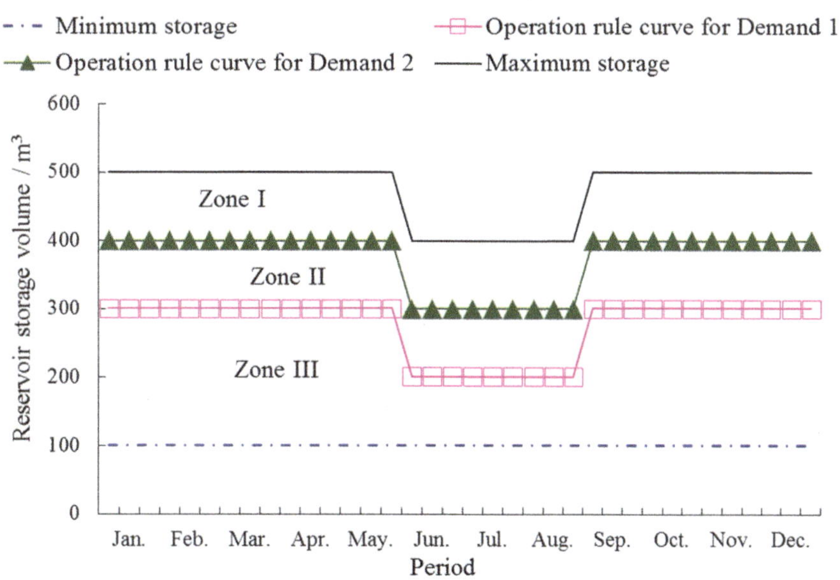

Figure 2. Reservoir operational rule curves.

In this paper, the active water storage of reservoir is divided into three parts, zone I, zone II and zone III, by two operation rule curves for industrial (D1) and agricultural (D2) demands. Each of the water demands has a related rule curve and a rationing factor. Because different kinds of water demands require different reliabilities and different degrees of priority in practice, the rationing factors usually have different values. For example, assume water rationing factor is α_1 for the operation rule curve for industrial demand (D1) in Figure 2. When the water storage of the reservoir lies in zone I or zone II, water demand

D1 is fully met, *i.e.*, the amount of water supply is D1; when the water storage of reservoir is in zone III, the amount of water supply is calculated by applying the water rationing factor, *i.e.*, $\alpha_1 \times$ D1.

Table 2. Operation rule implied by rule curves.

Reservoir Storage	Water Supply for Each Demand	
	Demand 1 (D1)	Demand 2 (D2)
Zone I	D1	D2
Zone II	D1	$\alpha_2 \times$ D2
Zone III	$\alpha_1 \times$ D1	$\alpha_2 \times$ D2
Rationing factor	α_1	α_2

Multi-Reservoir Operation Model

The multi-reservoir optimal operation problem is formulated as a single-objective optimization problem, where decision variables are the water storages at different periods on different operation curves, including industrial and agricultural curves of XN-2 (the virtual reservoir which GYG and SW reservoirs are aggregated into) and DHF reservoirs. Meanwhile, the operation for the entire system should meet the guaranteed rate of each water demand, *i.e.*, the guaranteed water-supply rate of industry and agriculture are above 95% and 50%, respectively. The guaranteed rate of each water demand can be calculated by statistics from the operation result, shown in Equation (10):

$$GR_i(x) = \frac{NTG_i}{NT_i + 1} \tag{10}$$

where x is the vector of decision variables, *i.e.*, the water storages at different periods on an operation rule curve; GR_i is the guaranteed rate for water demand i (industrial water demand when $i = 1$, agricultural water demand when $i = 2$), which measures the frequency of water demand being satisfied during entire time periods and is used as an indicator to reflect water supply capacity of the system; NT_i is the entire operation period for demand i (industrial demand when $i = 1$, agricultural demand when $i = 2$); similar to NT_i, NTG_i is the time period when water demand i is being satisfied.

Additionally, the number of depth damages, which indicate that the reservoir water storage has been lowered to dead water storage, should be as few as possible during the operation for the entire system. Similarly, the number of depth damages could be calculated by statistics from the operation result.

In this paper, the objectives are set to minimize the number of depth damages, ans improve the guaranteed rate of each water demand during the operation for the entire water supply system:

$$\begin{gathered} minf(x) = 1 - \sum_{i=1}^{m} \sum_{j=1}^{n} GR_{i,j} + \sum_{i=1}^{m} \sum_{j=1}^{n} ND_{i,j} \\ s.t.\, GR_{1,j}(\boldsymbol{x}) \geq 95\% \\ GR_{2,j}(\boldsymbol{x}) \geq 50\% \end{gathered} \tag{11}$$

where $GR_{i,j}$ is the guaranteed rate for water demand i in subsystem j (Sub A when $j = 1$, Sub B when $j = 2$, and Sub C when $j = 3$); similarly, $ND_{i,j}$ is the number of depth damages for water demand i in subsystem j.

For the multi-reservoir operation optimization problem, the mass balance equations are:

$$S_{t+1} - S_t = I_t - R_t - SU_t - E_t \tag{12}$$

$$R_t = g(\boldsymbol{x}),\ SU_t = k(\boldsymbol{x}), E_t = e(\boldsymbol{x}) \tag{13}$$

$$ST_t^{\min} \leq S_t \leq ST_t^{\max}, ST_t^{\min} \leq \boldsymbol{x} \leq ST_t^{\max} \tag{14}$$

where S_t is the initial water storage at the beginning of period t; S_{t+1} is the ending water storage at the end of period t; I_t, R_t, SU_t, and E_t are inflow, delivery for water use, spill and evapotranspiration loss, respectively; and ST_t^{max} and ST_t^{min} are the maximum and minimum storage, respectively.

3.2. Study Area

3.2.1. Tang-Wang River Basin

The Tang-Wang River is the first level tributary of the left bank of Song-Hua River, China. The total length of the Tang-Wang River is approximately 509 km. Its basin drains an area of 20,383 km^2. The study area is limited to the upper-middle stream region of the Tang-Wang River above Yixin hydrologic station, herein still referred to as Tang-Wang River Basin. The river basin drains an area of 11,376 km^2. The basin boundary and associated SWAT model sub-basin boundaries are presented in Figure 3. There are 87 sub-basins defined in the Tang-Wang River Basin, where 16 rain gauges and 1 streamflow gauge are located. Daily Meteorologic data (temperature, solar radiation, weed speed, relative humidity), daily precipitation data and stream flow data used in SWAT Model Calibration for Tang-Wang River Basin are listed in Table 3.

Figure 3. Tang-Wang River Basin.

Table 3. Hydrometeorological data for Tang-Wang River Basin.

Time Scale	Hydrologic/Meteorologic Element	Station	Period
Daily	Precipitation	16 gauges, such as Dongsheng	1979–2001
	Streamflow	Yixin	1979–2001
	Temperature, relative humidity, weed speed and solar radiation	Yichun	1979–2001

To evaluate the HDDS-S algorithm by comparing its performance with that of the original DDS algorithm on river basin model calibration, two scenarios are constructed: (1) model auto-calibration with the HDDS-S algorithm; and the (2) model auto-calibration with the DDS algorithm. The Yixin streamflow gauge was chosen as calibration station, and streamflow data from the 1979–2001 period were used to calibrate the model parameters. For each scenario, the first two months (January and February) are used as a warm up period for model simulation. The rest of the time periods in each scenario are used to assess the model's performance in the parameter calibration process.

3.2.2. DHF-GYG-SW Multi-Reservoir

This paper use thes DHF-GYG-SW multi-reservoir located in Huntai Basin as a case study, which consists of the Dahuofang reservoir (DHF), Guanyinge reservoir (GYG), and Shenwo reservoir (SW). Huntai basin, including Hun River and Taizi River, is located in a rapidly urbanizing region of Northeast China (Figure 4).

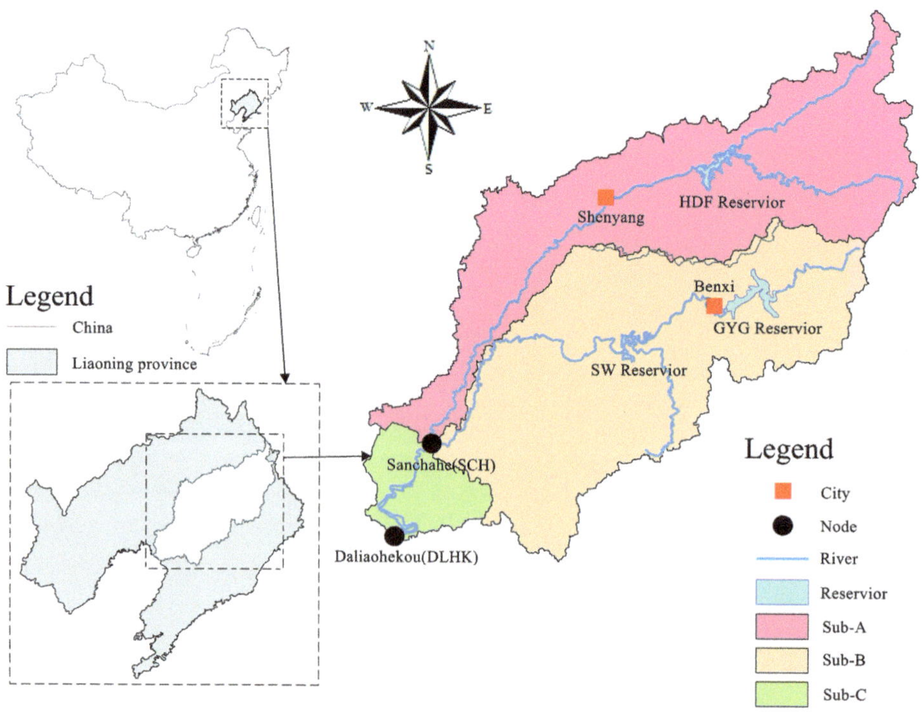

Figure 4. The location of the adopted water supply multi-reservoir.

DHF is a carry-over storage and is located in the trunk Stream of Hun River. It drains an area of 5437 km^2, and the total length of the Hun River is approximately 169 km within the basin. GYG and SW are cascade reservoirs and located in the trunk stream of the Taizi River. GYG is a carry-over storage and drains an area of 2795 km^2, while SW drains an area of 6175 km^2 with incomplete year regulation performance. The reservoir characteristics, annual average inflow and water supply tasks are illustrated in Table 4.

Table 4. Reservoir characteristics and tasks of water supply.

Reservoir	Minimum Capacity (10^8 m^3)	Active Capacity (10^8 m^3)		Annual Average Inflow (10^8 m^3)	Tasks of Water Supply
		Drought Season	Flood Season		
DHF	1.34	14.30	10.00	15.70	(2) (5)
GYG	0.35	14.20	14.20	11.10	(1) (3) (4) (5)
SW	0.35	5.43	2.14	12.80	(4) (5)

Notes: Indicate: (1) industrial water supply from reservoir directly; (2) agricultural water supply for DHF-SCH (DHF-Sanchahe) interval; (3) industrial water supply for GYG-SW interval; (4) agricultural and industrial water supply for SW-SCH (SW-Sanchahe) interval; (5) agricultural and industrial water supply for SW-DLHK (SW-Daliaohekou) interval.

The whole multi-reservoir system consists of three parts: subsystem A (Sub-A), which contains DHF and DHF-Sanchahe (SCH) interval; subsystem B (Sub-B), which contains GYG, GYG-SW interval, SW reservoir, and SW-SCH interval; and subsystem C (Sub-C), which contains SCH-Daliaohekou (DLHK) interval. The corresponding topological structure is generated, as shown in Figure 5a. Sub-A is supplied with water from DHF and uses DHF operation rules to decide how to supply water to each user in this subsystem. Sub-B is supplied with water from GYG and SW, which constitute cascade reservoirs and so are aggregated into one virtual reservoir (XN-2 reservoir), as shown in Figure 5b. Sub-C is supplied with water from the parallel reservoir, *i.e.*, XN-2 and DHF, the common tasks in Sub-C should be allocated to XN-2 and DHF. The water ratio method is employed to allocate the common tasks in Sub-C, which is a simple and efficient method shown in Figure 5c,d.

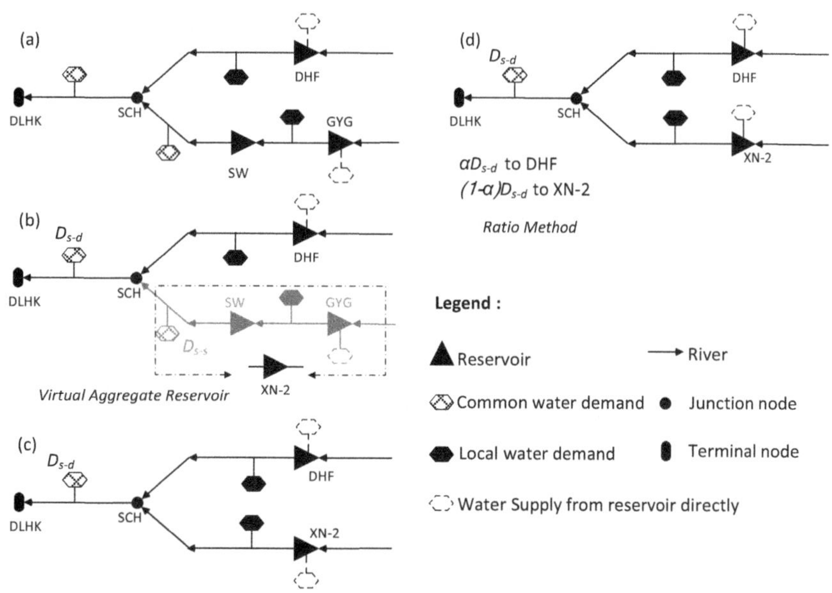

Figure 5. Reservoir aggregation and common water supply task allocation, (**a**) initial topological structure; (**b**) formation of virtual aggregate reservoir; (**c**) final topological structure after reservoir aggregation; (**d**) common water supply task allocation in the final topological structure.

The calculation of water division coefficient α is shown in Equations (15)–(17):

$$\alpha = SID_{DHF,t}/\left(SID_{DHF,t} + SID_{XN-2,t}\right) \quad (15)$$

$$SID_{DHF,t} = S_{DHF,t} + I_{DHF,t} - D_{DHF,t} \quad (16)$$

$$SID_{XN-2,t} = S_{XN-2,t} + I_{XN-2,t} - D_{XN-2,t} \quad (17)$$

where $S_{DHF,t}$ is the initial water storage of DHF at the beginning of period t; $I_{DHF,t}$ and $D_{DHF,t}$ are the inflow into DHF and water supply from DHF during the period t, respectively; $S_{XN-2,t}$ is the initial water storage of virtual aggregate reservoir XN-2 at the beginning of period t, *i.e.*, the sum water storage of GYG and SW; $I_{XN-2,t}$ and $D_{XN-2,t}$ are the inflow into XN-2 and water supply from XN-2 during the period t, *i.e.*, the sum inflow into GYG and SW, the sum water supply from GYG and SW, respectively.

4. Applications of the HDDS-S Algorithm and Results

4.1. River Basin Model Calibration

The river basin model calibration problem for the Tang-Wang River Basin is used to validate effectiveness of the new HDDS-S algorithm. The Yixin hydrologic station, located at the outlet of Tang-Wang River Basin, is chosen as the calibration station, and the streamflow data from the 1979–2001 periods are used to calibrate model parameters. Based on the two scenarios S0 and S1, *i.e.*, model auto-calibration with the HDDS-S and DDS algorithms respectively, performances of the two algorithms in the river basin model parameter auto-calibration are evaluated on Intel Core™ 2 Duo, with a 2.66-GHz CPU. In different scenarios, average required computation time under the same number of function evaluations, *i.e.*, $NFE = 1000$, or the same simulation accuracy, *i.e.*, $SSE = 370{,}000$, through 30 optimization trials are shown in Figure 6.

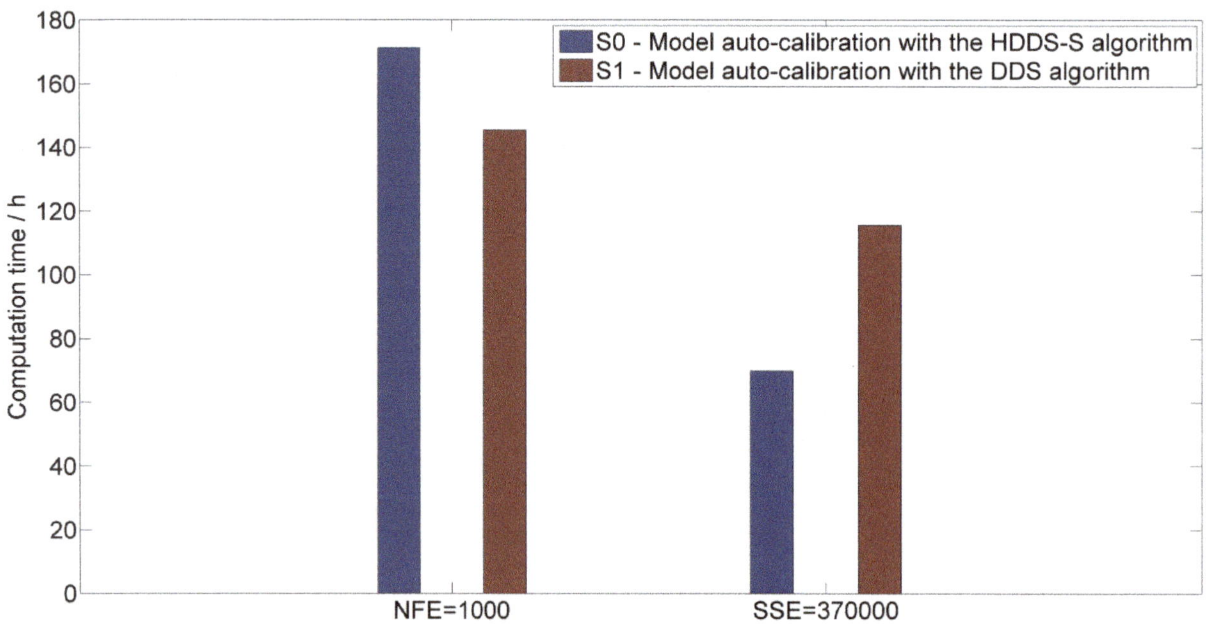

Figure 6. Average required computation time under the same number of function evaluations ($NFE = 1000$) and the same simulation accuracy ($SSE = 370{,}000$) in different scenarios through 30 optimization trials.

In different scenarios, average simulation results at the Yixin hydrologic station under the same number of function evaluations, *i.e.*, $NFE = 1000$, through 30 optimization trials is shown in Figure 7.

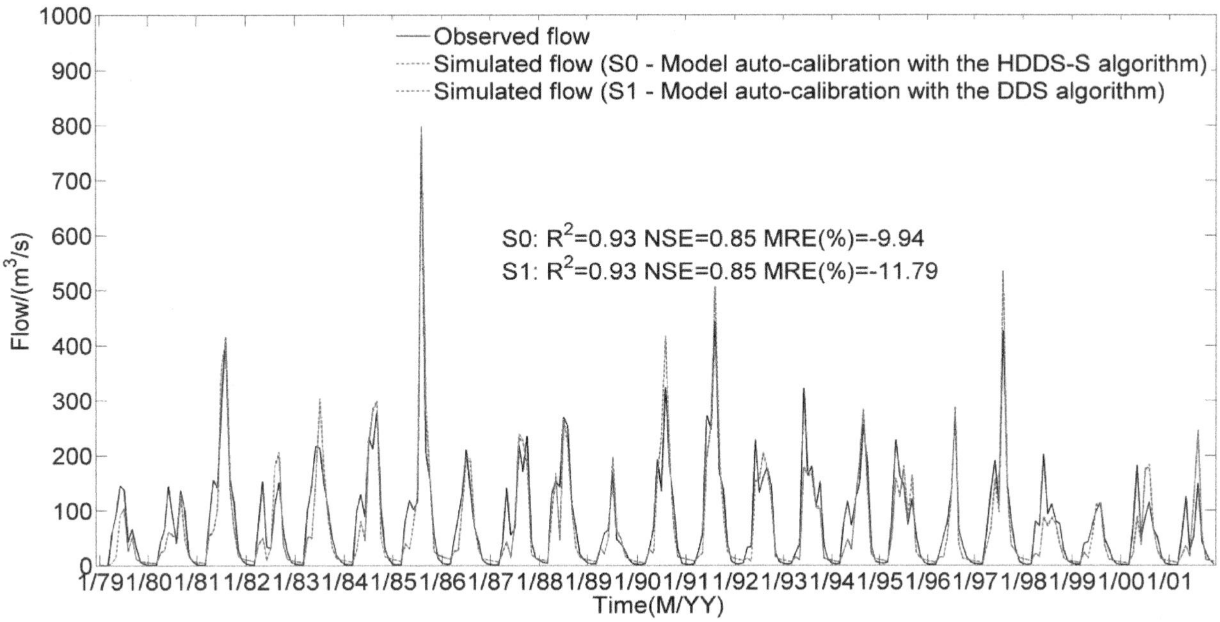

Figure 7. Observed and average simulated monthly streamflows during the period from 1979 to 2001 at the Yixin hydrologic station through 30 optimization trials.

Seen from Figures 6 and 7, under the same number of function evaluations ($NFE = 1000$), the average required computation time increases from 145.40 h in the case of the DDS algorithm to 171.32 h in the HDDS-S algorithm. Although the model performance improves from 11.79% (MRE) in the case of the DDS algorithm to 9.94% in the HDDS-S algorithm at the Yixin hydrologic station, there is little difference of model performances between the DDS algorithm and the HDDS-S algorithm in terms of the other two evaluation criterions because the simulation results will approximate their optimal values after 1000 function evaluations to a large extent. Under the same simulation accuracy ($SSE = 370{,}000$), the average required computation time reduces from 115.74 h ($NFE = 796$) in the case of the DDS algorithm to 69.90 h ($NFE = 408$) in the HDDS-S algorithm through 30 optimization trials. Although the average computation time in the case of the HDDS-S algorithm increases due to the additive sensitivity computation of each dimension decision variable compared to the DDS algorithm under the same iteration number, the average computation time in the case of the HDDS-S algorithm reduces greatly compared to the DDS algorithm under the same simulation accuracy.

The performances of the two algorithms in river basin model calibration are evaluated not only from average required computation time under the same number of function evaluations, average required computation time under the same simulation accuracy, and average simulation results at Yixin hydrologic station under the same number of function evaluations, but also from average convergence process of the objective function and average calibration process of model parameters under the same number of function evaluations. Convergence processes of the objective function under the same number of function evaluations ($NFE = 1000$) through 30 optimization trials in different scenarios are shown in Figure 8. It is further indicated that due to the consideration of the changing sensitivity information of each dimension decision variable, under the same number of function evaluations, convergence rate of the objective function is improved in the HDDS-S algorithm compared to the DDS algorithm. Additionally, there are few differences of the HDDS-S algorithm convergence processes among different optimization trials, which indicate the reliability and stability of the HDDS-S algorithm in solving model auto-calibration problem.

Under the same number of function evaluations ($NFE = 1000$), the average cumulative varied times for each parameter during model auto-calibration process in different scenarios through 30 optimization trials are shown in Figure 9, where the average cumulative varied times in S0 indicate the cumulative sensitivity of each parameter.

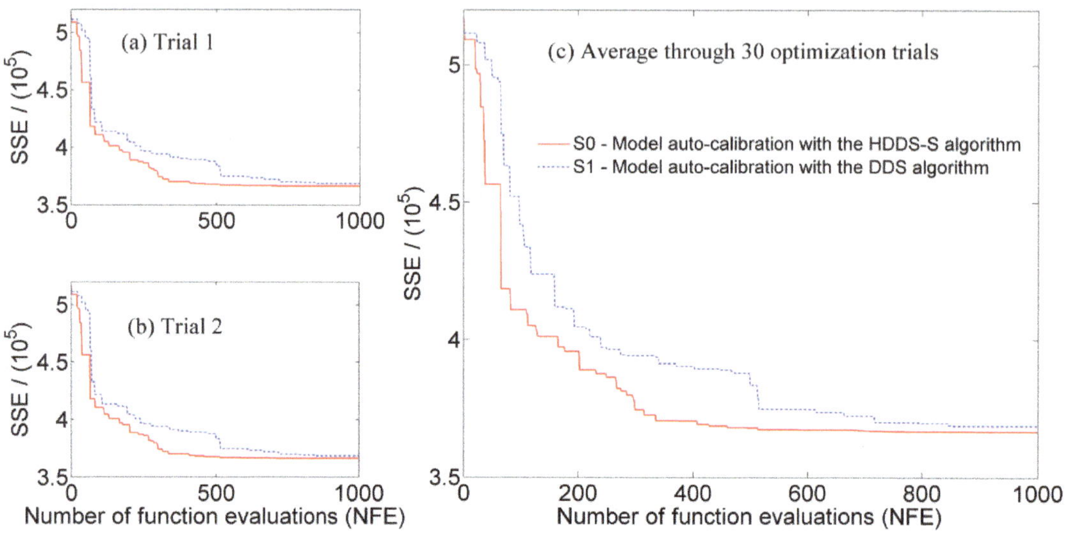

Figure 8. Convergence processes of the objective function under the same number of function evaluations ($NFE = 1000$) in different scenarios through 30 optimization trials, (**a**) and (**b**) convergence processes in two random optimization trials; (**c**) average convergence process through 30 optimization trials.

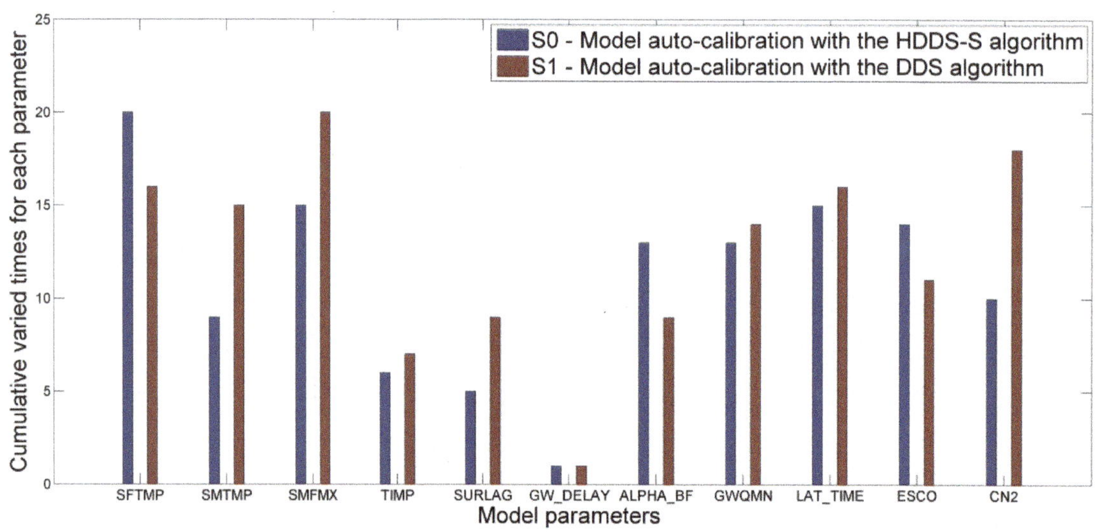

Figure 9. Average cumulative varied times for each parameter during model auto-calibration process under the same number of function evaluations ($NFE = 1000$) in different scenarios through 30 optimization trials, where the description of each parameter in the x axis could be found in Table 1.

The average cumulative varied times for each parameter during model auto-calibration process under the same number of function evaluations ($NFE = 1000$) through 30 optimization trials are different between the two scenarios, *i.e.*, the DDS algorithm, which selects dimension for perturbation completely at random with a uniform probability, could not reflect the sensitivity information of each parameter. The details are indicated from two aspects.

(1) As indicated in the cumulative varied times in the case of the HDDS-S algorithm, some parameters need to be varied many times due to strong sensitivities, e.g., SFTMP, ALPHA_BF and ESCO. However, the cumulative varied times of these parameters in the DDS algorithm are less.

(2) Similarly, some parameters need to be varied fewer times due to weak sensitivities, e.g., SMTMP, SMFMX, SURLAG and CN2. However, their cumulative varied times in the DDS algorithm are greater than.

Under the same number of function evaluations ($NFE = 1000$) through 30 optimization trials, average calibration processes of model parameters SFTMP, SMFMX, SURLAG and ESCO, in different scenarios are shown in Figure 10.

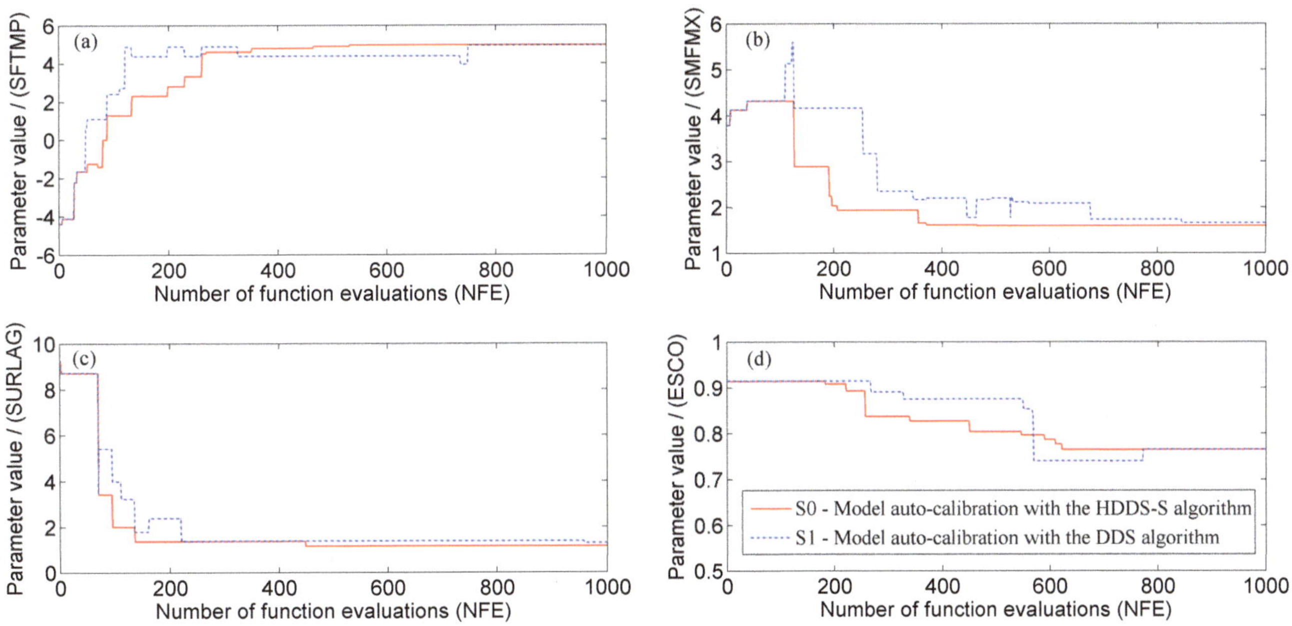

Figure 10. Average calibration process of model parameters under the same iteration number ($NFE = 1000$) through 30 optimization trials in different scenarios, (**a**) SFTMP; (**b**) SMFMX; (**c**) SURLAG and (**d**) ESCO.

It is indicated from Figures 9 and 10 that the value of the most sensitive parameter SFTMP needs to be varied through the whole calibration process, while its calibration is stopped when reaching some iteration number in the DDS algorithm. Similarly, the value of some more sensitive parameters, such as ESCO also needs to be varied many times, while it is calibrated with fewer variation times, scatter variation moments and obvious stagnation phenomenon in the DDS algorithm. However, less sensitive parameters SURLAG and SMFMX needs to be varied fewer times, while it is calibrated with more variation times, scatter variation moments and obvious stagnation phenomenon in the DDS algorithm.

Therefore, no matter how sensitive these parameters are, they could be calibrated to their optimal values much faster with the HDDS-S algorithm than the DDS algorithm.

River basin model calibration results show that the computation time and efficiency could be improved in the HDDS-S algorithm, including the enhanced convergence rate of objective function, concentrated varied moments, reduced blind variation and stagnation phenomenon, compared to the DDS algorithm. Therefore, the HDDS-S algorithm is more efficient than the DDS algorithm in the river basin model calibration for the specific catchment.

4.2. Multi-Reservoir Optimal Operation

The optimal operation problem for DHF-GYG-SW multi-reservoir is used to evaluate the universality of the HDDS-S algorithm in river basin management. The input data are inflow and water demand data, including the industrial and agricultural water demands in the future planning year, *i.e.*, 2030, and the history inflow from 1956 to 2006. Each calculation year is divided into 24 time periods (with ten days as scheduling horizon from April to September, and a month as scheduling horizon in the remaining months). Since the industrial water supply occurs through the whole year, there are twenty-four decision variables for industrial water supply. The agricultural water supply occurs only during the periods from the second ten-day of April to the first ten-day of September, thus there are fifteen decision variables for agricultural water supply. Therefore, the entire operation system has four rule curves and $(24 + 15) \times 2 = 78$ decision variables in total.

Similarly, based on the two scenarios S0 and S1, *i.e.*, multi-reservoir optimal operation in the case of the HDDS-S algorithm and multi-reservoir optimal operation in the DDS algorithm respectively, average performances of the two algorithms in multi-reservoir optimal operation are evaluated under the same iteration number ($NFE = 10{,}000$) through 30 optimization trials from three aspects, including average required computation time, average objective function value, and average guaranteed water-supply rate on Intel Core™ 2 Duo, with a 2.66-GHz CPU.

First of all, the rationality of operation rules derived after 10,000 function evaluations in one random optimization trial in the case of the HDDS-S algorithm is evaluated, and the joint scheduling charts are presented in Figure 11.

As industrial water demand is more urgent than agricultural water demand, and the agricultural water would be limited primarily when reservoir storage is insufficient, the water supply rule curve for agriculture is above the operation rule curve for industry, as shown in Figure 10. In the scheduling cycle duration, each rule curve decreases before the flood season and then has a different degree of uplift after the flood season. In order to prepare for the large inflow of the summer flood season and to reduce surplus water, the reservoir active storage before the flood season should be lowered and then the number of limited water supply is reduced, *i.e.*, the operation rule curves are decreased before the flood season. After the flood season, when the inflow decreases gradually, in order to ensure that no depth damage happens in dry years, the curves are raised to fill reservoirs as early as possible and the number of limited water supply is increased. In addition, the largest agricultural water use happens from April to May, thus the operation rule curve of agriculture is higher in the duration to conserve water and mitigate potential future shortages for industrial demand in some dry years. Specifically, referred to DHF reservoir, its water supply rule curve for industry falls in the duration from March to April due to spring floods produced by melting snow, meanwhile, its water supply rule curve for industry raises in May because the largest agricultural water use happens from April to May and much water has been supplied to agriculture.

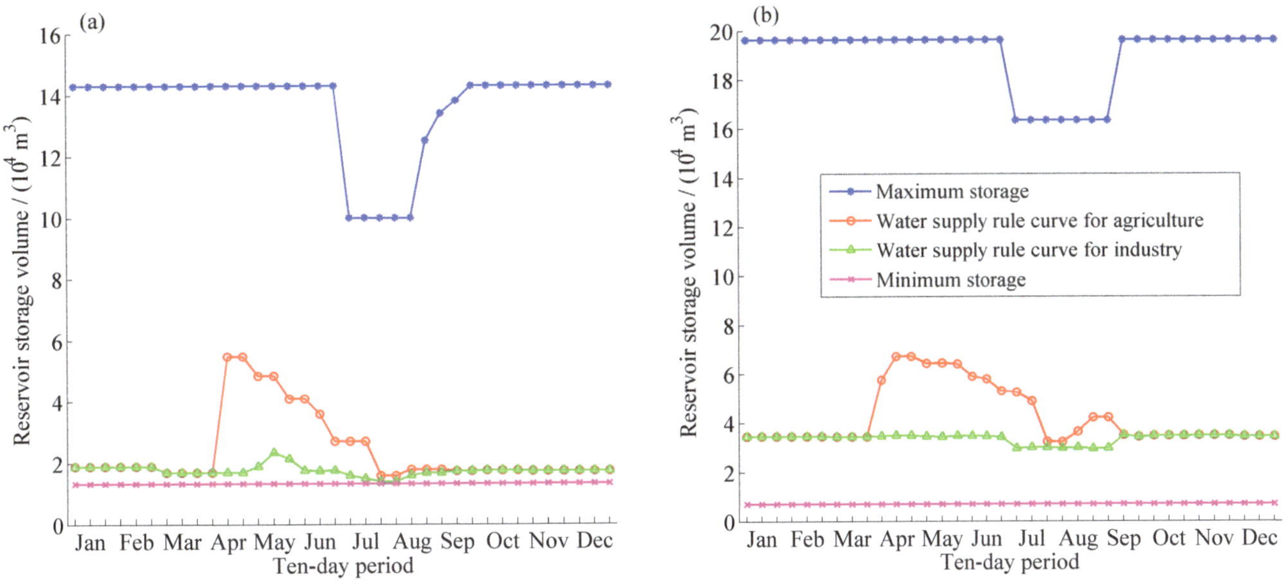

Figure 11. The derived operational rule curves with the HDDS-S algorithm after 10,000 function evaluations in one random optimization trial, (**a**) DHF reservoir operation rule curve; (**b**) XN-2 reservoir operation rule curve.

Additionally, the average scheduling results determined by operation rule curves derived with the HDDS-S algorithm after 10,000 function evaluations in 30 random optimization trials are: (1) In Sub A, the guaranteed rate for industrial and agricultural water demands are 95.10% and 75.35% respectively; (2) In Sub B, the guaranteed rate for industrial and agricultural water demands are 95.81% and 75.46% respectively; (3) In Sub C, the guaranteed rate for industrial and agricultural water demands are 96.08% and 62.56% respectively. The average guaranteed rates in all three subsystems meet the requirement, *i.e.*, guaranteed rate for industrial and agricultural water demands exceed 95% and 50% respectively.

Therefore, as mentioned above, the derived operational rule curves of DHF reservoir and XN-2 reservoir in the HDDS-S algorithm and the scheduling results are reasonable.

In different scenarios, under the same number of function evaluations ($NFE = 10{,}000$), average required computation time and objective function value through 30 random optimization trials are shown in Table 5.

Table 5. Average computation time and objective function value under the same iteration number ($NFE = 10{,}000$) through 30 random optimization trials in different scenarios.

Scenario	Objective Function Value	Computation Time/Second
S0	9.13	723.73
S1	25.68	294.37.2

Under the same number of function evaluations ($NFE = 10{,}000$), though larger differences of average required computation time exist in different scenarios, average objective function value decreases from 25.68 in the DDS algorithm to 9.13 in the HDDS-S algorithm, which improve obviously.

Average guaranteed water-supply rates for all water supply area under the same number of function evaluations ($NFE = 10{,}000$) through 30 random optimization trials in different scenarios are shown in Table 6. Under the same number of function evaluations ($NFE = 10{,}000$), though few differences of average guaranteed rate for industrial water demand exist in different scenarios, *i.e.*, the guaranteed rate for industrial water demand in all three subsystems exceed 95%, the average guaranteed rates for agricultural water demand in all three subsystems increase from 67.52%, 64.78% and 50.94% in the DDS algorithm to 75.35%, 75.46% and 62.54% in the HDDS-S algorithm in Sub A, Sub B and Sub C respectively, which improved obviously.

Table 6. Average guaranteed water-supply rate under the same iteration number ($NFE = 10{,}000$) through 30 random optimization trials in different scenarios.

Water Supply Area		Guaranteed Water-Supply Rate (%)			
		Industry		Agriculture	
		S0	S1	S0	S1
Sub-system	Sub A (DHF subsystem)	95.10	95.06	75.35	67.52
	Sub B (GYG-SW subsystem)	95.81	96.00	75.46	64.78
	Sub C (SCH-DLHK subsystem)	96.08	96.25	62.54	50.94

Optimal operation of DHF-GYG-SW multi-reservoir results show that the derived operational rule curves of DHF reservoir and XN-2 reservoir and the scheduling results from one random optimization trial are reasonable in the case of the HDDS-S algorithm, while average guaranteed water-supply rates in all subsystems could be improved in the HDDS-S algorithm under the same number of function evaluations ($NFE = 10{,}000$) through 30 random optimization trials compared to the DDS algorithm. Therefore, the HDDS-S algorithm is able to derive good solutions and thus is effective in the specific large multi-reservoir optimal operation.

5. Discussions and Conclusions

The DDS algorithm could automatically refine and update optimal search space, *i.e.*, searches globally at the start of the search and conducts a more localized search as the number of iterations approaches the maximum allowable number of function evaluations, and automatically adjust and exhibit good performance within the user's timeframe for calibration to obtain a quick and approximate result, which is achieved by dynamically and probabilistically reducing the number of dimensions in the neighborhood and the reducing probability is distributed to each decision variable equally [45,46]. With multiple optimization test functions as well as real and synthetic SWAT2000 model automatic calibration formulations, Tolson and Shoemaker [32] have proved that the DDS algorithm could return good solutions faster than the SCE algorithm within the maximum allowable number of function evaluations.

In this paper, the DDS algorithm is improved with reference to the changing sensitivity information of multi-dimension decision variables, and the HDDS-S algorithm is proposed based on the sensitivity analysis. The HDDS-S algorithm consists of three steps: firstly, the changing sensitivity matrix of multi-dimension parameters and cumulative sensitivity of each parameter are updated dynamically during the process of optimization according to the relationship between dimension selection and optimizing efficiency; secondly, the probability of choosing each parameter is calculated during the process of optimization according to its cumulative sensitivity; finally, the optimization strategy with non-uniform probability of choosing parameters during the process of parameter optimization is realized to improve convergence rate and optimizing efficiency. The HDDS-S algorithm is tested extensively against the DDS algorithm on the river basin model calibration problem for Tang-Wang River Basin in Jilin province, China, and the optimal operation problem for DHF-GYG-SW multi-reservoir in the northeast of China.

In the river basin model calibration problem for Tang-Wang River Basin, although there is little difference of model performances between the DDS algorithm and the HDDS-S algorithm because the simulation results will approximate their optimal values after 1000 function evaluations to a large extent, the average required computation time reduces from 115.74 h ($NFE = 796$) in the case of the DDS algorithm, to 69.90 h ($NFE = 408$) in the HDDS-S algorithm, through 30 optimization trials under the same simulation accuracy ($SSE = 370{,}000$). Meanwhile, due to the consideration of the changing sensitivity information of each dimension decision variable, under the same iteration number, the convergence rate of the objective function is improved in the HDDS-S algorithm compared to the DDS algorithm, and there are few differences of the HDDS-S algorithm convergence processes among different optimization trials, which indicate the reliability and stability of the HDDS-S algorithm in solving the model auto-calibration problem. Additionally, in the optimal operation problem for the DHF-GYG-SW multi-reservoir, under the same number of function evaluations ($NFE = 10{,}000$), though few differences of average guaranteed rate for industrial water demand exist in different scenarios, *i.e.*, the guaranteed rate for industrial water demand in all three subsystems exceed 95%, the average guaranteed rates for agricultural water demand in all three subsystems increase from 67.52%, 64.78% and 50.94% in the DDS algorithm to 75.35%, 75.46% and 62.54% in the HDDS-S algorithm in Sub A, Sub B and Sub C respectively. Overall, results show that the HDDS-S algorithm could be more efficient and effective than the DDS algorithm in the two specific problems.

Since one algorithm may be better for calibration of the river basin model for only one catchment or one specific multi-reservoir optimal operation, we will choose other catchments and multi-reservoir operation systems with different characteristics including climate, topography and human activities, to examine the performance of the HDDS-S algorithm under different conditions in future studies.

Acknowledgments

This study was supported by the China Postdoctoral Science Foundation Grants (2014M561231), and National Natural Science Foundation of China Grants (51409043).

Author Contributions

Jinggang Chu designed the heuristic dynamically dimensioned search with sensitivity information (HDDS-S), conceived the experiments to evaluate the new global optimization algorithm, performed the experiments, analyzed the results, and wrote the first draft of the manuscript. Yong Peng, Wei Ding and Yu Li provided their support and guidance throughout the manuscript writing. Their timely suggestions have greatly improved the manuscript. All authors reviewed the final manuscript.

Appendix

Pseudo-Code of HDDS-S

The complete HDDS-S algorithm pseudo-code for minimization is provided in the following:

STEP 1. Define HDDS-S inputs as follows:

- maximum number of objective function evaluations, N,
- neighborhood perturbation size parameter, r (0.2 is default),
- vectors of lower, x^{min}, and upper, x^{max}, bounds for all D decision variables,
- Set initial solution within the bond of each decision variable randomly, $x^0 = [x_1, \cdots, x_D]$,
- Initial decision variable changing sensitivity matrix, $(s_{ij})_{N\times D} = (0)_{N\times D}$.

STEP 2. Set objective function evaluation counter i to 1, *i.e.*, $i = 1$, and evaluate objective function F at initial solution, $F(x^0)$:

- $F_{best} = F(x^0)$, and $x^{best} = x^0$.

STEP 3. Referred to all decision variables, according to their cumulative sensitivities from beginning, select K of the D decision variables to search, *i.e.*, to be included in neighborhood, $\{M\}$, take decision variable $x_i (i = 1, \cdots, D)$ as an example to illustrate the process:

- Calculate the sensitivity on iteration j: $s_{i,j}$.
- Calculate the cumulative sensitivity from beginning to iteration j: $sc_{i,j}$.
- Calculate the probability of choosing x_i to search based on its cumulative sensitivity $sc_{i,j}$ from beginning to iteration j: $P(i, j+1)$.
- Add x_i to $\{M\}$ with probability $P(i, j+1)$.
- If $\{M\}$ empty, select all decision variables to $\{M\}$.

STEP 4. Referred to each decision variable in $\{M\}$, perturb x_k^{best} $(k = 1, \cdots, M)$ using a standard normal random variable, $N(0,1)$, reflecting at decision variable bounds if necessary:

- $x_k^{new} = x_k^{best} + \sigma_k N(0,1)$, where $\sigma_k = r(x_k^{max} - x_k^{min})$.
- If $x_k^{new} < x_k^{min}$, reflect perturbation:
  - $x_k^{new} = x_k^{min} + (x_k^{min} - x_k^{new})$,
  - If $x_k^{new} > x_k^{max}$, set $x_k^{new} = x_k^{min}$.
- If $x_k^{new} > x_k^{max}$, reflect perturbation:
  - $x_k^{new} = x_k^{max} - (x_k^{new} - x_k^{max})$,
  - If $x_k^{new} < x_k^{min}$, set $x_k^{new} = x_k^{max}$.

STEP 5. Evaluate $F(x^{new})$ and update current best solution if necessary:

- If $F(x^{new}) \le F_{best}$, update new best solution:

- $F_{best} = F(x^{new})$ and $x^{best} = x^{new}$.

STEP 6. Update objective function iteration counter, $i = i + 1$, and check stopping criterion:

- If $i = N$, STOP, save (e.g., F_{best} & x^{best}).
- Else, go to STEP 3.

Conflicts of Interest

The authors declare no conflict of interest.

References

1. Labat, D.; Godderis, Y.; Probst, J.L.; Guyot, J.L. Evidence for global runoff increase related to climate warming. *Adv. Water Resour.* **2004**, *27*, 631–642.
2. Barnett, T.P.; Pierce, D.W.; Hidalgo, H.G.; Bonfils, C.; Santer, B.D.; Das, T.; Dettinger, M.D. Human-induced changes in the hydrology of the western United States. *Science* **2008**, *319*, 1080–1083.
3. Zhang, A.; Zhang, C.; Chu, J.; Fu, G. Study on the human-induced runoff change in Northeast China. *J. Hydrol. Eng.* **2014**, doi:10.1061/(ASCE)HE.1943-5584.0001078.
4. Zhang, A.; Zhang, C.; Fu, G.; Wang, B.; Bao, Z.; Zheng, H. Assessments of impacts of climate change and human activities on runoff with SWAT for the Huifa River Basin, Northeast China. *Water Resour. Manag.* **2012**, *26*, 2199–2217.
5. Zhang, C.; Shoemaker, C.A.; Woodbury, J.D.; Cao, M.; Zhu, X. Impact of human activities on stream flow in the Biliu River basin, China. *Hydrol. Process.* **2013**, *27*, 2509–2523.
6. Zhang, C.; Zhu, X.; Fu, G.; Zhou, H.; Wang, H. The impacts of climate change on water diversion strategies for a water deficit reservoir. *J. Hydroinform.* **2013**, doi:10.2166/hydro.2013.053.
7. Zhang, C.; Wang, D.; Wang, G.; Yang, W.; Liu, X. Regional differences in hydrological response to canopy interception schemes in a land surface model. *Hydrol. Process.* **2014**, *28*, 2499–2508.
8. Wang, F.; Zhang, C.; Peng, Y.; Zhou, H. Diurnal temperature range variation and its causes in a semiarid region from 1957 to 2006. *Int. J. Climatol.* **2014**, *34*, 343–354.
9. Labadie, J.W. Optimal operation of multireservoir systems: State-of-the-art review. *J. Water Resour. Plan. Manag.* **2004**, *130*, 93–111.
10. Draper, A.J.; Lund, J.R. Optimal hedging and carryover storage value. *J. Water Resour. Plan. Manag.* **2004**, *130*, 83–87.
11. Sadegh, M.; Mahjouri, H.; Kerachian, R. Optimal inter-basin water allocation using crisp and fuzzy Shapley games. *Water Resour. Manag.* **2010**, *24*, 2291–2310.
12. Zhang, C.; Wang, G.; Peng, Y.; Tang, G.; Liang, G. A negotiation-based multi-objective, multi-party decision-making model for inter-basin water transfer scheme optimization. *Water Resour. Manag.* **2012**, *26*, 4029–4038.
13. Zhao, T.; Zhao, J.; Yang, D. Improved dynamic programming for hydropower reservoir operation. *J. Water Resour. Plan. Manag.* **2014**, *140*, 365–374.

14. Xu, W.; Zhang, C.; Peng, Y.; Fu, G.; Zhou, H. A two stage Bayesian stochastic optimization model for cascaded hydropower systems considering varying uncertainty of flow forecasts. *Water Resour. Res.* **2014**, *50*, doi:10.1002/2013WR015181.
15. Fu, G.; Kapelan, Z.; Reed, P. Reducing the complexity of multi-objective water distribution system optimization through global sensitivity analysis. *J. Water Resour. Plan. Manag.* **2012**, *138*, 196–207.
16. Fu, G.; Kapelan, Z.; Kasprzyk, J.R.; Reed, P. Optimal design of water distribution systems using many-objective visual analytics. *J. Water Resour. Plan. Manag.* **2013**, *139*, 624–633.
17. Wang, Q. The genetic algorithm and its application to calibrating rainfall-runoff models. *Water Resour. Res.* **1991**, *27*, 2467–2471.
18. Simpson, A.R.; Dandy, G.C.; Murphy, L.J. Genetic algorithms compared to other techniques for pipe optimization. *J. Water Resour. Plan. Manag.* **1994**, *120*, 423–443.
19. Savic, D.A.; Walters, G.A. Genetic algorithms for least-cost design of water distribution networks. *J. Water Resour. Plan. Manag.* **1997**, *123*, 67–77.
20. Wu, Z.; Boulos, P.F.; Orr, C.H.; Ro, J.J. Using genetic algorithms to rehabilitate distribution systems. *Am. Water Works Assoc. J.* **2001**, *93*, 74–85.
21. Tolson, B.A.; Maier, H.R.; Simpson, A.R.; Lence, B.J. Genetic algorithms for reliability-based optimization of water distribution systems. *J. Water Resour. Plan. Manag.* **2004**, *130*, 63–72.
22. Fu, G.; Butler, D.; Khu, S.T. Multiple objective optimal control of integrated urban wastewater systems. *Environ. Model. Softw.* **2008**, *23*, 225–234.
23. Fu, G.; Kapelan, Z. Fuzzy probabilistic design of water distribution networks. *Water Resour. Res.* **2011**, *47*, doi:10.1029/2010WR009739.
24. Suribabu, C.R.; Neelakantan, T.R. Design of water distribution networks using particle swarm optimization. *Urban Water J.* **2006**, *3*, 111–120.
25. Montalvo, I.; Izquierdo, J.; Pérez, R.; Iglesias, P.L. A diversity-enriched variant of discrete PSO applied to the design of water distribution networks. *Eng. Optim.* **2008**, *40*, 655–668.
26. Duan, Q.; Sorooshian, S.; Gupta, V. Effective and efficient global optimization for conceptual rainfall-runoff models. *Water Resour. Res.* **1992**, *28*, 1015–1031.
27. Duan, Q.; Gupta, V.K.; Sorooshian, S. Shuffled complex evolution approach for effective and efficient global minimization. *J. Optim. Theory Appl.* **1993**, *76*, 501–521.
28. Maier, H.R.; Simpson, A.R.; Zecchin, A.C.; Foong, W.K.; Phang, K.Y.; Seah, H.Y.; Tan, C.L. Ant colony optimization for design of water distribution systems. *J. Water Resour. Plan. Manag.* **2003**, *129*, 200–209.
29. Zecchin, A.C.; Simpson, A.R.; Maier, H.R.; Leonard, M.; Roberts, A.J.; Berrisford, M.J. Application of two ant colony optimization algorithms to water distribution system optimization. *Math. Comput. Model.* **2006**, *44*, 451–468.
30. Zecchin, A.C.; Maier, H.R.; Simpson, A.R.; Leonard, M.; Nixon, J.B. Ant colony optimization applied to water distribution system design: Comparative study of five algorithms. *J. Water Resour. Plan. Manag.* **2007**, *133*, 87–92.
31. Duan, Q.; Sorooshian, S.; Gupta, V.K. Optimal use of the SCE-UA global optimization method for calibrating watershed models. *J. Hydrol.* **1994**, *158*, 265–284.

32. Tolson, B.A.; Shoemaker, C.A. Dynamically dimensioned search algorithm for computationally efficient watershed model calibration. *Water Resour. Res.* **2007**, *43*, doi:10.1029/2005WR004723.
33. Hansen, N.; Ostermeier, A. Completely derandomized self-adaptation in evolution strategies. *Evol. Comput.* **2001**, *9*, 159–195.
34. Hansen, N.; Müller, S.D.; Koumoutsakos, P. Reducing the time complexity of the derandomized evolution strategy with covariance matrix adaptation (CMA-ES). *Evol. Comput.* **2003**, *11*, 1–18.
35. Vrugt, J.A.; Robinson, B.A. Improved evolutionary optimization from genetically adaptive multimethod search. *Proc. Natl. Acad. Sci. USA* **2007**, *104*, 708–711.
36. Zhang, X.; Srinivasan, R.; Van Liew, M. On the use of multi-algorithm, genetically adaptive multi-objective method for multi-site calibration of the SWAT model. *Hydrol. Process.* **2010**, *24*, 955–969.
37. Liu, P.; Guo, S.; Xu, X.; Chen, J. Derivation of aggregation-based joint operating rule curves for cascade hydropower reservoirs. *Water Resour. Manag.* **2011**, *25*, 3177–3200.
38. Liu, P.; Cai, X.; Guo, S. Deriving multiple near-optimal solutions to deterministic reservoir operation problems. *Water Resour. Res.* **2011**, *47*, doi:10.1029/2011WR010998.
39. Zhu, X.; Zhang, C.; Yin, J.; Zhou, H.; Jiang, Y. Optimization of water diversion based on reservoir operating rules—A case study of the Biliu River reservoir, China. *J. Hydrol. Eng.* **2014**, *19*, 411–421.
40. Arnold, J.G.; Srinivasan, R.; Muttiah, R.S.; Williams, J.R. Large area hydrologic modeling and assessment part I: Model development. *J. Am. Water Resour. Assoc.* **1998**, *34*, 73–89.
41. Neitsch, S.L.; Arnold, J.G.; Kiniry, J.R.; Williams, J.R. *Soil and Water Assessment Tool User's Manual-Version 2000*; Texas Water Resources Institute: College Station, TX, USA, 2002; p. 412.
42. Hao, F.; Cheng, H.; Yang, S. *Non-Point Source Pollution Model*; Environmental Science Press: Beijing, China, 2006.
43. Moriasi, D.N.; Arnold, J.G.; van Liew, M.W.; Bingner, R.L.; Harmel, R.D.; Veith, T.L. Model evaluation guidelines for systematic quantification of accuracy in watershed simulations. *Trans. ASABE* **2007**, *50*, 885–900.
44. Luo, Y.; He, C.; Sophocleous, M.; Yin, Z.; Hongrui, R.; Ouyang, Z. Assessment of crop growth and soil water modules in SWAT2000 using extensive field experiment data in an irrigation district of the Yellow River Basin. *J. Hydrol.* **2008**, *352*, 139–156.
45. Behrangi, A.; Khakbaz, B.; Vrugt, J.A.; Duan, Q.; Sorooshian, S. Comment on "Dynamically dimensioned search algorithm for computationally efficient watershed model calibration" by Bryan A. Tolson and Christine A. Shoemaker. *Water Resour. Res.* **2008**, *44*, doi:10.1029/2007WR006429.
46. Tolson, B.A.; Shoemaker, C.A. Reply to comment on "Dynamically dimensioned search algorithm for computationally efficient watershed model calibration" by Ali Behrangi *et al. Water Resour. Res.* **2008**, *44*, doi:10.1029/2008WR006862.

7

Water Information as a Tool to Enhance Sustainable Water Management—The Australian Experience

James Horne

The Australian National University, Canberra ACT, 2600, Australia;
E-Mail: jameshorne@iinet.net.au;

Academic Editors: Markus Disse and Davide Viaggi

Abstract: Many countries and regions have struggled to put in place adequate water information systems to assist with sustainable water management. This article describes and assesses the key components of Australia's water information and data systems, with particular reference to rural and regional Australia, focusing on progress with strengthening these systems at a national level since 2007. Through the early part of the period, much of Australia was experiencing a crisis in water availability. The article concludes with ongoing challenges for Australia and lessons from the Australian experience for other countries embarking on upgrading their water information and data systems. Upgrading a nation's water information systems is a long-term task, but an important one in a world of climate change and increased climate variability. Substantial progress is likely to take five to 10 years to materialize. From the outset, upgrading information systems needs to be focused on data series that will facilitate answering key policy questions, assist water users in making significant decisions more effectively, and allow businesses and government to better address risks from water-related events. As always, political support matters. To sustain investments in information, its coverage must facilitate illuminating key questions and issues. Custodians of information systems must ensure that the value proposition is clear to all.

Keywords: water information; sustainable water management; water regulations; science; regulatory framework; water policy; Australia; Water Act; Bureau of Meteorology

1. Introduction

Over the past decade there has been a massive upsurge in interest in water management regimes on every continent [1,2].

Managing our water resources effectively requires appropriate information across many areas, in part depending on the use of the water resource, the role of water markets in determining or regulating allocation decisions, the priority accorded the environment, and the governance arrangements in place. Demand for better, quality-assured information also comes from a myriad of users covering water market participants, resource managers and operators, regulators, policy makers, and analysts. Many countries and regions have struggled to put in place adequate water information systems [3–7].

The collection, dissemination, quality, and transparency (both in reporting and analysis) of water information and data are not ends in themselves. In the face of increasing water scarcity and value in some areas, and an increasing incidence of floods in others, adequate information is critical to understanding the essence of the "water problem". It should assist to provide a firm foundation for crafting solutions, or understanding progress towards outcomes sought. It assists water managers (be they businesses, households, or environmental water managers) to optimize their use of water, as well as businesses that are involved in assessing or managing risks related to water resources. Overall, there is a strong public good element in funding good basic information covering our water resources and markets.

It goes to ensuring we:

- Use our resources as we intend—in mining and agriculture, and increasingly the environment—in a transparent way, and that we make important decisions about resource use with the benefit of basic knowledge and information, not solely on the basis of "if a problem emerges we will handle it then".
- Manage risks to the environment purposefully in developing our resources and industries. If we intend to manage the environment in line with "sustainability benchmarks" we should be able to articulate what this means, and ought to be able to track outcomes to ensure we are achieving them.
- Invest our scarce public sector resources (for example, in both hard and soft infrastructure) wisely, understanding and making transparent the implications of our decisions.
- Provide a transparent understanding of the drivers of water-related outcomes, helping to identify problems or issues as and when they arise, and to drive productivity and efficiency [3,7].

However, there will always be a legitimate and important discussion concerning how much information is needed to respond to user demands for water information. Do existing data collections and underlying science facilitate policymaking, allow testing of compliance regimes, or help to establish accountability? Do existing information collections facilitate economic decisions by industries, farmers, and irrigators, or facilitate assessment of the sustainability of ecosystem services? Is the existing state of understanding of our water resources likely to change with the application of new information technology, or does new technology allow data collection that was hitherto impossible? In a rapidly changing economy and society, is the available information adequate to answer key questions of users in an enduring fashion? An ongoing task is to ensure the relevance and cost-effectiveness of data collection, and at the same time not allow collection and dissemination to be stymied by vested interests, or groups that may be adversely affected by the availability of transparent information. Information will not replace the role society and governments play in determining future boundaries on water use, but it will go a long way toward ensuring that decisions are taken with "eyes open".

In this article, Section 2 describes the key components of Australia's water information and data collections with particular reference to rural and regional Australia, focusing on national efforts to strengthen these collections since 2007. Section 3 assesses the progress in strengthening the Australian water information system (as of the end of 2014), and the concerns that continue to exist. Section 4 outlines the ongoing challenges for Australia and the lessons from the Australian experience for other countries embarking on upgrading their water information and data systems. The assessment is qualitative, and the judgments drawn based on the referenced literature and the author's direct involvement as a senior national policy advisor.

2. Key Components of Australia's Water Information System

A country's water information system will at least in part reflect the roles and responsibilities set out in its legislative and institutional framework. In Australia's case, state governments have historically played a key role in managing its water resources, in the context of a federal system. State water agencies continue to collect and disseminate much local level water information, including important information in regards to annual water allocations to holders of water entitlements.

However, over the past two decades this role has changed significantly, with the national government and national agencies playing an increasingly important role. The need for improved water information in Australia was acknowledged at some length in its 2004 National Water Initiative (NWI), an overarching national intergovernmental water-planning document. The NWI set out the importance of additional knowledge and information, and where extra work was required to assist in delivering improved management outcomes [3,8,9]. At that time, the national government provided substantial resources to improve "water management, capacity, knowledge, skills and innovation." Initiatives included development of a nationally compatible water accounting system, a suite of groundwater projects, a national hydrological modeling tool, the CSIRO sustainable yields project (at the direct behest of the national government), and established national data sets for commencing monitoring and reporting of water markets [10,11]. This program commenced operation in June 2006. Issues with data collections were also raised but not immediately progressed [12,13].

Focused holistic attention ensuring the adequacy of water information commenced in late 2006, as a part of preparatory work for the 2007 national government's National Plan for Water Security. This 10-year plan sought "to improve water efficiency and address overallocation of water in rural Australia," from which some 70% of Australian consumption emanates. A catchphrase of the January 2007 announcement was "You can't manage what you don't measure" [14].

The major driver for this new holistic approach to establish an appropriate and effective national water information system was the Commonwealth Water Act of 2007. It was passed with bipartisan support in the national parliament at a time of crisis in water availability. The Act reflected a widely-held view that state-based information systems were not delivering an effective platform for decision-making. It provided the authority to certain national agencies (particularly the Australian Bureau of Meteorology (BOM), the Murray Darling Basin Authority (MDBA), and the Australian

Competition and Consumer Commission (ACCC)) to compel the provision of certain types of information from designated bodies. In practice, national institutions sought to gather information in a cooperative, transparent framework, but the mandatory powers conferred through legislation and regulation provided the teeth to ensure good compliance. Agencies sought to implement their new responsibilities cooperatively, focusing on the value proposition that improved information will facilitate enhanced economic and environmental outcomes, and a need to change the culture to one that was more open and transparent.

Prior to 2007, the Australian Bureau of Statistics (ABS) had published its Water Account Australia on two occasions (2000–2001 and 2004–2005) and Water Use on Australian Farms (from its Agricultural Survey, taken annually since 2002–2003) [15–17]. Overall, these series were less than satisfactory, being unable to answer most key questions of policy makers, river operators, and businesses, particularly irrigators. Bodies at the local or state level collected water information for local purposes, but it was far from transparent and consistent, was often not comparable state to state or basin to basin, and, as documented in the NWI and elsewhere, left much to be desired [9,18].

2.1. Water Information Services

A key part of the national government's 2007 National Plan for Water Security (later rebranded as "Water for the Future") was announced as a A$480 million 10-year program to provide consistent, improved water services for all users [14]. This included large investments in modernizing state based streamflow, groundwater, and storage monitoring systems, and the automation and standardization of data collection [7,19]. Coupled with the work already in the pipeline, the new water information program was the central vehicle for building a new fit for purpose water information framework. The concept of "fit for purpose" information was well recognized. Information needed to facilitate national (or basin-wide, interjurisdictional) and state-based policymaking, and decision-making at the local level by farmers and irrigators. It also needed to provide more timely and accurate information to businesses in general. Historical issues of inconsistent standards and measurement needed to be rectified, access to all users improved, and levels of transparency increased dramatically [3]. The driver for the national government's intentions was the Water Act 2007.

The overall national water information program would be implemented in large part by BOM [19] building on and in collaboration with state agencies. Other institutions that played important supporting roles included the ABS, the competition policy regulator (the Australian Competition and Consumer Commission (ACCC), the Murray Darling Basin Authority (MDBA), and the National Water Commission (NWC). A myriad of state agencies and state or regional water service providers that had previously collected information for their own purposes would now provide existing information streams to the BOM.

The national Water Regulations of 2008 prescribed the water data to be provided to the BOM, pursuant to the provisions of the Water Act 2007. They list the parties from whom the BOM would gather data sets, limited to existing information held in electronic form. Additionally, the Regulations require organizations to give metadata and contextual information about water information. The regulations specify the information particular organizations must provide the BOM. They also specify the terms under which this information passes to the central repository, to minimize transaction costs.

Aspects of this new program included making local or state level data available on a national basis, bringing to bear much more rigorous standards for collection to ensure consistency and comparability, and supplementing this data with new information series to allow analysis of issues that had been neglected. Some of this would require development of new tools, better collection infrastructure, or better application and development of technology and scientific analysis.

The information covered by the national Water Regulations includes:

- surface water resource information (level and flow of surface water in watercourses, excluding water storages) (category 1);
- ground water resource information (level and pressure of groundwater and aquifer recharge volumes) (category 2);
- water storage information (water storage level, volume, inflows and outflows, and the location, capacity, and ownership of major storages) (category 3);
- meteorological information (meteorological phenomena that affect the availability of water. Includes: rainfall, wind speed, humidity, evaporation, temperature, pressure, vapor pressure deficit, and solar exposure and irradiance) (category 4);
- water use information (water takes, supplies, and returns) (category 5), and information about water rights, allocations, and trades (water access rights, irrigation rights, water trades, water allocation announcements, and permits) (category 6);
- information about urban water management (urban water resources and management. Includes information on water takes, supplies, sewage, urban stormwater, and recycled water (category 7), and information about water restrictions (announcements of water use restrictions) (category 8);
- water quality information (electrical conductivity, suspended solids, turbidity, nutrient content, acidity, and temperature of surface water. Includes electrical conductivity and acidity of ground water) (category 9); and
- water information for flood warning purposes (level and flow of surface water in watercourses, and rainfall information used for flood warning purposes) (category 11) [20].

Vertessy noted that these new services needed to be "enduring, repeatable, robust, trusted and tailored to end user need" [7].

Before much of this work could yield direct benefits to users, detailed preparatory work was required—for example, agreeing standards, building information systems, and ensuring data sets would assist answering key questions. The Australian Water Resources Information System (AWRIS) was to be the vehicle to underpin many of the BOM's water information products and services, get data suppliers on board, and develop ways to minimize transaction costs [19]. AWRIS was to be a powerful information system capable of receiving, standardizing, organizing, and interpreting water data from across the nation [21]. As with many novel IT systems, serious implementation issues arose, delaying the delivery of some new products and services [18]. However, an initial implementation of AWRIS (known as AWRIS 1) was completed, with the BOM making important new "value adding" products available to users by alternative means (for example, Water Data Online [22] provides access to surface water data in a way not seen previously in state-based portals). Direct access to data is currently more limited than originally envisaged [18,19].

AWRIS 2 is scheduled to be operational in 2015, with the ambition that it will be more capable and more independently searchable, which will assist delivery of water information to users more in line with what was originally expected. The original capability sought for AWRIS was the access it would give to all users to independently tap into a national, compatible, reliable, independent, and searchable database, dramatically improving transparency. This is still the objective, but its potential to enhance decision-making and monitoring is yet to be fully realized.

The BOM is now publishing much of the "progress thus far" through its website, including new dashboards and forecasting services. Nearly all the published or searchable products involve intensive collaboration with state and territory governments, water service providers, and other Commonwealth agencies. In the context of water regulation information categories presented above, key areas of progress include:

- Design rainfall—updated intensity-frequency data information as a part of a major revision of Engineer Australia's guide to flood estimation (category 4) [23,24].
- Development of the Australian Hydrological Geospatial Fabric (Geofabric)–enabling a detailed understanding of how water is "stored, transported and used through the landscape," and is being used in water accounting, assessment, forecasting, and prediction [25].
- Groundwater—a suite of groundwater-specific products is available, allowing access to data at 800,000 bore locations, and includes a Groundwater Dependent Ecosystems Atlas. Groundwater information is included in the Geofabric, Australian Water Resources Assessment and the National Water Account (categories 2, 5, and 6).
- Modernization and extension of hydrologic monitoring systems of state agencies and water service providers resulting in improved measurements and data management systems, enabling higher quality and more standardized national reporting across many information categories [26].
- Water Data Online—bringing together nationally consistent water information at a local level (category 1) [22].
- Australian Water Resource Assessments—provides a national-scale assessment structured around 13 regions, answering questions about water availability (categories 1, 2, 3, 5, 7, and 9) [19].
- Streamflow forecasts—seasonal stream flow forecasts are now available at over 70 locations and a new short term streamflow forecasting service is available to registered users for the first time (categories 1, 4, and 11) [19,27].
- Standards—significant progress. Version 1.2 of the Water Data Transfer Format available and Australian Water Accounting Standard 2 released [28].
- The National Water Account—the release provides an annual report on the significant water regions in Australia encompassing the capital cities and their hinterlands, and the MDB, Ord, and Daly systems [29]. As can be seen from Figure 1 below, while the report only covers a small proportion of the Australian continent, it covers well over 70% of the population and 70%–80% of national water consumption (depending on the year). Its limited area of coverage underscores the focus on cost effectiveness, as the regional accounts contained detailed line items on assets, liabilities, and flows of water (categories 1, 5, 6, and 7).

A key feature of the information series produced by national bodies is the routine use of the "Creative Commons" data licensing approach. This "open data" approach is in contrast to previous state-based approaches. It greatly simplifies access to the information and *inter alia* allows development of apps and other products by third parties to enhance the utilization of water information in decision-making. At this point some state governments have not adopted fully this transparent approach. Some agencies still require special permissions for access to, for example, underlying water modeling. In a real sense, an open national approach to information can provide much enhanced access information by all information users.

The above progress by the BOM amounts to a massive step forward in terms of information availability and transparency compared with a decade ago. However, the job is not complete. Areas where less or little progress has been achieved include water quality and water prices, and some remaining definitional issues in key data series between national and state agencies, and even between national agencies. These issues are discussed below.

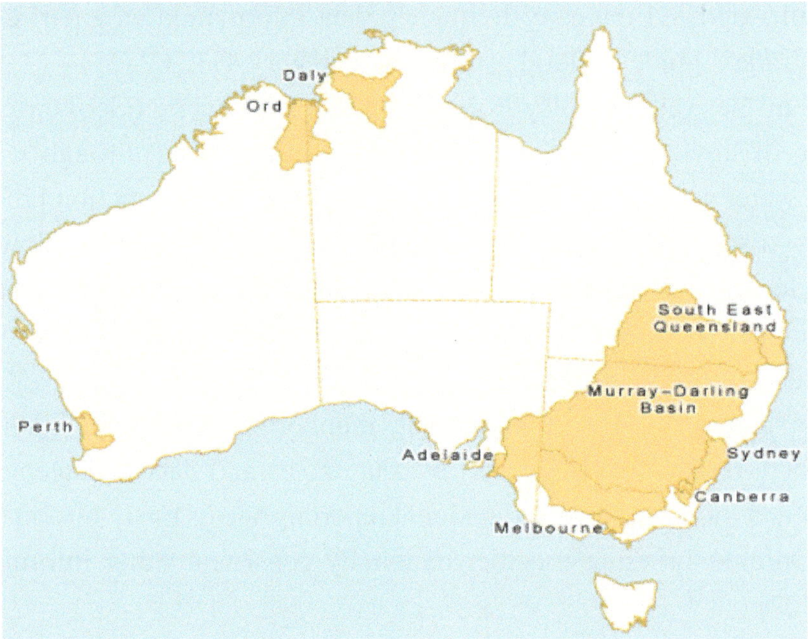

Figure 1. Coverage of National Water Account, 2013. Source: [29].

The BOM's water information program is not only about improving data collection and availability to users but importantly focuses on developing new tools for water users. For example, under a joint program between the BOM and Australia's preeminent science organization, the CSIRO, under the Water Information Research and Development Alliance (WIRADA) framework, tools have been developed that allow enhanced stream flow forecasting, new data standards, and more transparent, detailed, and reliable water availability modeling in many key areas [30]. While the suite of available services continues to evolve, there is little doubt that significant improvements have been made in data availability and its timing, answering many key questions of water information users [19]. For example, data on water storages, streamflow, and groundwater conditions are now transparent, and provide much improved quality, timeliness, and comparability. The data are not yet searchable in complex ways [18], but it is intended that this function be added in the context of completion of AWRIS 2 in 2015. This "searchable" functionality will allow many water information users further opportunities to add value.

Observation points underpinning streamflow forecasting have increased significantly following a major BOM-led investment program and, when combined with new tools, enabled "improved forecasting of inflows into reservoirs and management of environmental water reserves and provided valuable preparation for flood forecasting" ([19], p. 2). Geoscience Australia (GA) continues to develop its path-breaking "data cube" technology, which will substantially aid managing flood risk [31], and in time may be able to address other large risks to water resource sustainability, such as the expansion of farm dams in the MDB.

Of course, much of this can only take place because of recent developments in information technology, and the assistance from a large number of collection agencies making this information widely and transparently available. This has been part of what has been a central problem in the implementation and development of the water information system.

Two large projects—the AWRIS project noted above, and a project to bring together all state water registries to enable users to maximize benefits from water markets that cross state boundaries—involved large information technology elements. Both these projects have had serious delivery problems. In the case of the AWRIS, the BOM was initially forced to reduce the scope of the project in the short term as data management issues proved insurmountable, significantly affecting its initial value to users [18]. In the case of the second, the project has been terminated [32]. It remains to be seen if key aspects relating to accurate and timely price and quantity data will be completed.

Some users and data collectors (mainly water service providers and state-based regulators) have not so warmly supported the increased attention from the broadly based information program. Some of the submissions to the review of the Water Act 2007, from state governments, irrigation companies, and shire councils provide evidence of this attitude [33]. In bureaucratic politics terms, the move to improve the quality, transparency, and usability of data collections is viewed as a threat. In these submissions, there is a tendency to omit mention of what is useful, forget the service that was being provided a decade ago, and largely ignore the type of service users demand today, where online access is seen as a minimum starting point. However, there is a valid point being made: a lesson from these submissions is that it is essential that the benefits and costs of enhanced data collection, and information that is subsequently made freely and publicly available, be carefully assessed. It is also important to increase awareness of the progress being made, and what can be achieved relative to what could previously be achieved. Good design of collection methods is an important part of keeping costs low.

2.2. Economic Information

2.2.1. Australian Bureau of Statistics Data

The ABS began producing (irregularly) its "Water Account, Australia" (WAA), in accordance with the System of Environmental-Economic Accounting. The ABS notes WAA presents information on the physical and monetary supply and use of water in the Australian economy [17]. Data covering agricultural water use were largely collected through periodic censuses and more regular surveys directed at farmers and irrigators [15], but there are significant questions relating to its adequacy in a contemporary water management framework. The ABS has not delved much into water prices, trading, or environmental water use, all key economic questions.

Much of the discussion on water information when the NWI was being negotiated in 2003–2004, and later in late 2006 prior to the announcement of the National Plan for Water Security, recognized that the data collected and the method of collection by the ABS were unable to address many of the key policy questions being discussed in the Australian water sector. The data series for water use on farms, largely based on an Agricultural Survey/Census, was not adequate to examine issues around scarcity, sustainability, quality, and value that were at the forefront of the NWI in rural and regional Australia. Its concept of "use" is on the one hand narrow (for example, not engaging with how much conveyance water was required to underwrite on-farm use) and from an irrigator/policymaker perspective unusual (attributing large volumes of "use" to hydroelectric power generation when the net water lost to the system from this activity was small). It was difficult for many water information users to engage with. The ABS sought to examine "physical water supply and use" and "monetary water supply and use" but (at least in part reflecting limited resourcing and the decision to widen the BOM's water information remit) key technical water resource issues were not engaged with. The "Water Account Australia" helps us to start addressing questions around urban water use in Australia, and useful research continues on valuation of water resources [34]. However, by design it was left to BOM and others to develop accessible products that enabled information to be readily accessed. BOM, CSIRO, and GA focused on the issues, skillsets, and technology needed to address mainstream water policy and user questions.

One of the weaknesses of the ABS data a decade ago (and it remains) was that the environment was not recognized as a "water user", and none of the water data looked at water quality or sustainability issues. Moreover, issues around farm dams, groundwater use, water required to deliver water consumed on farms, and return flows all detracted from the published data set. For example, understanding the change in return flows over time will assist in accurately understanding the benefits and costs of on-farm infrastructure subsidy programs to bolster environmental flows. The issues with the ABS data in part reflect the collection methodology (built around the national farm census and periodic surveys), and the fact that the original data collection did not have the environment as a focus of attention or need. It underscores the need for data collections to be reassessed periodically to respond to need. Estimates addressing some of these issues are now available. These estimates have been produced from work originally commissioned by the former National Water Commission and now overseen by the BOM, through the BOM's developing AWRIS framework and through information collected by the ACCC. In the future, drawing together new technologies (see below) seems likely to enable more effective management of ground and surface water resources, including getting a much more accurate grasp of the risks involved in managing resources.

2.2.2. Water Market Data

Water markets play an important role in ensuring that available water is used productively in some important areas of rural Australia. An important new player in these markets, particularly in the MDB, is the ACCC, which took on new regulatory functions under the Water Act of 2007 and the Competition and Consumer Act of 2010 [35]. Its Water Monitoring Report outlines progress on many of the economic reforms being pursued in the rural water markets, covering regulated water charges, water entitlement transformation, and compliance with water charge rules and water market rules [36]. This has greatly increased transparency of interjurisdictional water management across the MDB.

For example, the ACCC now publishes, *inter alia*, data on the extent of transformation of water rights by water entitlement holders (and the time taken by irrigation infrastructure operators (IIOs) to undertake this task), the impact of old and new restrictions on trade imposed by state governments, the adherence to water charging rules by IIOs (measuring the number of cases and seriousness of overcharging by IIOs), and the levels of water planning and management charges of state authorities. This information has increased the level of transparency around water trading rules and behavior of irrigation infrastructure operators (IIOs), and of itself places ongoing pressure on them and state authorities to behave appropriately. The data also makes transparent to irrigators how their IIO operates compared with other IIOs. For example, in 2012–2013, the data illustrate how the size of the "casual (water) user fee" significantly affects the volume of water delivered on a casual basis [37]. For more background see [3,38].

A "National Water Market System" was to be a cornerstone of data provision on both quantity and price for Australia's water markets. However, as with many information technology projects, resistance to change from state agencies and lack of commitment at the national level led to this project being abandoned, and future of the progress made uncertain [39,40].

As a result, an ongoing substantial weakness in Australian water information on water trading continues to be the absence of accurate (full price disclosure) and timely (better than weekly) price data, organized in a manner useful to market participants (through a single portal). At one level, it is surprising that comprehensive availability and transparency of water price data has not improved dramatically as water markets from allocation water and entitlement water in the MDB in particular have grown. All the necessary data are capable of being collected within existing state-based administrative frameworks when individual transactions are registered. Such an information service would be a major plus for all participants in the water markets, from irrigators to investors and brokers, and environmental water holders that are permitted to trade. The water trading rules of the Basin Plan, which commenced mid-2014, require states to publish trading prices.

2.3. Importance of Basic Science and Technology to Water Information

A series of one-off projects since 2007 has considerably enhanced Australia's basic understanding of its water resources. Three such projects are outlined below to illustrate the role of basic science and new analytical tools to the water information system; the role of independent expert science agencies undertaking the work; and the importance of this work being undertaken at a basin level, putting aside state-based political and other interjurisdictional issues.

2.3.1. CSIRO "Sustainable Yields" Study

In November 2006, in the face of rapidly worsening water availability in parts of the Murray Darling Basin (MDB), the Australian government commissioned the CSIRO to undertake a long-term comprehensive assessment of water availability in the MDB. CSIRO was the preeminent [41], independent national research organization and the only organization at the time capable of undertaking basin-wide research with the support of the governments of the four states and one territory involved in the MDB.

Later this work was extended, with the CSIRO undertaking water availability studies covering other major Australian water systems, to allow a consistent analytical framework for water policy decision-making [42,43]. These studies, undertaken between 2007 and 2012, covered both surface water and groundwater, and included modeling of the effects of climate change on water availability to 2030. The areas covered (shown in Figure 2) focused on "at risk" water basins, and areas where further expansion of activities involving water use was being proposed.

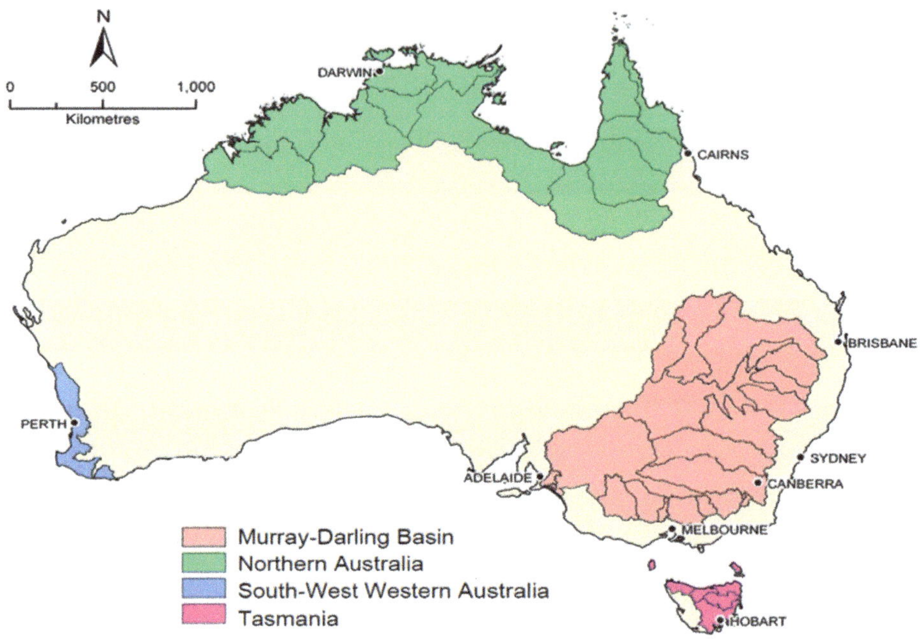

Figure 2. CSIRO "Sustainable Yield" Studies. Source: [44].

The studies were detailed desktop scientific analyses, involving cutting-edge modeling technology, and provided water information users with a much clearer understanding of water availability in key areas, and likely impacts of climate change in the medium term [38]. This was the first time such an analysis had been undertaken on a basin-wide basis. The MDB study was developed further as the MDB Plan was drafted, as issues such as appropriate sustainable diversion limits were considered [45]. Each study used technical skills not present at the state level, and a consistent rigorous methodology to examine catchment development, groundwater extraction, and climate change [46]. This has led to some fundamental reassessing of forward-looking water management options and risks. For example, the Northern Australia study expressed caution over the viability of new dams in Northern Australia [47], while the Tasmanian study provided detailed insights into the impacts of the expected drier climate on proposed irrigation scheme developments [48]. Two projects involved cross border investigations that had been hitherto unable to be undertaken. It was done by the respected independent national science "brand" and the methodology was repeatable across the country.

2.3.2. Broken Hill Managed Aquifer Recharge

Again, in an effort to provide improved basin-wide solutions to water problems, the national government commissioned Geoscience Australia, a respected national leader in groundwater modeling, to undertake research on groundwater resources in the Menindee Lakes area. The study has not only transformed understanding of groundwater around the Menindee Lakes, but has proven technology that will assist in transforming assessment of groundwater resources around the world [49,50]. While previous work had been undertaken at the state level, this study provided the basis for fundamental reappraisal of surface and groundwater management options in the study area itself, with basin-wide implications.

2.3.3. Bioregional Assessments

A third project (currently in progress) involves the national government undertaking detailed basin-wide, cross-border "bioregional assessments" in relation to water resources that may be affected by large-scale coal and unconventional gas mining activities. The regions covered by these assessments are set out in Figure 3.

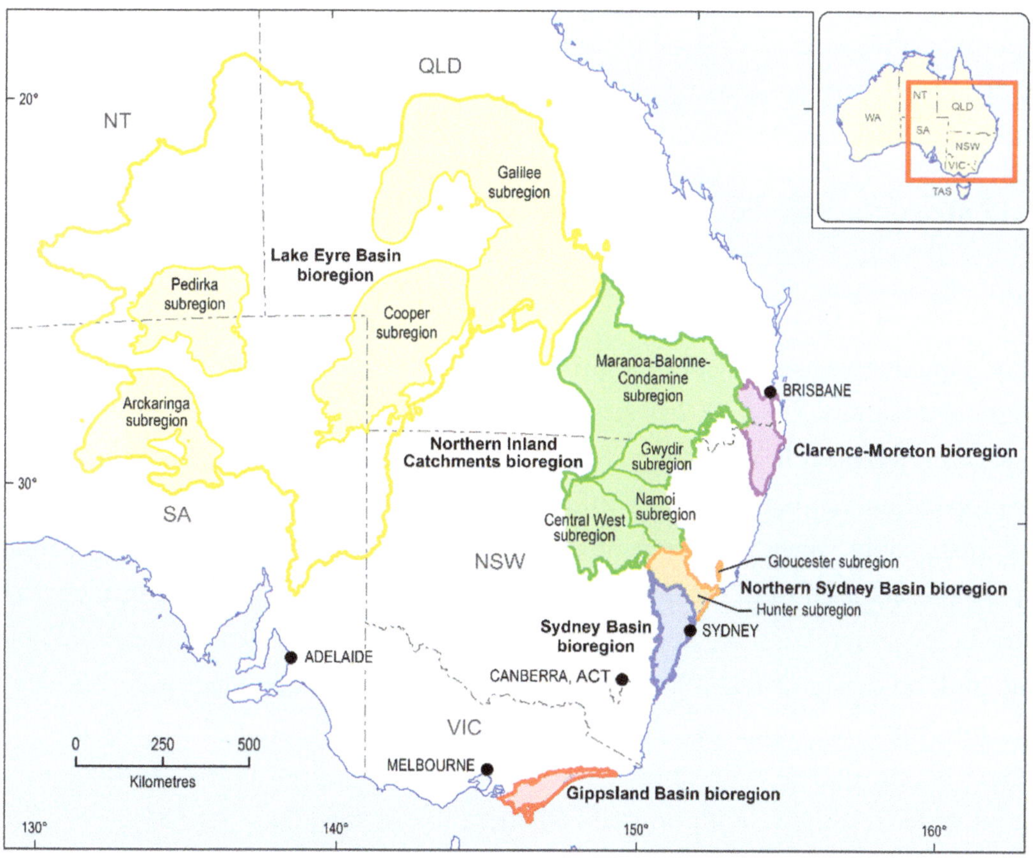

Figure 3. Bioregional Assessment Regions. Source: [51].

According to the Australian government, "bioregional assessments are world-first science-based studies that will develop detailed, multi-layered records of the environment and provide all interested stakeholders with a comprehensive, customized suite of baseline information and analysis about water resources in their local area. The assessments will examine the possible impacts of coal seam gas and large coal mining developments on water-related resources" ([52], p. 1).

The bioregional studies are supported by communities that want an assessment independent of the state governments that have backed large-scale resource projects. The studies should provide a sound basis on which to examine the cumulative impact of individual large-scale mining and agricultural development projects. Some of these studies are being undertaken "after the fact"; even then they will be useful to suggest whether given consents need to be altered in light of the future deterioration of a water resource. Ideally, these studies should have been undertaken before decisions on individual projects were made. For the other bioregional assessment regions, where the studies are preceding decisions on development (either water resources or other natural resources including unconventional gas and coal mining), the studies should provide sound baseline information as inputs to further decision-making.

Taken together, these three groups of studies form an important part of Australia's water information system. The studies illustrate the importance of science and the importance of institution selection. They have all produced or are producing information of quite profound scientific and economic significance that will assist in making well-based public policy decisions and private investment decisions. These are all detailed studies with potentially profound long-term resource management implications (for an interesting article on this subject see Hirsch *et al.* [53]). How they are used in the public policy arena will depend very much on the regional, state, and national politics at the time, but at the very least decisions can be based on a materially improved baseline understanding of existing water resources.

3. Progress in Strengthening the Australian Water Information System

3.1. Qualitative Assessment of Improvements in Water Information

This author's qualitative assessment of improvements in the availability of useful water information for rural and regional Australia since 2007 is set out in Table 1. These comments are grounded in published information series referenced in this article. Readers are encouraged to view this source material. Table 1 supports the proposition that national water information is playing an important role driving water reform and assisting sustainable water management and decision-making. It is now hard to imagine a world without useful data sets in most of the areas set out in this table. Yet a decade ago, many rows in Table 1 were close to being empty sets or, at least, data availability was at best piecemeal, not transparently available, and built on standards that often differed between states or individual water service providers.

However, information availability does not always settle questions. For example, the Australian political debate is still replete with untested assertions on the opportunities to develop new frontiers utilizing extant water resources, with limited consideration of sustainability [54].

Another case where enhanced science-based water information has been critical is in the MDB. The debate over the sustainable diversion limits in the MDB Basin Plan was underpinned by solid science, with the limits themselves being appropriately societal decisions [55]. Even now it is clear that farming bodies do not agree with the outcome [56]. In addition, mining companies still argue that any broad rules on water management should be altered to meet their needs [57]. These arguments will continue, but at least society can take basic decisions with a better understanding of some of the ramifications of that decision.

Table 1. Water Information for Sustainable Water Management—Progress in Rural and Regional Australia since 2007.

Type	Administering Agency	Authority	Comment	Beneficiary	Benefit
Water availability, supply and use	BOM; MDBA; ABS State water managers (e.g., New South Wales Office of Water)	Commonwealth Water Act; Water regulations; Basin plan ABS Act; State water legislation	Seasonal forecasting now routinely available. Short-term forecasting being trialed. Significantly improved timeliness streamflow data, volumes in storages, trade data. Overarching national water account in place but not well accepted. New technology (Geofabric, data cube, AWRIS)	Irrigators; environmental water managers; river operators; regional water providers; policy makers	Improved flood control and flood risk assessment; improved irrigator water ordering; improved river operations; improved demand management; improved environmental water use; better policy responses
Basic groundwater information	BOM State water managers	Commonwealth Water Act State water legislation	Groundwater levels and aquifer characteristics available online, along with national atlas of groundwater dependent ecosystems. New technology and investments in upgraded collection system (e.g., automated bore technology) providing significant improvement in basic data.	Small urban populations; irrigators; mining interests; NRM bodies	Better management of risk; draws attention to sustainability issues
Market regulation (MDB focus)	ACCC; BOM; MDBA State water managers State water regulators	Water charge and market rules; Water regulations; State-based regulations	Significant improvement in transparency; rules combined with publication of annual data on monitoring has led to improved market behavior.	Water entitlement holders; water users	More effective water portfolio/input management; increased market pressure on IIOs to perform
Market data—MDB focus (price and quantity)	Dept. of Environment State water registries BOM Private brokers	Water regulations State Water legislation	Entitlement and trade volumes reported but key benefit not yet realized as price data still poor. National water market system project abandoned, reflecting state intransigence and poor Commonwealth management.	Irrigators; environmental water managers; river operators; regional water providers; policy makers	More timely, cost-effective irrigator interventions; transparency; efficiency of water market

Table 1. *Cont.*

Type	Administering Agency	Authority	Comment	Beneficiary	Benefit
Basic water information (water access and irrigation rights, trades, allocation announcements)	BOM (until recently NWC and NWMS) State water managers	Water Act 2007; Water regulations; State Water legislation	Significantly improved basic data, but overall usefulness limited by inadequate price data and working common registry interface.	Irrigators; environmental water managers; river operators; regional water providers; policy makers	Strengthens market fairness
Environmental Watering	BOM, National and state environmental water managers	Various	Improved data on planned actions and ongoing activity.	Ecosystem	Optimizing ecosystem benefit from water flows; assuring other users of benefit from actions
Meteorological data affecting water resource availability	BOM	Water Act 2007, BOM Act, Water Regulations 2008	Issue of further refining outputs to meet user needs.	Irrigators and farmers; water managers; river operators; environmental water managers; businesses	More effective irrigator use of water resources; Improved irrigator planning, more effective environmental water use
Water Quality	BOM State water managers	Water Regulations 2008 State Water legislation	Little progress outside of data collection on salinity. Little data collected on environmental water quality.	Irrigators and farmers; environmental water managers	Clear understanding of "fit for purpose" implications

Sources: [3,19,21–31,33,37–40,45,49,50,58,59].

To buttress this ongoing debate, Australia now has rapidly developing information systems and databases that allow many more statements and propositions to be tested than was the case a decade ago. There is now a much better appreciation of the extent and characteristics of surface water assets, the rights governing their use, and the likely risks that might arise over time, and this helps to underpin and enhance water resource planning and management. There are still important shortcomings in some data sets, and important shortcomings in how users can access data that is already collected. Further progress is required, with specific attention to user requirements and whether all information being collected is required. However, many assertions that there is widespread duplication between information collections do not ring true; rather, they suggest that particular interests do not welcome being put under the microscope.

Understanding of ground water assets remains much more limited, and management of the resources is much less transparent and less uniform. The BOM's online efforts here are encouraging [60].

3.2. Water "Abstractions" and "Use" in the MDB

In any upgrading process, care is clearly needed to avoid overlapping data series and effort and definitional differences collected by different agencies. As needs change, a difficult task is to ensure that data collections remain relevant and vital, to minimize regulatory and transaction costs, and maximize benefit to users. In the case of Australian water information, few national data series existed a decade ago that exist today. However, one important area that remains puzzling is data published on water use in the MDB, for which there are currently three published data series.

Currently the MDBA, BOM, and ABS publish data series on annual water use/water abstracted/water diverted in the MDB. The MDBA data initially came from state obligations to monitor and report water use in the MDB after the cap on extractions was introduced in 1996. The BOM publishes a series in its Water Account largely based on MDBA data, but this currently contains some definitional differences with the MDBA data, resulting in material differences in aggregate numbers. Finally, the ABS publishes survey/census data reflecting self-reported water used on farms that is quite different from the above series [15,29,61].

Two points can be made about these three data series. First, differences in total annual volumes are significant. For example, the BOM calculates water abstracted from water resources in the MDB in 2012–2013 at 11,896 GL. The ABS calculates water used in the MDB in 2012–2013 to be 8574 GL. The MDBA 2012–2013 data is not yet available, some 18 months after the completion of the reporting period. Definitional differences are confusing to both casual and serious users. The ABS farm "use" figures are some 27% less than BOM abstraction numbers. Definitional differences appear to explain much of the difference but there is no concordance available between the series to test their respective veracity. Published data does not allow this to be carried out. This is necessary from a public policy perspective to test what each series contributes to understanding.

Second, it is not that there has not been discussion over resolving differences, as there has, but that it takes considerable time to overcome administrative inertia when introducing change, particularly when new institutions are asked to play a key ongoing role. These data series illustrate how, when change occurs, considerable effort is required to reexamine the role played by the original information collections. As the quality bar is lifted, what previously was "useful" may need to be carefully reexamined.

4. Challenges and Lessons

This article has sought to describe and assess the impact of an increased national effort on water information and data collections, with a focus on rural and regional Australia (and particularly the MDB) over the decade to 2014.

The article shows that the recent decade of considerable resource investment in water information has produced real results and a much better understanding of rural and regional water resources, and resulted in an enhanced ability to manage them in a sustainable way. Increased resourcing and a political commitment to serious water reform have been important elements in progress. Three ongoing challenges are noteworthy. First, data collection needs to closely reflect the current and prospective needs of water information users. Effective communication with those stakeholders is critical. New information technology and scientific understanding allow new information products to be developed for users—cost and use are two key considerations for ongoing assessment of the content of Australia's water information needs. Water information products and services need to focus strongly on user needs and delivering value to all data users, but have one eye to the future and cost *versus* benefits for data collection that may not appear to have a clear use today.

Second, there are interest groups that would prefer data to be collected at a state level, thus reducing standardization and comparability. Governments should strongly endorse standardization and transparency, rather than use "regulatory cost burdens" as reasons not to collect and disseminate necessary data. The BOM has stature as an independent non-evaluative source of data, unaffected by political interference, which has been present at the state level. It has put much effort into encouraging standardization of data, much like the ABS in the area of economic and social data (and, incidentally, is in the vanguard of developing standards for international use).

Third, some key data series, such as price and quantity of transactions in the water market in the MDB, could be improved dramatically and quickly if governments agreed there was value, with virtually no additional regulatory cost. This should be a priority for the MDBA, the BOM, and their information partners. The MDBA will need to ensure that states meet their commitments under the Basin Plan in this regard.

The applicability of the Australian approach of increasing the resourcing of national water information to other countries that perceive a need to upgrade their water information systems will depend in large part on where they are starting from, the legal framework in which water information is collected, the level of resourcing, the institutional and governance frameworks that are in place, the professional skill base, and the political and cultural forces that are either promoting or stifling reform.

The Australian experience suggests nine lessons for other countries:

- Upgrading a nation's water information systems is a long-term task, but an important one in a world of climate change and increased climate variability. Substantial progress is likely to take five to 10 years to materialize. In the current political landscape anywhere in the world, this is a long time, so attention needs to be given continuously to the purpose of the exercise, and the benefits from it. As there is a tendency to forget the original baseline (particularly as crises dissipate), attention needs to be given to reiterating the value proposition, and ensuring that data collected covers issues relevant to users and in a form that is readily accessible to them.

- The process of upgrading is likely to require substantial new resources. This underscores the need to examine the case to continue existing data series as well as introduce new ones, and to take opportunities to lower collection and regulatory costs.
- Effective data series tend to be long ones, so each series should reflect a careful initial consultation and development process. From the outset, upgrading information systems needs to be focused on data series that will facilitate answering key policy questions, assist water users in making significant decisions more effectively, or equip businesses and government to better address risks from water-related events.
- As always, political support matters. Politicians generally have little sustained interest in these issues, so drawing out the key "good news" stories, and the benefits to users and the economy and environment will be vital to sustain the reform process.
- Developing large new information systems involving large IT systems is fraught with high project risk, which will require careful project management. The IT skill base and the skill base of the water professionals managing and delivering the new water information system will be critical in transforming it. Developing new tools often goes hand in hand with data collection and data dissemination to users. Understanding the user group and its needs, and the optimal way to make the data base easily accessible to that group, will play a critical factor in delivering the project.
- New information may radically alter the way both groundwater and surface water resources are viewed. New technology is resulting in radically new approaches becoming cost effective. Flexibility and, on occasion, risk taking will be required to ensure that new approaches can be introduced into information collections. New approaches require good governance arrangements, particularly if several levels of government are involved.
- Any new national data collection and dissemination exercise will require new approaches and relationships between water information managers and users.
- Information collection and dissemination should be put in the hands of bodies that can best produce the required products. Entrenched interests are very likely to argue why a new approach should not be followed, as new information may be detrimental to their interests. Their case should be given no more or less weight; each new series needs to be fully justified on its own merits.
- The success or failure of a new information system will in part be determined by how it is used. For many users, their acceptance will be based around such factors as the ability of the system to enhance analysis and understand risk, reduce transaction costs, and boost business viability. At a policy level, for governments, the yardstick for acceptance is somewhat different: it is more that governments and public sector managers can develop policy based on a good understanding of the country or region's water resources. How it is used depends on the judgments and decisions of the decision-making body.

Upgrading Australian water information began in the middle of a crisis, and long after it was well known that existing water information was inadequate to make the decisions required. Seven years later, with the crisis having receded, governments appear to be giving the importance of information less weight. To sustain investments in information, its coverage must facilitate illuminating key questions and issues. Custodians of information systems must ensure that the value proposition is clear to all. Selling the case is critical. Communicating effectively with stakeholders is a key part of implementing and sustaining any successful upgrading program.

Acknowledgments

The author would like to thank colleagues and a referee for many useful comments on an earlier draft of this paper.

Conflicts of Interest

The author declares no conflict of interest.

References

1. Maestu, J. *Water Trading and Global Water Scarcity*; RFF Press: Milton Park, Abingdon, UK, 2013.
2. Grafton, R.Q.; Hussey, K. *Water Resource Planning and Management*; Cambridge University Press: Cambridge, UK, 2011.
3. Horne, J. Economic approaches to water management in Australia. *Int. J. Water Resour. Dev.* **2013**, *29*, 526–543.
4. Maestu, J.; Gomez, C. Drivers of Economic Information in River Basin Planning. *Int. J. Water Resour. Dev.* **2011**, *27*, 665–676.
5. OECD. In Proceedings of the OECD Workshop on Improving the information base to better guide water resource management decision making, Zaragoza, Spain, 4–7 May 2010. Available online: http://www.oecd.org/greengrowth/oecdworkshoponimprovingtheinformationbasetobetterguidewaterresourcemanagementdecisionmakingzaragozaspain4-7may2010.htm (accessed on 25 January 2015).
6. Parris, K. Improving the Information Base to Better Guide Water Resource Management Decision Making. *Int. J. Water Resour. Dev.* **2011**, *27*, 625–632.
7. Vertessy, R.A. Water information services for Australians. *Aust. J. Water Resour.* **2013**, *16*, 91–106.
8. Crase, L. *Water Policy in Australia: The Impact of Change and Uncertainty*; RFF Press: London, UK, 2008.
9. Australian Government. *Intergovernmental Agreement on a National Water Initiative*; Australian Government: Canberra, Australian, 2004. Available online: http://www.nwc.gov.au/nwi (accessed on 25 January 2015).
10. Raising National Water Standards Program—Evaluation Overview. Available online: http://www.nwc.gov.au/__data/assets/word_doc/0005/25349/RNWS-evaluation-overview.doc (accessed on 25 January 2015).
11. Raising National Water Standards Program. Available online: http://archive.nwc.gov.au/rnws (accessed on 25 January 2015).
12. Sinclair Knight Merz. The Need for Improved Water Data and Water Sharing. In *Waterlines Occasional Paper No. 4*; National Water Commission: Canberra, Australian, January 2008. Available online: http://archive.nwc.gov.au/__data/assets/pdf_file/0004/11020/improved-water-data-body-waterlines-0108.pdf (accessed on 25 January 2015).

13. Cordery, I.; Weeks, B.; Loy, A.; Daniell, T.; Knee, R.; Minchin, S.; Wilson, D. Water Resources Data Collection and Water Accounting. *Aust. J. Water Resour.* **2007**, *11*, 257–266.
14. Howard, J. *A National Plan for Water Security*; Australian Government: Canberra, Australian; pp. 1, 15.
15. Australian Bureau of Statistics. 4618.0—Water Use on Australian Farms, 2012–2013. Available online: http://www.abs.gov.au/ausstats/abs@.nsf/Latestproducts/4618.0Main%20Features72012-13?opendocument&tabname=Summary&prodno=4618.0&issue=2012-13&num=&view (accessed on 25 January 2015).
16. Australian Bureau of Statistics. 4655.0—Australian Environmental-Economic Accounts, 2014. Available online: http://www.abs.gov.au/AUSSTATS/abs@.nsf/Lookup/4655.0Explanatory%20Notes12014?OpenDocument#(accessed on 25 January 2015).
17. Australian Bureau of Statistics. 4610.0—Water Account, Australia, 2012–2013. Available online: http://www.abs.gov.au/AUSSTATS/abs@.nsf/second+level+view?ReadForm&prodno=4610.0&viewtitle=Water%20Account,%20Australia~2012-13~Latest~27/11/2014&&tabname=Past%20Future%20Issues&prodno=4610.0&issue=2012-13&num=&view=& (accessed on 25 January 2015).
18. Australian National Audit Office. Audit Report No.18 2013-14—Administration of the Improving Water Information Program. Available online: http://anao.gov.au/~/media/Files/Audit%20Reports/2013%202014/Audit%20Report%2018/AuditReport_2013-2014_18.pdf (accessed on 25 January 2015).
19. Bureau of Meteorology. Publications—Improving Water Information Programme Progress Report 2013. Available online: http://www.bom.gov.au/water/about/publications/ (accessed on 25 January 2015).
20. Bureau of Meteorology. Categories of Water Information. Available online: http://www.bom.gov.au/water/regulations/categoriesWaterAuxNav.shtml (accessed on 25 January 2015).
21. Bureau of Meteorology. Publications–Information Sheets–3 Australian Water Resources Information System (AWRIS). Available online: http://www.bom.gov.au/water/about/publications/ (accessed on 25 January 2015).
22. Bureau of Meteorology. Water Data Online. Available online: http://www.bom.gov.au/waterdata/ (accessed on 25 January 2015).
23. Bureau of Meteorology. Design Rainfalls. Available online: http://www.bom.gov.au/water/designRainfalls/index.shtml (accessed on 25 January 2015).
24. Engineers Australia. Australian Rainfall and Runoff Discussion Paper: An Interim Guideline for Considering Climate Change in Rainfall and Runoff Draft, November 2014. Available online: http://www.arr.org.au/wp-content/uploads/2013/Projects/Draft_ARR_interim_guidance_Format.pdf (accessed on 25 January 2015).
25. Bureau of Meteorology. Australian Hydrological Geospatial Fabric (Geofabric). Available online: http://www.bom.gov.au/water/geofabric/index.shtml (accessed on 25 January 2015).
26. Bureau of Meteorology. Modernisation and Extension of Hydrologic Monitoring Systems Program. Available online: http://www.bom.gov.au/water/regulations/fundingProgram/index.shtml_(accessed on 25 January 2015).
27. Bureau of Meteorology. Seasonal Streamflow Forecasts. Available online: http://www.bom.gov.au/water/ssf/ (accessed on 25 January 2015).

28. Bureau of Meteorology. Assurance Engagements on General Purpose Water Accounting Reports. Available online: http://www.bom.gov.au/water/standards/wasb/gpwareports.shtml (accessed on 25 January 2015).
29. Bureau of Meteorology. National Water Account. Available online: http://www.bom.gov.au/water/nwa/2013/ (accessed on 25 January 2015).
30. CSIRO. WIRADA Annual Reports. WIRADA Annual Report 2013–2014. Available online: http://www.csiro.au/Organisation-Structure/Flagships/Water-for-a-Healthy-Country-Flagship/WIRADA-annual-reports.aspx (accessed on 25 January 2015).
31. Geoscience Australia. Data Cube—The Future of Earth Observation Data Management and Analysis (narrated by Dr Adam Lewis). Available online: https://www.youtube.com/watch?v=AYB5Sw80Vfs&index=25&list=UUgglOEaXbvCJCrIj6R2EIBw (accessed on 25 January 2015).
32. McCormick, B. Water: Budget Review 2014–2015 Index. Parliament of Australia. Available online: http://www.aph.gov.au/About_Parliament/Parliamentary_Departments/Parliamentary_Library/pubs/rp/BudgetReview201415/Water (accessed on 25 March 2015).
33. Report of the Independent Review of the Water Act 2007. Available online: https://www.environment.gov.au/water/publications/report-of-the-independent-review-water-act-2007 (accessed on 25 January 2015).
34. Speers, R.A. Report on upcoming experimental estimates of water resource stocks for Australia. In Proceedings of the 20th Meeting of the London Group on Environmental Accounting, New Delhi, India, 15–17 October 2014.
35. Australian Competition and Consumer Commission. ACCC Role in Water. Available online: https://www.accc.gov.au/regulated-infrastructure/water/accc-role-in-water#monitoring-and-reporting (accessed on 25 January 2015).
36. Grafton, R.Q.; Horne, J. Water Markets in the Murray-Darling Basin. *J. Agric. Water Manag.* **2014**, *145*, 61–71.
37. Australian Competition and Consumer Commission. ACCC Water Monitoring Report 2012–2013, April 2014. Available online: https://www.accc.gov.au/publications/accc-water-monitoring-report/
accc-water-monitoring-report-2012-13 (accessed on 25 January 2015).
38. National Water Commission. *Australian Water Markets: Trends and Drivers 2007–2008 to 2011–2012*; Commonwealth of Australia: Canberra, Australia, 2013.
39. Australian Government. The National Water Market. Archived page 28 February 2014. Available online: http://web.archive.org/web/20140228235833/http://www.nationalwatermarket.gov.au/site-information/index.html (accessed on 25 January 2015).
40. Australian Government. The National Water Market. Available online: http://www.nationalwatermarket.gov.au (accessed on 10 February 2015).
41. CSIRO. Community attitudes towards science and technology in Australia. Available online: https://publications.csiro.au/rpr/download?pid=csiro:EP145330&dsid=DS1 (accessed on 1 April 2015).
42. CSIRO. The Sustainable Yields Projects. Available online: http://www.clw.csiro.au/publications/waterforahealthycountry/sustainable-yields.html (accessed on 25 January 2015).
43. Horne, J. Australian water policy in a climate change context: Some reflections. *Int. J. Water Resour. Dev.* **2013**, *29*, 137–151.

44. South-West Western Australia Sustainable Yields Project. PowerPoint presentation. Available online: http://www.clw.csiro.au/publications/waterforahealthycountry/swsy/pdf/SWSY-stakeholder-presentation-Bunbury.pdf (accessed on 25 January 2015).
45. Young, W.; Bond, N.; Brookes, J.; Gawne, B.; Jones, G. *Science Review of the Estimation of an Environmentally Sustainable Level of Take for the Murray-Darling Basin*; A Report to the Murray-Darling Basin Authority from the CSIRO Water for a Healthy Country Flagship; CSIRO: Canberra, Australian, 2011.
46. Petheram, C.; Rustomii, P.; Vleeshouwer, J. *Rainfall-Runoff Modelling Across Northern Australia*; A report to the Australian Government from the CSIRO Northern Australia Sustainable Yields Project; Water for a Healthy Country Flagship: Canberra, Australia, 2009.
47. CSIRO. Water in Northern Australia: Summary of Reports to the Australian Government from the CSIRO Northern Australia Sustainable Yields Project. Available online: http://www.csiro.au/Portals/Publications/Research--Reports/NASY-Summary-report.aspx (accessed on 25 January 2015).
48. CSIRO. Water Availability for Tasmania: CSIRO Tasmania Sustainable Yields Project. Available online: http://www.clw.csiro.au/publications/waterforahealthycountry/tassy/ (accessed on 25 January 2015).
49. Lawrie, K.C.; Brodie, R.S.; Tan, K.P.; Gibson, D.; Magee, J.; Clarke, J.D.A.; Halas, L.; Gow, L.; Somerville, P.; Apps, H.E.; *et al. Broken Hill Managed Aquifer Recharge (BHMAR) Project: Securing Broken Hill's Water Supply: Assessment of Conjunctive Water Supply Options Involving Managed Aquifer Recharge and/or Groundwater Extraction at Menindee Lakes: Data acquisition, processing, analysis and interpretation method*; Record 2012/011; Geoscience Australia: Canberra, Australia, 2012. Available online: http://www.ga.gov.au/metadata-gateway/metadata/record/gcat_73819 (accessed on 25 January 2015).
50. Geoscience Australia. Broken Hill Managed Aquifer Recharge. Available online: http://www.ga.gov.au/about/what-we-do/projects/water/broken-hill-managed-aquifer-recharge (accessed on 25 January 2015).
51. Australian Government. Bioregional Assessments. Available online: http://www.bioregionalassessments.gov.au (accessed on 25 January 2015).
52. Australian Government. Factsheets—Bioregional Assessments. Available online: http://www.bioregionalassessments.gov.au/factsheets.shtml (accessed on 25 January 2015).
53. Hirsch, R.M.; Miller, T.L.; Hamilton, P.A. Using Today's Science to Plan for Tomorrow's Water Policies. *Environment* **2001**, *43*, 8–17.
54. Campbell, A.; Turnour, J. We need a smarter debate on developing northern Australia. *The Conversation*, 13 August 2013. Available online: https://theconversation.com/we-need-a-smarter-debate-on-developing-northern-australia-16846 (accessed on 16 February 2015).
55. Australian Government. *Australian Government Response to the Senate Rural and Regional Affairs and Transport References Committee Report: Management of the Murray-Darling Basin System: Second Interim Report: The Basin Plan*; Australian Government: Canberra, Australian, March 2013.
56. National Farmers' Federation. Review of Water Act 2007 Submission. Available online: http://www.environment.gov.au/minister/birmingham/2014/mr20141219.html (accessed on 25 January 2015).

57. Minerals Council of Australia. Submission to the 2014 Review of the Water Act 2007. Available online: http://www.environment.gov.au/water/legislation/water-act-review (accessed on 25 January 2015).
58. Extended Hydrologic Prediction Services for Improved Water Resources Management: An Australian Perspective. Global Water Forum. Available online: http://www.globalwaterforum.org/2015/03/02/extended-hydrologic-prediction-services-for-improved-water-resources-management-an-australian-perspective/ (accessed on 20 March 2015).
59. Docker, B.; Robinson, I. Environmental water management in Australia: Experience from the Murray-Darling Basin. *Int. J. Water Resour. Dev.* **2014**, *30*, 164–177.
60. Bureau of Meteorology. Australian Groundwater Explorer. Available online: http://www.bom.gov.au/water/groundwater/explorer/map.shtml (accessed on 9 February 2015).
61. Murray Darling Basin Authority. Water Audit Monitoring Reports. Available online: http://www.mdba.gov.au/what-we-do/managing-rivers/the-cap/wam (accessed on 25 January 2015).

8

Quantity *versus* Quality in China's South-to-North Water Diversion Project: A System Dynamics Analysis

Yu Yang [1,*], Lei Yin [2] and Qingzong Zhang [2]

[1] Department of Public Administration, Southeast University, Nanjing 210096, China

[2] School of Economics and Management, Southeast University, Nanjing 210096, China; E-Mails: 101011694@seu.edu.cn (L.Y.); 101001820@seu.edu.cn (Q.Z.)

* Author to whom correspondence should be addressed; E-Mail: yangyu.seu@gmail.com;

Academic Editor: Miklas Scholz

Abstract: The South-to-North Water Diversion (SNWD) Project has entered the operational stage. The infrastructure decreases water shortages and promotes economic growth for North China, whereas the environmental effects have yet to be fully examined. In this paper, an analysis of the interaction between water quantity and water quality is conducted using the system dynamics method. In a specific province covering areas that provide and receive water, simulation results for economic and environmental indicators at different levels of transferred water and pollution rates are obtained. Certain conclusions can be drawn from the findings. First, the SNWD Project can provide a sufficient quantity of water for economic growth in receiving regions. Second, an efficient quantity of transferred water exists, which demonstrates the growth limitation for both agriculture and the groundwater funnel. Third, upgrades in water quality can mutually reinforce the quantity effect and greatly promote economic growth. An environmental limitation exists for all economic activities. Therefore, water quantity and water quality are equally crucial in the SNWD Project to provide an opportunity for sustainable development.

Keywords: South-to-North Water Diversion (SNWD) Project; system dynamics modeling; water quantity; water quality; growth limitation

---

1. Introduction

The South-to-North Water Diversion (SNWD) Project (official translation), or the South-to-North Water Transfer Project, is an infrastructure project designed to divert water from flood-prone southern rivers to China's water-poor northern region [1]. According to the project plan, the annual volume of diverted water is 44.8 billion cubic meters. The size is close to the total water volume of China's Yellow River, which is the third longest river in Asia [2]. Although approximately 253 billion Yuan (Chinese currency, 1 Yuan = 0.16 U.S. dollar) has been invested in its construction [3], from a cost-benefit point of view, the SNWD Project is a feasible choice for North China and the rest of the country [4,5]. A study on water vulnerability and demand-supply balance dynamics indicates that the incremental water supply from the project would help the receiving region keep pace with the development of the entire country [6].

Similarly to other mega-projects in China and worldwide [7,8], however, controversies surrounding the SNWD Project have continuously been debated by numerous stakeholders and researchers [9–11]. Technological, economic, social, ecological and even political disputes are constant [12]. One-third of the provinces in China are included in the project, and over 200 million people would benefit from the water diversion, making the endeavor more than a simple infrastructure project. To ensure that the project better serves households, firms and farmers, among the disputes, the central concern should focus on the basic demands of those affected. Water demands can be interpreted from two perspectives: efficiently tackling the increasing water gap for human water users and effectively meeting drinking and nondrinking water requirements. Finally, a sustainable water management pattern that not only guarantees a sufficient amount of water but also satisfies a certain standard of water quality should be achieved. Thus, a water transfer project should attain water quantity and water quality goals simultaneously, which is the primary focus of this article.

With decades of studies on the planning and construction of the SNWD Project, the amounts of diverted water and related effects have been well addressed [13,14]. Nonetheless, for a project that virtually violates the laws of nature, the indirect effects of the water diversion have yet to be fully discussed. At the pre-construction stage, some scholars and experts took interest in the environmental uncertainties associated with the project [15,16]. Moreover, cooperation among regions has been obstructed by the activities of the primary polluters [17]. Some authors have proposed that the ultimate objective of the SNWD Project is to improve the water environment rather than meet the water demand gap [12]. The costs of the SNWD Project include furthering environmental deterioration through rapid rural transition [18]. Evidence from virtual water calculations shows that North China exports over 50 billion m^3 of virtual water (amounting to over one-third of the supply) to South China annually. A majority of the virtual water flow occurs in the form of agriculture products, and brings stress for the water-exporting provinces. [19,20]. Thus, diverted high-quality water is critical for the receiving region as well as the supplier, which is the actual receiver of the "water". In December 2014, President Xi Jinping urged stronger management to protect water quality and conserve water [21]. With the SNWD Project entering the formal operational phase, controversies should be analyzed from a dynamic perspective. In this paper, therefore, we consider the future and investigate the intrinsic link between water quantity and water quality during the upcoming operational stage instead of studying water quantity during the pre-construction and construction stage, as has been done in the existing literature.

Regarding the dilemma of quantity and quality in the SNWD Project, we must provide answers to certain unanswered questions. Which is comparatively more essential, water quantity or water quality? Would an emphasis on water pollution treatment introduce extra costs and offset growth in certain regions? To gain better knowledge on potential outcomes, the research community has sought to apply mathematical models in the analysis. Typical quantitative decision-making tools and the computable general equilibrium model have been employed [6,22,23]. Using established simulation methods, data output under various conditions have been obtained to provide information for comparison and further policy design. In most previous studies, the two variables, water quantity and water quality, have been treated separately, which is partly a result of the complexity of the economic-social-ecological analytical framework and is also related to the difficulty of the model description. With the goal of systematically characterizing the inherent relationship between water quantity and water quality, this paper utilizes a methodology of system dynamics analysis to investigate the effects and the trends of the SNWD Project and discusses the implications for sustainable development based on the results of the quantity-quality interaction. Furthermore, these findings may contribute to research and policy applications in other similarly marginal environments around the world. The remainder of this paper is structured as follows. We first describe the study area, the modeling method and data sources used. Then, through a well-designed system dynamics model, we analyze the effects of water quantity and quality in various scenarios. The results of the simulation and prediction are then reviewed and further discussed. Finally, we summarize the study and present policy suggestions for sustainable water management.

2. Methods and Data

2.1. Study Area

The SNWD Project consists of three canal systems, the eastern route, central (or middle) route and western route. As a result of varying biophysical conditions and social-economic characteristics [24], a single study could not cover them all. Among the three routes, the western route, drilling through the Tibetan Plateau to divert water to the Yellow River, is the most ambitious and is still in the planning phase [25,26]. Thus, Figure 1 does not accurately depict this route. The central route is a closed system in which the water is pumped through canals, pipes or tunnels (e.g., under the Yellow River), thus offering relatively good water quality able to meet Class II standards. However, the construction of this canal requires the large-scale relocation of 330,000 people in Henan and Hubei Provinces [27], and the potential disruptions are primarily social rather than economic or environmental [28]. Unlike the central route, the eastern route mostly uses an existing channel. The open system passes through the Beijing-Hangzhou Grand Canal, the world's longest man-made river, developed in the 13th century. Thus, the investment in and displacement of the eastern route are far less than those associated with the central route [29]. However, the eastern route endangers the quality of diverted water because it collects a large amount of pollutants and sediment through the industry-heavy areas, including Jiangsu and Shandong Provinces [30].

The comparison suggests that the eastern route is the most appropriate to research. Although water quality is a general issue for all routes, the eastern route faces the most serious emerging risks, particularly because it covers 71 counties (or cities, districts), directly benefiting a population of 100 million people. Furthermore, the eastern coastal region is more advanced in economic growth than the central provinces, such as Henan and Hebei. In 2013, Jiangsu and Shandong ranked among the top three nationwide in terms of GDP (gross domestic product), and the per capita GDP of Tianjin has remained the highest in the nation over the last two years [31]. These features make research on water quantity and quality, as well as the economic and environmental consequences of constructing the eastern route, well worth examining. The first stage of the eastern route was completed in December 2013. The route starts in southern Jiangsu Province, the downstream of Yangtze River, then transfers water to northern Jiangsu, Shandong and Tianjin. In other words, Jiangsu Province not only provides water, but receives water as well. Furthermore, Jiangsu is a critical source of polluted water. Its total, wastewater discharge was the nation's second highest in 2013, accounting for approximately 9% of that of the entire country [31]. Such conditions would naturally spur fear among people in water-receiving cities. Therefore, along the eastern route, Jiangsu is an ideal study area that sufficiently encapsulates the research concerns and content of this paper.

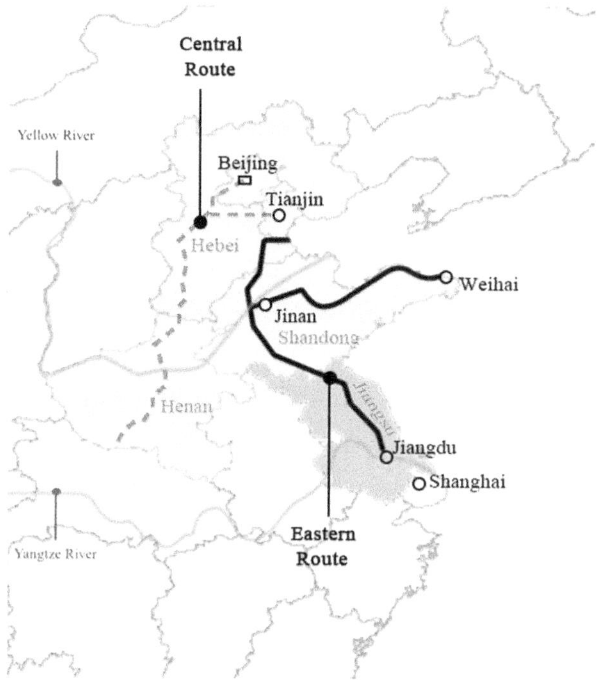

Figure 1. Map of the study area.

2.2. System Dynamics Modeling

We implemented a system dynamics methodology by building an analytical framework for modeling complex systems, such as the SNWD Project [32]. This method has been applied to a wide range of water resource management and ecological studies [33–35]. With visualized simulation techniques, a system dynamics platform, Vensim PLE Version 6.3 in this study, can allow for a better understanding of the interrelationships among the variables in a large system.

With respect to conflicts between water supply and demand along the southern and northern cities, the water diversion project creates direct or indirect impacts on urban or regional economic and ecological systems. The interaction mechanism is rather complex, but with a detailed model based on the principles of system dynamics, the architecture becomes apparent. As shown in Figure 2, the entire system is described from the perspective of water demand and water supply (the two circles). To capture the dynamic relationships within the system, the model includes over 50 variables, 14 specified constants, and 13 lookup functions. Five subsystems have been framed, *i.e.*, water, agriculture, industry, population and the environment. Each subsystem is mutually interrelated to any of the other subsystems. Connectors establish relationships between the elements in the water diversion system, and information transferred among the connected elements is indicated by the direction of the arrows. All flows will contribute to the stocks of the five main outcome variables (the five rectangles). We select five specific subsystems based on three considerations. First, the agricultural, industrial and residential sectors occupy the vast majority of the overall water demand/supply. Second, the effects of water quantity are actually economic effects, which are reflected by the growth in agriculture and industry. Third, the environmental effects of water quality can be further divided into two parts: the usage of surface resources and the usage of groundwater resources. The former produces wastewater pollution associated with rapid agricultural and industrial development, whereas the latter leads to overexploitation as urbanization accelerates. Both results have severely impacted the environment [36].

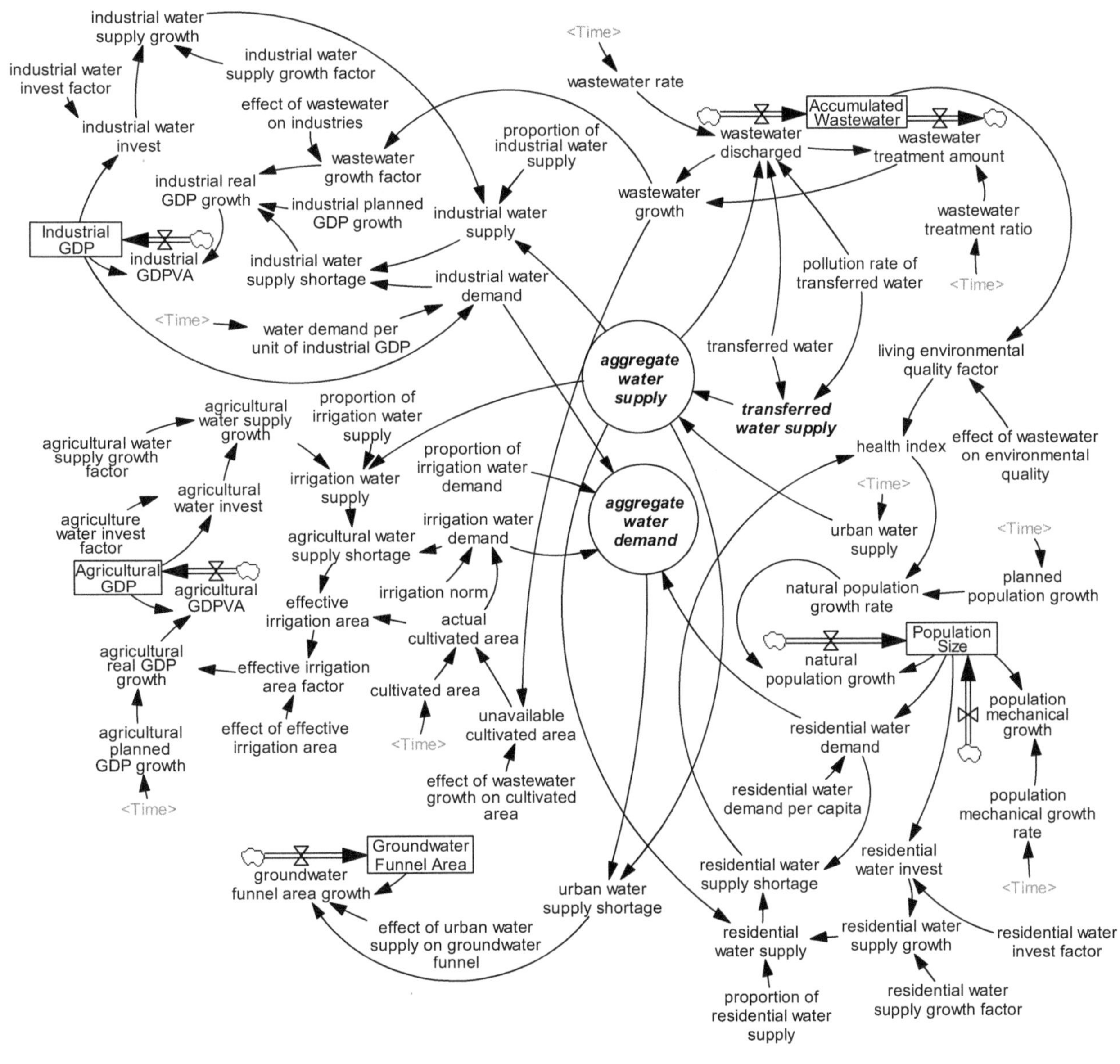

Figure 2. Causal pathways in the system dynamics model.

This model also includes the effects of accumulations and time delays, as well as impacts from resource constraints and behavioral feedback. The model helps analyze the dynamics caused by the causal relationships between elements in the system and provides insight about reinforcing and balancing feedback loops. Consider the groundwater funnel as an example. The funnel is a certain area where the ground is subsiding due to the excessive pumping of groundwater. North China is the largest subsidence area with the most funnels on the planet, and some regions have even fallen below the sea level. The urban water supply shortage would aggravate the funnel area with an increasing depletion of underground water. If the system receives transferred water, which alleviates the shortage, the groundwater funnel area decreases. From a macro-level perspective, the urban water supply is determined by factors associated with other subsystems, e.g., agricultural demand or the pollution rate of transferred water. The next step of the system dynamics analysis is assigning variables and obtaining the results related to the economic and environmental effects of the SNWD Project.

2.3. Data

Data form the major cities located along the diversion route were collected from various official databases and key references, e.g., the Jiangsu Statistical Yearbook and the Jiangsu Water Resources Bulletin. The cities include Yangzhou, Taizhou, Huaian, Yancheng, Suqian, Xuzhou and Lianyungang. Although the formal operation of the eastern route began in late 2013, data between 2002 and 2012 were gathered and calculated. One reason is that the construction of the eastern route began in 2002, and it utilized the existing Grand Canal as the main waterway for water transport. Thus, the impacts from the construction stage are as significant as those from the operational stage, a situation unlike that of the central route. The other reason is the consideration of the accuracy of the data. Model validation can be achieved by comparing the performance of the model to a real system [37]. Before the simulation over the operational period, the system dynamics model was calibrated with the data in the construction period. We obtained desirable results from the comparison between the simulation and historical data, with a bias that was approximately within 4% for the economic (GDP) data, 0.7% for the population data and 6% for the environmental (wastewater, groundwater funnel) data. The validation ensured the data quality indicated in Table 1. It needs to be clarified that the sum of proportions does not equal 1 because only sectors directly connected with the subsystems in the system dynamics model are counted. Water consumptions from some sectors such as transportation, public facilities and social services (e.g., urban afforestation, firefighting) are far fewer and could be omitted. The proportions listed here have covered nearly 90 percent of the overall water demand or supply.

Some other variables not listed above were also calculated with the latest data available. The two central variables in this research, water quantity and quality, are indicated by the amount of transferred water (TW) and the pollution rate (PR), respectively. The baseline is determined from the integration of official publications. In the feasibility study of the SNWD Project, the annual amount of diverted water before construction was completed was 4.97 billion m^3 [38], whereas after the formal operation began, an additional 3.60 billion m^3 was diverted, which amounts to an aggregate of 8.57 billion m^3 [39]. According to China's National Development and Reform Commission, the actual amount of diverted water in the eastern route is fixed at 14.80 billion m^3 [40]. Therefore, we analyzed three water levels: 5, 10 and 15 billion m^3. Furthermore, we also considered two levels, 30 and 50 billion m^3, for the scenarios in which the water demand significantly increases with expanding urbanization and economic activities in the near future. With respect to the pollution rate, there is only one reliable source. The People's Daily officially reported that the overall water quality compliance rate was 89% before the project began its formal operation in 2012 [41]. The data indicate that the pollution rate is approximately 10%.

Table 1. Initial values and sources of some key variables.

Variables	Initial Values (Year = 2012)	Units and Data Sources
proportion of irrigation water supply	66.45	%, Jiangsu Water Resources Bulletin
irrigation norm	649.50	million m^3 per thousand hectare, Jiangsu Water Resources Bulletin
proportion of irrigation water demand	88.64	%, Jiangsu Water Resources Bulletin
proportion of industrial water supply	14.00	%, Jiangsu Water Resources Bulletin
proportion of residential water supply	7.00	%, Jiangsu Water Resources Bulletin
agricultural GDP	154.58	billion Yuan, Jiangsu Statistical Yearbook
industrial GDP	483.52	billion Yuan, Jiangsu Statistical Yearbook
population size	3400.82	ten thousand, Jiangsu Statistical Yearbook
accumulated wastewater	0	billion m^3, authors' assumption
groundwater funnel area	150	thousand hectare, Jiangsu Water Resources Bulletin

To assess the effects of changes in water quantity and quality, we run simulations through the year 2020. The results offer stakeholders and policy-makers new perspectives for understanding the interaction of water quantity and quality, as well as the trajectory of sustainable water management.

3. Results

This section reports on four outcome variables: agricultural and industrial GDP as two economic indicators, and accumulated wastewater and the groundwater funnel area as two environmental indicators. The results concerning population are omitted, because the annual changes are minor (within approximately 0.2%). Although the model provides results through 2020, its purpose is not to predict exact forecasts, but to offer insight into how the economic and environmental systems tend to respond to water quantity and quality.

3.1. Different Transferred Water Levels

As explained in the previous section, we chose five different levels of transferred water amounts to simulate the effects of quantity on economics and the environment. The results obtained for the four variables are listed in Figure 3. Five curves of accumulated wastewater can be observed. A higher level of water diversion accumulates more pollution, which is clearly attributed to the open water system along the eastern route. The effects on the three other variables are inconspicuous, which is partially a result of the unit setting. If the billions unit is replaced by hundred millions (cubic meters or Yuan), which is the official unit in China, the gap between the fives curves will become more distinct. To better present the results, specific simulated values for the three variables from 2015 to 2020 are listed in Table 2.

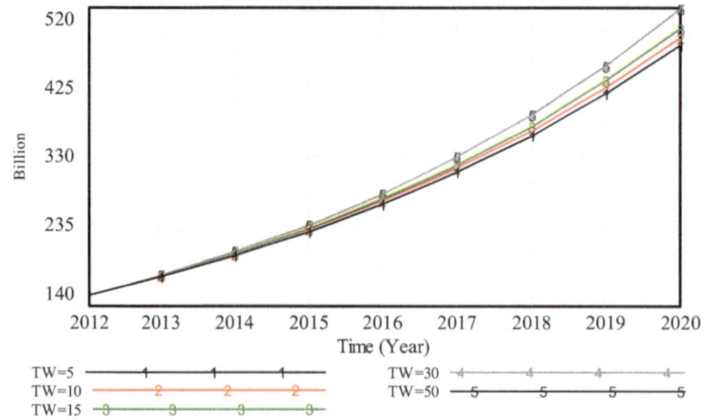

(a)

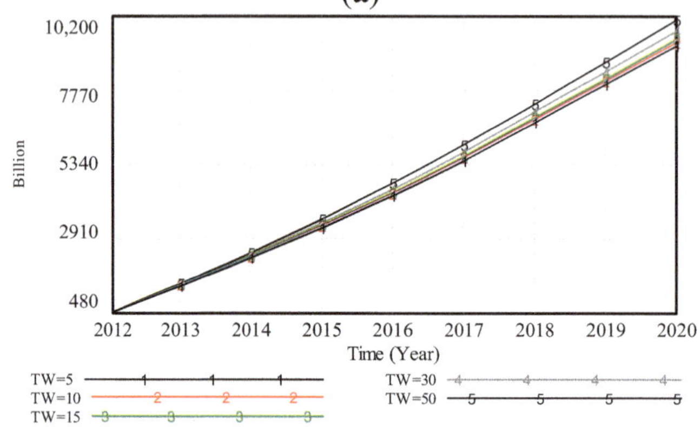

(b)

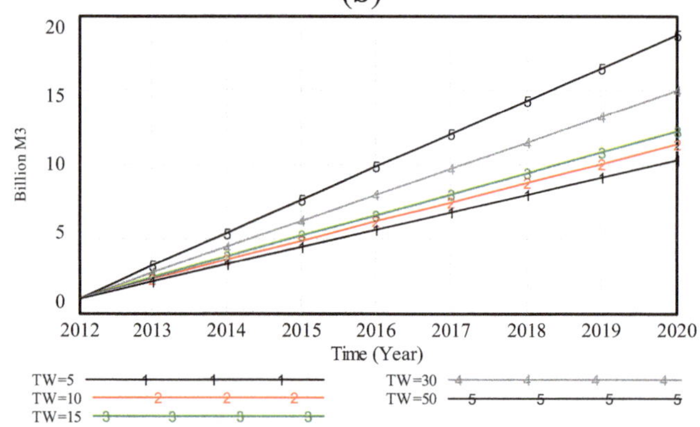

(c)

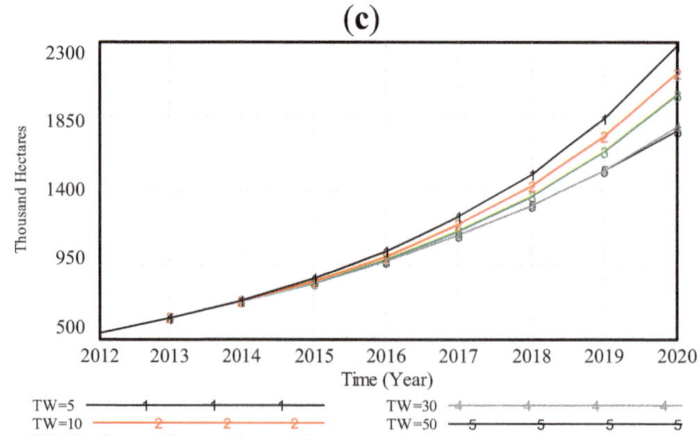

(d)

Figure 3. Economic and environmental effects respond to different transferred water (TW) levels. (**a**) Agricultural GDP; (**b**) Industrial GDP; (**c**) Accumulated Wastewater; (**d**) Groundwater Funnel Area.

Table 2. Outcomes as water quantity changes, 2015–2020.

Outcome Variables	Transferred Water (TW)	2015	2016	2017	2018	2019	2020
Agricultural GDP (billion Yuan)	TW = 5	234.02	269.18	309.68	356.34	410.10	472.10
	TW = 10	235.64	271.65	313.20	361.14	416.47	480.36
	TW = 15	237.84	275.03	318.06	367.87	425.53	492.30
	TW = 30	242.62	282.38	328.66	382.53	445.23	518.20
	TW = 50	242.62	282.38	328.66	382.53	445.23	518.20
Industrial GDP (billion Yuan)	TW = 5	3253.95	4309.78	5451.21	6692.87	7940.09	9192.91
	TW = 10	3288.38	4357.32	5512.86	6769.82	8032.35	9300.49
	TW = 15	3321.49	4403.01	5572.11	6843.77	8121.01	9403.85
	TW = 30	3412.82	4529.04	5735.50	7047.63	8365.33	9688.63
	TW = 50	3532.34	4693.98	5949.36	7314.47	8685.17	10061.50
Groundwater funnel area (thousand hectare)	TW = 5	861.45	1027.99	1236.27	1496.83	1830.78	2271.35
	TW = 10	843.18	994.75	1183.14	1417.19	1715.61	2097.92
	TW = 15	836.75	975.81	1147.70	1359.88	1629.20	1972.49
	TW = 30	836.75	970.63	1125.93	1306.08	1515.55	1779.62
	TW = 50	836.75	970.63	1125.93	1306.08	1515.05	1757.46

The vertical comparison reveals a similar pattern for the agricultural GDP as well as the groundwater funnel area. As indicated by the grey shading areas in Table 2, the agricultural GDP is constant after the amount of transferred water reaches 30 billion m^3, implying that a growth limitation may have been reached. The decreasing trend of the groundwater funnel area would end at the same amount of transferred water. The termination begins even earlier in 2015, at a level of 15 billion m^3, and exhibits a slower pace in 2019 and 2020. The similarity is primarily caused by the natural characteristics of these two variables. Given the current technology used for agricultural cultivation and groundwater exploitation, there is limited potential for improvements despite the massive water diversion [42,43]. These results also validate the rationale for the planned amount of diverted water [38,39].

The situation is different for industries. The industrial GDP would increase by 50–100 billion Yuan with an extra five billion m^3 of water supply and increase by 100–300 billion Yuan with an additional 15–20 billion m^3 of water supply. Because the natural restrictions on industries are not considerable, industrial growth always produces a majority of the pollution, which should not continue without limit [44]. In determining whether to avoid the negative environmental aftereffects, there exists a conflict that is precisely the central issue addressed in this study. What effects would the interaction between the changes in water quantity and water quality generate? The next step of the calculation will attempt to offer additional solutions.

3.2. Different Pollution Rate Levels

The adjustments to transferred water suggest a rational amount not to exceed 30 billion m^3 to prevent inefficiency. Thus, the following simulation will examine scenarios that change the pollution rate for a certain level of transferred water. In each particular case, we test the water quality effects of a pollution rate at the current 10% condition, and for two additional scenarios: 5% as an improvement in water quality and 15% as deterioration. As shown in Figure 4, with a pollution rate ranging from 5% to

10% and 15%, there is an increasing accumulation of wastewater at various amounts of transferred water. The detailed simulation results of the three other outcome variables are also reorganized in Table 3 for reference.

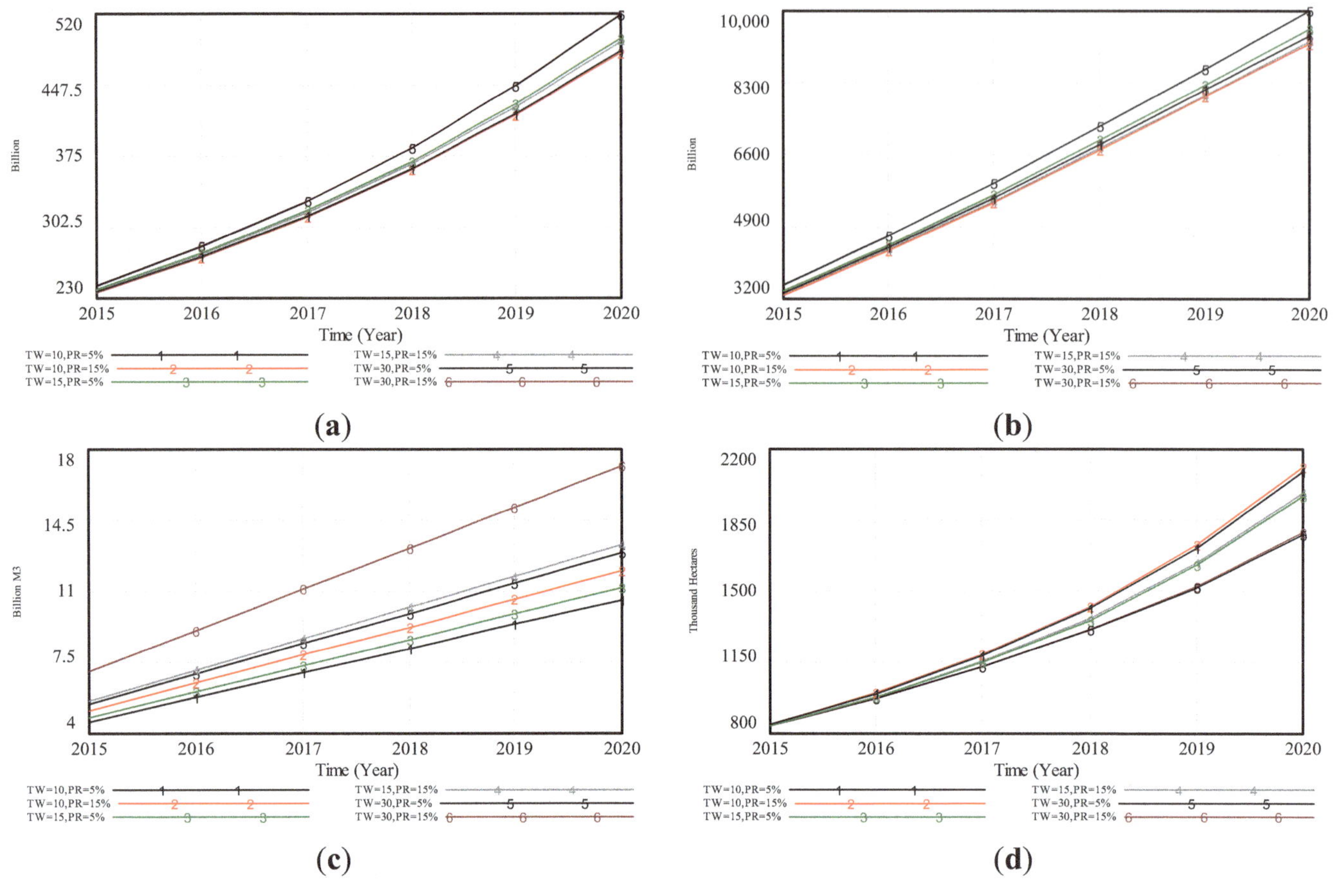

Figure 4. Economic and environmental effects response to different pollution rate (PR) levels (**a**) Agricultural GDP; (**b**) Industrial GDP; (**c**) Accumulated Wastewater; (**d**) Groundwater Funnel Area.

Table 3. Outcomes as water quality changes, 2015–2020.

Outcome Variables	Transferred Water (TW), Pollution Rate (PR)	2015	2016	2017	2018	2019	2020
	TW = 10, PR = 15%	235.40	271.27	312.64	360.36	415.41	478.96
	10%	+0.25	+0.39	+0.56	+0.78	+1.06	+1.40
	5%	+0.26	+0.41	+0.59	+0.81	+1.10	+1.44
Agricultural	TW = 15, PR = 15%	238.19	275.58	318.86	368.97	427.02	494.27
GDP	10%	+0.36	+0.55	+0.79	+1.10	+1.48	+1.96
(billion Yuan)	5%	+0.36	+0.55	+0.79	+1.10	+1.49	+1.97
	TW = 30, PR = 15%	242.62	282.38	328.66	382.53	445.23	518.20
	10%	0	0	0	0	0	0
	5%	0	0	0	0	0	0

Table 3. *Cont.*

Outcome Variables	Transferred Water (TW), Pollution Rate (PR)	2015	2016	2017	2018	2019	2020
Industrial GDP (billion Yuan)	TW = 10, PR = 15%	3258.98	4316.61	5459.90	6703.49	7952.55	9207.10
	10%	+29.40	+40.71	+52.96	+66.33	+79.80	+93.39
	5%	+29.52	+40.87	+53.17	+66.60	+80.13	+93.76
	TW = 15, PR = 15%	3276.78	4341.11	5491.58	6742.91	7999.67	9261.87
	10%	+44.71	+61.90	+80.53	+100.86	+121.34	+141.98
	5%	+44.97	+62.26	+81.00	+101.45	+122.05	+142.82
	TW = 30, PR = 15%	3321.10	4402.08	5570.35	6840.82	8116.55	9397.56
	10%	+91.72	+126.96	+165.15	+206.81	+248.78	+291.07
	5%	+94.08	+130.24	+169.44	+212.20	+255.29	+298.71
Groundwater funnel area (thousand hectare)	TW = 10, PR = 15%	844.02	996.62	1186.28	1421.93	1722.33	2107.11
	10%	−0.84	−1.86	−3.14	−4.74	−6.72	−9.19
	5%	−0.84	−1.86	−3.12	−4.71	−6.66	−9.11
	TW = 15, PR = 15%	836.75	977.09	1150.53	1364.61	1636.23	1982.33
	10%	0	−1.28	−2.83	−4.73	−7.03	−9.84
	5%	0	−1.27	−2.81	−4.7	−6.96	−9.72
	TW = 30, PR = 15%	836.75	970.63	1125.93	1306.08	1517.77	1784.2
	10%	0	0	0	0	−2.22	−4.58
	5%	0	0	0	0	−0.50	−2.48

Notes: "+"/"–"denotes the relative increase/decrease of the results above; "0" indicates no change.

It should be noted that the agricultural GDP and groundwater funnel area would stabilize after reaching a certain level of transferred water. This situation is mostly consistent with the results in Section 3.1, even though the pollution rate varies. This result further verifies the potential limits on the water demand from agriculture. Below a fixed level (TW = 30), these two variables would change at a different pollution level. Less-polluted water could promote agricultural growth. The water quality is increasingly critical in the long run (2018–2020), when a 5% improvement in the pollution rate could produce an additional 1–2 billion Yuan in agricultural GDP and reduce approximately 5–10 thousand hectares of the groundwater funnel area.

An exception to the case of TW = 30 is the groundwater funnel area in the long term, *i.e.*, 2019 and 2020, when a lower pollution rate results in a lower funnel area. This change might be correlated with the interactions of population growth, residential water demand and supply shortage. In the long run, the residential population will increase to a significantly high level as a result of urbanization. It is assumed that the population size will have a positive impact on the residential water demand. Thus, the effective water supply will be negatively correlated with the pollution rate, whereas the groundwater funnel area, which depends on a water supply shortage, will exhibit a notable decrease.

The most important results are the impacts due to both water quantity and water quality on the industrial GDP. The simulation updates the understanding of the unsolved issue discussed in Section 3.1. It demonstrates that more investment in pollution treatment would stimulate industrial growth. The industrial GDP would increase by approximately 30–150 billion Yuan with a 5% upgrade in water quality between 2015 and 2018 and increase by as much as 300 billion Yuan from 2017 to 2020.

Furthermore, the effect from the additional water quantity could reinforce the water quality impacts. The simulation has also indirectly demonstrated the growth limitation of industries, although no obvious natural restrictions exist, such as those on agriculture or groundwater. If we were to compare the gains from better water quality and the costs of pollution treatment, the findings would be more meaningful. According to a public data source, 15.4 billion Yuan has been devoted to water pollution control [45]. This investment could easily be offset by the economic gains from the extra growth in the industrial GDP.

4. Discussions

4.1. Trade-Offs in the SNWD Project

The simulations in the previous section have provided definite answers to the central question raised in this paper. There is no contradiction between the quantity and quality of diverted water. It is clear that economic and social development in regions with water shortages requires that more water be made available and that the SNWD Project could vastly improve the situation. Furthermore, the quality of transferred water is equally essential to the sustainable growth of the water-receiving areas. Environmental limitations exist in all economic activities, including agriculture and industry, as well as the urban expansion that creates more groundwater funnels. Therefore, the water diversion program should address the economic dimension and also avoid the imminent conflict with the water environment.

The preliminary findings have shown that the water quantity-quality trade-off could obtain a point of equilibrium, while this trade-off is not the only one in the SNWD Project. In the simulations of various subsystems, despite the minor changes in population, another potential trade-off exists between agriculture and industry. This trade-off is essentially a comparison between marginal costs and marginal benefit. As for the agricultural sector, it is related to the safeguarding of the national food security in China [46]. The simple calculation on the water usage and the economic output is not sufficient for the social optimization, especially when its production and yields would decrease due to urbanization [47]. As for the industrial sector, it has contributed the majority in the GDP accumulation, yet with even more pollution and violating social welfare [44]. Therefore, there exists a certain side effect in the water diversion, *i.e.*, the results show no limitation for industrial growth in Section 3.1. For the purpose of realizing synergy among economic, social and environmental goals, it is necessary to build a multi-objective coordination model for the SNWD supply system. The trade-off between water resource allocation and quality protection is a fundamental issue. Meanwhile, the full consideration of the production decision-making and related by-products in agricultural, industrial and other sectors is a prerequisite as well. Without this mechanism, there might be some obstructions to achieving the ideal results of water quantity and quality in Section 3.

The research findings presented above are straightforward, one reason possibly being the study area of focus. We focused on a single province, so that certain other factors could be ignored. If we were to include other provinces, the analysis and simulation model would be more complicated. Because the conflicts and coordination between provincial boundaries are increasingly significant [48], the political dimension factors are difficult to describe in the quantitative model. Furthermore, in this case, we should also consider that the central government would be involved, making the entire conflict hierarchical [49]. Thus the third trade-off between provincial boundaries, or between central and local governments, would make a further renegotiation and possibly change the designed water quantity and quality proposals.

4.2. Further Stages under Planning

The SNWD Project has yet to be fully accomplished. As described in Section 2.1, the completed project in late-2013 was the first stage of the eastern route. Thus, the quantitative analysis in the sections above was actually based on the conditions and data in that early stage. Findings from the simulation did not take the new project construction into consideration.

When discussing the upcoming second stage that remains in the planning process, a more comprehensive thinking is required. Though some explicit predictions have been achieved, they need to be combined with some other real-world policies. The most considerable one is from the State Council of China. In January 2013, the No. 2013-2 Document "Assessment Methods for the Most Strict Water Resource Management System" was issued. The policy outlines the principles of China's water resource management from a macro perspective. It sets the overall national water consumption on a limitation of 700 billion m^3 per year, as well as the control targets for every province. Among them, Jiangsu Province is given a quota restriction of 50.8 billion m^3 in 2015, and 52.4 billion m^3 in 2020 [50]. The limitation for Shandong Province is 25.1 billion m^3 (2015) and 27.7 billion m^3 (2020), while for Tianjin, it is 2.8 billion m^3 and 3.8 billion m^3, respectively. Comparing with the results in Section 3.1, the amount of transferred water is equivalent to nearly 60 percent of Jiangsu's total water supply. It would significantly relieve stresses for the water consumption in both Shandong and Tianjin. However, the water quantity-controlling regime has brought more significant pressure for Jiangsu. Related evidence is the No. 2011-29 Document "The Water Resource Planning in Jiangsu Province" from the Water Resources Department of Jiangsu Province. The planning goal of water usage in 2020 is 59 billion m^3 [51]. However, the State Council cut this standard down 6.6 billion m^3 two years later in the No. 2013-2 Document.

The solution for this rigorous situation is quite complicated, while the analysis in Section 3.2 might put forward a workable suggestion. Since the water quality has a specific connection with the water diversion amount, the control of the pollution rate could be the breakthrough point. A reduced pollution level leads to a higher efficiency of water consumption. For example, in 2020, the industrial GDP for the case of TW = 15 and PR = 10% (9403.85 billion) is greater than the case of TW = 30 and PR = 15% (9397.56 billion). If the water quality could be improved in the following years, the targeted water supply goal would not be a problem for Jiangsu. Therefore, the outputs of the first stage are the inputs for the second stage. With an increasing demand in the second stage, it is necessary for water providing regions to establish a clear objective system on water quality upgrading so as to enhance the current usage efficiency and to satisfy the further needs from the northern provinces and even the capital.

4.3. Some Neglected Factors

The research on the water quantity-quality interaction has covered several factors including economic, social and ecological dimensions. Meanwhile, the system dynamics model has its unique advantage in describing the effects of water diversion from various perspectives. However, some key influential factors are still missing.

This study has only predicted the case for the time horizon up to the year 2020, which is based on an implicit assumption that no eventful technological or structural change will occur. Thus, all of the results pertain to the current technological conditions. While it is increasingly emphasized in China that local governments put huge efforts into the research and development of techniques with water efficiency and pollution control [52], the results in the next five years would possibly change. In view of this surmise, it could be optimistic about the potential difficulties in further project stages under planning.

The structural characteristics of the various economic sectors have also been considered to remain unchanged. This means that the initial proportion settings in Table 1 are relatively stable during the eight-year simulation period. Most of the graphs indicate exponential growth curves, corresponding to the ongoing and upcoming high-speed growth in China. The growth rate is extraordinarily high for Jiangsu Province, especially for its north and central regions (the cities included in this study), which have annual GDP growth rates over 12 percent and industrial growth rates of nearly 14 percent [53]. However, we could expect that in the long run, more significant technological progresses will occur, with a transformed industrial structure, and the flow and stock of the outcome variables would differ. Currently, the transforming degrees in the agriculture (first) sector are relatively low, since the changes of its share in GDP were less than 0.2 percent in the recent five years. The changes are more clear and satisfying in the industrial (second) sector, with the annual declining rates between one and 1.4 percent [53]. Thus, a transition to an s-shaped curve would occur a few decades later, when the parameters given in this study fundamentally change. This trend will bring a remission to the water consumption pressure, especially for the providing regions. Meanwhile, the central and local governments would definitely push forward the acceleration of economic structural transformation [54]. This is a factor of policy and will be discussed in the next paragraphs.

For a great national infrastructure project, the influence of the policy dimension is crucial, as well. Most of this paper presented an engineering perspective by setting different water transfer and pollution rate levels, which is independent of policy. All of the initial parameters have been partially determined by central or local policies and stabilized after years of evolution. Governments have the ability to co-opt environmental policies to pursue other strategic objectives [55]. As the SNWD Project has completed its construction stage and begun formal operations, the influence from policy-makers, rather than engineers, should be allotted more significance.

There are three priority factors related to policies. The first is the water price mechanism. Although the Jiangsu Water Source Company Ltd., on behalf of the provincial government, is charged with the water market businesses, including water price setting, water is a special product that is a necessity for the basic needs of residents and farmers. Water consumption could not be handled totally under commercial principles. Water-providing areas have complained about the costs they bear as part of the SNWD water treatment and other issues. The central government is requested to impose an "SNWD tax" on beneficiaries of the project [56]. If similar tax or subsidy policies are introduced into the water price mechanism, an ideal condition that makes the marginal costs approximate the marginal revenue could be formed. In an economic view, the efficiency of water usage would increase significantly under a better price mechanism. As a consequence, the results of the water transfer amounts in Section 3.1 could reduce accordingly. The second is the water pollution control mechanism. Prior to the operation of the SNWD Project, over 400 pollution prevention projects, with a total investment over 15 billion Yuan, have been implemented in the water quality programs of the eastern route [57]. However, the decentralized regime in China and in this project has weakened local environment agencies, and the pollution fees from firms are often too low [58]. As a national infrastructure project, local officials should receive powerful support from the central governments. With the incentives of the pilot program on Water Ecological Civilization Cities [59], greater efforts in monitoring and detecting water pollution, advocating for responsive policies in local levels will lead to an improvement in water quality. Last, but not least, social participation is substantial. It is critical for both water quantity and water quality. As for the water consumption, policies focusing on social concerns with education programs implemented for water conservation could reduce individual daily water usages. Meanwhile, calling for stronger institutionalized public engagements would create a culture of conservation and provide a strong impetus for addressing water pollution [60].

Furthermore, we only treat the water quantity and quality as exogenous variables. Some endogenous factors, such as biological and chemical ones, are highly relevant to water quality, as well. Future climate conditions that might affect economic (especially agriculture) and environmental indicators are also unknown. Further research on those factors might be helpful to better understanding the findings in this study.

5. Conclusions and Policy Implications

This study uses the system dynamics method to analyze the interaction between water quantity and water quality in China's SNWD Project. The economic and environmental effects for various scenarios of transferred water and pollution rates are calculated. This research differs from other work in that we choose a specific province, covering both the water-providing and water-receiving areas. Although the system dynamics model and other mathematical tools have been used in other studies in the past, most of those studies have treated the factors of quantity and quality separately. In our model, five subsystems of water, agriculture, industry, population and the environment are included simultaneously, and data from 2002 to 2012 are collected. We develop simulations through the year 2020 to assess water quantity and quality effects and to obtain insight into sustainable water management in the related areas.

Based on the simulation results obtained for the economic and environmental indicators at the different levels of water quantity and water quality, certain conclusions can be drawn. First, the SNWD Project could provide a sufficient quantity of water for economic growth in the receiving regions. Second, an efficient quantity of transferred water exists, which demonstrates the growth limitation for both agriculture and the groundwater funnel. Third, upgrades in water quality could mutually reinforce the quantity effect and greatly promote economic growth, as well. Environmental limitation exists in all economic activities. Therefore, from a general perspective, both water quantity and water quality are equally crucial in the SNWD Project. The water diversion project is not only intended to reduce water constraints; it is also an opportunity to promote sustainable development and the establishment of an ecologically conscious civilization [61]. These findings also provide certain policy implications for the government and the public.

The source of the SNWD eastern route is the Yangtze River, the longest in Asia and the third longest in the world, and its downstream could provide almost a limitless water supply. This study shows that although the water source has no natural limitation, the use of it has environmental limitations for all economic activities. Therefore, an essential policy recommendation is that the operational stage of the SNWD Project requires a transition from efficient water diversion to effective water diversion. This means greater priority should be given to providing a sufficient amount of drinking and nondrinking water of high quality, rather than simply satisfying the water shortage with a specific amount of water. In practice, local governments also face the dilemma of low water prices and high water use efficiency, with the aim of satisfying economic needs, as well as environmental objectives [62]. Thus, certain regulations should be employed. Typical policy choices for some local governments include economic incentives, such as water quality permit trading or the use of environmental taxes (e.g., the "SNWD tax"). The central government needs to launch a water-quality-based control program with higher water quality criteria and stringent environmental legislation. Additionally, in terms of residential water usage, non-governmental participation and the promotion of water conservation awareness might also be necessary.

Acknowledgments

This research was financially supported by the Social Science Foundation (Youth Project) of Jiangsu Province (Grant No. 14ZHC001), the Social Science Foundation (Key Project) of Jiangsu Province (Grant No. 13ZHA001) and the Science and Technology Innovation Project on the South-to-North Water Diversion Project (Grant No. NSBDKJCX-2013-01). The authors would like to thank the reviewers for providing valuable comments that helped us to improve the manuscript.

Author Contributions

Yu Yang performed the research, wrote the paper and validated the results. Lei Yin collected original data and analyzed the results. Qingzong Zhang provided background materials and suggestions for reviewing the manuscript throughout the entire writing process. All authors read and approved the final manuscript.

Conflicts of Interest

The authors declare no conflict of interest.

References

1. Freeman, C. Quenching the Dragon's Thirst: The South-North Water Transfer Project—Old Plumbing for New China? Available online: http://www.wilsoncenter.org/publication/quenching-the-dragons-thirst-the-south-north-water-transfer-project8212old-plumbing-for (accessed on 30 September 2014).
2. China Daily. Flow Test for Water Project Gets Underway. Available online: http://usa.chinadaily.com.cn/epaper/2013–05/31/content_16552718.htm (accessed on 13 October 2014).
3. A Total Investment of 208.6 Billion in Central Route. Available online: http://finance.sina.com.cn/china/20140704/052719605525.shtml (accessed on 10 December 2014).
4. World Bank. *China 2020: Issues and Options for China*; World Bank: Washington, DC, USA, 1997.
5. World Bank. *Agenda for Water Sector Strategy for North China*; World Bank: Washington, DC, USA, 2001.

6. Feng, S.; Li, L.X.; Duan, Z.G.; Zhang, J.L. Assessing the impacts of South-to-North Water Transfer Project with decision support systems. *Decis. Support Syst.* **2007**, *42*, 1989–2003.
7. Le Mentec, K. The Three Gorges Dam and the demiurges: The story of a failed contemporary myth elaboration in China. *Water Hist.* **2014**, *6*, 1–19.
8. Barca, S. Telling the right story: Environmental violence and liberation narratives. *Environ. Hist.* **2014**, *20*, 535–546.
9. Chellaney, B. Water, power, and competition in Asia. *Asian Surv.* **2014**, *54*, 621–650.
10. James, K.R. Policy and Planning for Large Water Infrastructure Projects in the People's Republic of China, Doctoral Dissertation, Wesleyan University, Wesleyan, CT, USA, 2013.
11. He, C.; He, X.; Fu, L. China's South-to-North Water Transfer Project: Is it needed? *Geogr. Compass* **2010**, *4*, 1312–1323.
12. Chen, X.; Zhang, D.; Zhang, E. The south to north water diversions in China: Review and comments. *J. Environ. Plan. Manag.* **2002**, *45*, 927–932.
13. Liu, C.; Zheng, H. South-to-north water transfer schemes for China. *Int. J. Water Resour. Dev.* **2002**, *18*, 453–471.
14. Wang, W.; Gao, L.; Liu, P.; Hailu, A. Relationships between regional economic sectors and water use in a water-scarce area in China: A quantitative analysis. *J. Hydrol.* **2014**, *515*, 180–190.
15. Changming, L. Environmental issues and the south-north water transfer scheme. *China Quart.* **1998**, *156*, 899–910.
16. Kim, H.S. Sustainable development and the South-to-North Water Transfer Project in China. Doctoral Dissertation, Central Connecticut State University, New Britain, CT, USA, 2003.
17. Wei, S.; Yang, H.; Abbaspour, K.; Mousavi, J.; Gnauck, A. Game theory based models to analyze water conflicts in the Middle Route of the South-to-North Water Transfer Project in China. *Water Res.* **2010**, *44*, 2499–2516.
18. Berkoff, J. China: The South-North Water Transfer Project—Is it justified? *Water Policy* **2003**, *5*, 1–28.
19. Ma, J.; Hoekstra, A.Y.; Wang, H.; Chapagain, A.K.; Wang, D. Virtual *versus* real water transfers within China. *Philos. Trans. R. Soc. B* **2006**, *361*, 835–842.
20. Zhao, X.; Liu, J.; Liu, Q.; Tillotson, M.R.; Guan, D.; Hubacek, K. Physical and virtual water transfers for regional water stress alleviation in China. *Proc. Natl. Acad. Sci. USA* **2015**, *112*, 1031–1035.
21. China Daily. China's New Grand Canal Brings Water to Arid North. Available online: http://usa.chinadaily.com.cn/china/2014–12/16/content_19093457.htm (accessed on 18 December 2014).
22. Gu, W.; Shao, D.; Jiang, Y. Risk evaluation of water shortage in source area of middle route project for South-to-North Water Transfer in China. *Water Resour. Manag.* **2012**, *26*, 3479–3493.
23. Berrittella, M.; Rehdanz, K.; Tol, R.S. *The Economic Impact of the South-North Water Transfer Project in China: A Computable General Equilibrium Analysis*; Working Papers FNU-117; Research unit Sustainability and Global Change, Hamburg University: Hamburg, Germany, 2006.
24. Zhang, Q. The South-to-North Water Transfer Project of China: Environmental implications and monitoring strategy. *J. Am. Water Resour. Assoc.* **2009**, *45*, 1238–1247.
25. Liu, J.; Yang, W. Water sustainability for China and beyond. *Science* **2012**, *337*, 649–650.
26. Chen, H.; Wenger, R.B. Water Diversion Projects in China. In *Securing Water and Wastewater Systems: Global Experiences*; Clark, R.M., Hakim, S., Eds.; Springer International Publishing: Cham, Switzerland, 2014; pp. 213–232.

27. International Rivers. Resettlement in Action: An Eyewitness Report from the Middle Route of China's South-North Water Transfer Project. Available online: http://www.internationalrivers.org/files/attached-files/snwtreportenglish.pdf (accessed on 18 December 2014).
28. Lindalen, T. A New House or a New Life? A Case Study of Involuntary Resettlement in the South-North Water Transfer Project in Henan China. Master Thesis, University of Oslo, Oslo, Norway, 2012.
29. Magee, D. Moving the River? China's South–North Water Transfer Project. In *Engineering Earth*; Brunn, S.D., Ed.; Springer: Dordrecht, The Netherlands, 2011; pp. 1499–1514.
30. South-to-North Water Diversion Project, China. Available online: http://www.water-technology.net/projects/south_north/ (accessed on 8 December 2014).
31. National Bureau of Statistics of China. *China Statistical Yearbook 2013*; China Statistics Press: Beijing, China, 2013.
32. Forrester, J.W. Industrial dynamics: A major breakthrough for decision makers. *Harv. Bus. Rev.* **1958**, *36*, 37–66.
33. Niazi, A.; Prasher, S.O.; Adamowski, J.; Gleeson, T.P. A system dynamics model to conserve arid region water resources through aquifer storage and recovery in conjunction with a dam. *Water* **2014**, *6*, 2300–2321.
34. Winz, I.; Brierley, G.; Trowsdale, S. The use of system dynamics simulation in water resources management. *Water Resour. Manag.* **2009**, *23*, 1301–1323.
35. Beall, A.; Fiedler, F.; Boll, J.; Cosens, B. Sustainable water resource management and participatory system dynamics. Case study: Developing the Palouse basin participatory model. *Sustainability* **2011**, *3*, 720–742.
36. Liu, C.; Yu, J.; Eloise, K. Groundwater exploitation and its impact on the environment in the North China Plain. *Water Int.* **2001**, *26*, 265–272.
37. Pomar, C.; Harris, D.L.; Savoie, P.; Minvielle, F. Computer simulation model of swine production systems: III. A dynamic herd simulation model including reproduction. *J. Anim. Sci.* **1991**, *69*, 2822–2836.
38. Ministry of Water Resources. *A General Report on Feasibility Study of South-to-North Water Diversion Project*; Ministry of Water Resources: Beijing, China, 2005. (In Chinese)
39. Jiangsu Provincial Department of Water Resources. *Planning Report on Supporting Projects of South-to-North Water Diversion Project in Jiangsu Province*; Jiangsu Provincial Department of Water Resources: Jiangsu, China, 2011. (In Chinese)
40. The Overall Water Quality has Reached the Designed Goals in Eastern and Central Route. Available online: http://news.sohu.com/20141014/n405116497.shtml (accessed on 10 November 2014).
41. People's Daily. Water Quality Compliance Rate Rose from 3% to 89% in East Route after 9 Years of Pollution Treatment. Available online: http://paper.people.com.cn/rmrb/html/2012–07/26/nw.D110000renmrb_20120726_8–02.htm (accessed on 10 November 2014).
42. Hess, T.M.; Knox, J.W. Water savings in irrigated agriculture: A framework for assessing technology and management options to reduce water losses. *Outlook Agric.* **2013**, *42*, 85–91.
43. Caliman, F.A.; Robu, B.M.; Smaranda, C.; Pavel, V.L.; Gavrilescu, M. Soil and groundwater cleanup: Benefits and limits of emerging technologies. *Clean Technol. Environ. Policy* **2011**, *13*, 241–268.
44. Tang, D.; Yang, Y. Relationship between industrial structure and pollution density: Evidence from recent industrialization in China. *Int. J. Ecol. Dev.* **2011**, *19*, 30–43.

45. National Business Daily. Nansi Lake's Dilemma is to be Solved and the Water Rights are Not Clear. Available online: http://www.nbd.com.cn/articles/2014–08–12/855375.html (accessed on 17 November 2014).
46. Wong, J.L. The Food-Energy-Water nexus: An integrated approach to understanding China's resource challenges. *Harv. Asia Quart.* **2010**, *12*, 15–19.
47. Yan, T.; Wang, J.; Huang, J. Urbanization, agricultural water use, and regional and national crop production in China. *Ecol. Model.* **2015**, *2015*, doi:10.1016/j.ecolmodel.2014.12.021.
48. Just, R.E.; Netanyahu, S. *Conflict and Cooperation on Trans-Boundary Water Resources*; Kluwer Academic Publishers: Norwell, MA, USA, 1998.
49. He, S.; Hipel, K.W.; Kilgour, D.M. Water Diversion Conflicts in China: A hierarchical perspective. *Water Resour. Manag.* **2014**, *28*, 1823–1837.
50. State Council of China. Announcement on the Assessment Methods for the Most Strict Water Resource Management System. Available online: http://www.mwr.gov.cn/zwzc/zcfg/xzfghfgxwj/201301/t20130107_336155.html (accessed on 10 February 2015).
51. Water Resources Department of Jiangsu Province. The Water Resource Planning in Jiangsu Province. Available online: http://www.jswp.gov.cn/WebMain/Main/News.aspx?Id=3109 (accessed on 10 February 2015).
52. Matus, K.J.; Xiao, X.; Zimmerman, J.B. Green chemistry and green engineering in China: Drivers, policies and barriers to innovation. *J. Clean. Prod.* **2012**, *32*, 193–203.
53. Bureau of Statistics of Jiangsu. *Jiangsu Statistical Yearbook 2014*; China Statistics Press: Beijing, China, 2014.
54. Lin, J.Y.; Xu, J. The potential for green growth and structural transformation in China. *Oxford Rev. Econ. Policy* **2014**, *30*, 550–568.
55. Moore, S.M. Modernization, authoritarianism, and the environment: The politics of China's South–North Water Transfer Project. *Environ. Politics* **2014**, *23*, 1–20.
56. Moore, S. Water Resource Issues, Policy and Politics in China; Brookings Issue Brief. Available online: http://www.brookings.edu/research/papers/2013/02/water-politics-china-moore#_edn20 (accessed on 13 March 2015).
57. China Daily. Project Set to Deliver Clean Water Next Year. Available online: http://usa.chinadaily.com.cn/china/2012-07/26/content_15618298.htm (accessed on 13 March 2015).
58. Economy, E. *The River Runs Black: Environmental Challenge to China's Future*; Cornell University Press: Ithaca, NY, USA, 2004.
59. Li, J. *The Water Ecological Civilization City Policy Conception and Explore in China*; CEWP Scientific Research Cooperation Meeting: Jinan, China, 2013.
60. Benney, J.; Marolt, P. Introduction: Modes of activism and engagement in the Chinese public sphere. *Asian Stud. Rev.* **2015**, *39*, 88–99.
61. Magdoff, F. Ecological civilization. *Mon. Rev.* **2011**, *62*, 1–25.
62. Yang, H.; Zehnder, A.J. The South-North Water Transfer Project in China: An analysis of water demand uncertainty and environmental objectives in decision making. *Water Int.* **2005**, *30*, 339–349.

China's Policy on Dams at the Crossroads: Removal or Further Construction?

Chiyuan Miao [1,2,*], Alistair G. L. Borthwick [3], Honghu Liu [1] and Jigen Liu [1]

[1] Changjiang River Scientific Research Institute, Changjiang Water Resources Commission, Wuhan 430010, Hubei, China; E-Mails: liuhh@mail.crsri.cn (H.L.); jigenliu@163.com (J.L.)

[2] State Key Laboratory of Earth Surface Processes and Resource Ecology, College of Global Change and Earth System Science, Beijing Normal University, Beijing 100875, China

[3] School of Engineering, The University of Edinburgh, The King's Buildings, Edinburgh EH9 3JL, UK; E-Mail: Alistair.Borthwick@ed.ac.uk

* Author to whom correspondence should be addressed; E-Mail: miaocy@bnu.edu.cn;

Academic Editor: Sharon Megdal

Abstract: During the past century, the number and scale of reservoirs worldwide has grown substantially to meet the demand for water and hydropower arising from increased population, industrialization, and urbanization. This is particularly the case in China, where reservoir construction increased rapidly after the Chinese economic reform and the introduction of open-door policies. On average, 4.4 large reservoirs with a capacity greater than 0.1 km^3 were constructed per annum during the 1970s–1990s. This average reached 11.8 such reservoirs per annum in the 2000s. Considering the adverse impact of dams on rivers and riparian communities, various environmentalists and non-governmental organizations in China have begun to protest against the construction of dams. Now China's policy on dams is at a crossroads: Removal or further construction? In this paper, we systematically assess the construction of reservoirs in China and discuss the benefits and drawbacks of large-scale reservoir projects on several major rivers in China: The Yangtze River, the Yellow River and the Mekong River. Lastly, we provide a perspective on the future of reservoir development in China, taking into account natural conditions, renewable hydropower resources, and greenhouse gas emissions.

Keywords: reservoir; natural resource; eco-environment; China

---

1. Introduction

During the past century, the number and scale of reservoirs worldwide has grown substantially to meet the demand for water and hydropower arising from increased population, industrialization and urbanization. It is reported that an average of two dams with heights exceeding 15 m were completed each day throughout the latter half of the 20th Century [1]. To date, over 50,000 such dams have been constructed globally, contributing to a total storage capacity of more than 8300 km^3 [2]. At present, more than 60% of the world's rivers are dammed. According to the latest Global Reservoir and Dam (GRanD) database [3], the total capacities, by country, of the largest reservoirs capacity (>0.1 km^3) are: Canada (860.7 km^3), Russia (811.8 km^3), USA (767.5 km^3), Brazil (530.3 km^3), and China (451.0 km^3).

China currently has the largest agricultural economy in the world, as well as the largest population. Dams and reservoirs have played an important role in the recent socio-economic development of China, especially since the Chinese economic reform and the introduction of open-door policies in the late 1970s. Statistics show that reservoir construction in China developed rapidly over the past 30 years. For example, an average of 4.4 large reservoirs (of capacity greater than 0.1 km^3) were constructed per annum during the 1970s–1990s (Figure 1). The pace of construction accelerated after 2000, with an average of 11.8 large reservoirs constructed each year during the 2000s. By 2011, China possessed 552 large capacity (>0.1 km^3) reservoirs, 3269 medium capacity (0.01–0.1 km^3) reservoirs, and 84,052 small capacity (0.0001–0.01 km^3) reservoirs—with a total storage volume of 692.4 km^3 [4]. The total capacity grew at a rate of 4.43 km^3 per annum between 1973 and 1999, and the rate increased to 20.63 km^3 per annum after 2000 (Figure 1). Although almost all the rivers in China are now regulated by reservoirs, China still has considerable capacity for further exploitation of its residual water resources. According to the China Water Resources Bulletins from 2006 to 2010, on average about 22.34% of the total available water resources in China were being exploited and utilized. There are very significant differences among the river basins: The northern basins have >50% utilization and the Haihe river basin has even reached >90% utilization.

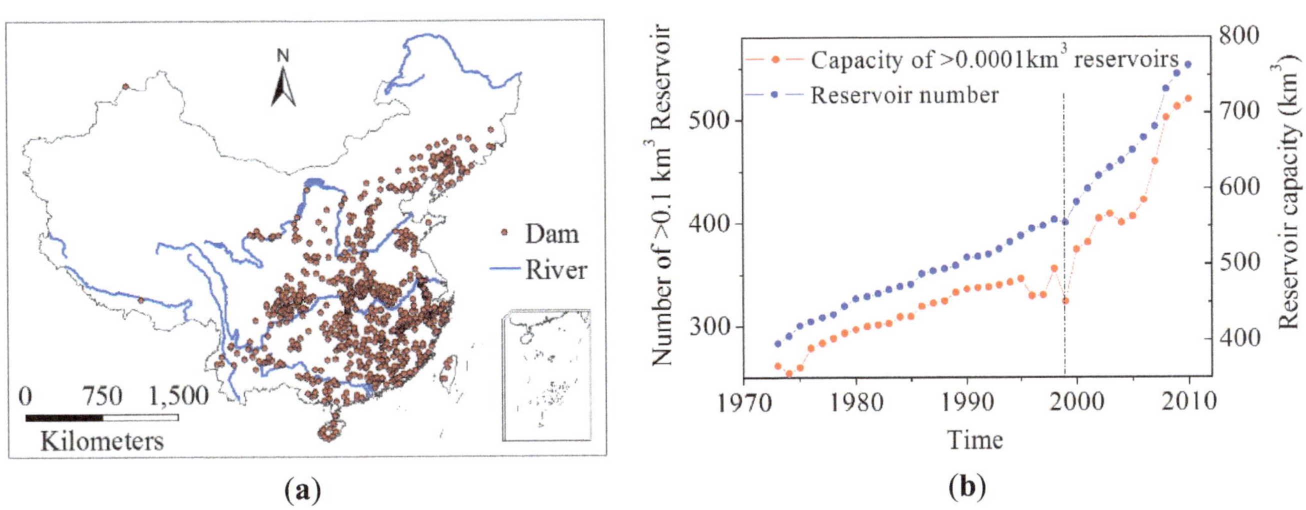

Figure 1. Distribution of dams throughout China in 2011 (**a**); and growth in the number of large reservoirs and the total capacity of reservoirs in China over the past 30 years (**b**).

2. Negative Impacts from Reservoir Development

Reservoirs provide considerable socio-economic benefits by increasing the water supply available for municipal, industrial, and agricultural uses, controlling floods, and generating hydropower. At the same time, reservoirs also cause severe negative impacts, including biodiversity depletion, ecosystem degradation, and population displacement, *etc.* Illustrative examples are provided by the large-scale reservoir projects on the Yangtze River, the Yellow River, and the Mekong River in China.

The Three Gorges Reservoir (TGR) on the Yangtze River is the largest river regulation and hydroelectric power station in the world [5]. It cost more than US $32 billion to construct and has an operational water level 110 m above the downstream river level. The TGR has a water surface area of 1080 km^2, and the Three Gorges Dam controls a watershed area of over 1 million km^2. The TGR has involved resettlement of more than 1.2 million local residents whose original habitations were located below the post-TGR upper water level. Although China has spent more than US $6.3 billion on resettlement programs [6], complicated social challenges have arisen in helping the resettled former residents to adapt successfully to their new surroundings [7]. Water retained in the TGR has submerged an estimated 1300 sites of cultural and archaeological interest [8]. Construction of dams along the Yangtze River has impeded the spawning migration of certain fish species, leading to considerable reductions in the numbers of Chinese sturgeon (*Acipenser sinensis*) and Dabry's sturgeon (*Acipenser dabryanus*), and the possible extinction of the Chinese paddlefish (*Psephurus gladius*) [9]. Following the 156 m impoundment in 2007, the phytoplankton diversity index was observed to have decreased from 85.52% to 32.32% in tributaries in the TGR region. River flow in the Yangtze is generally slower than it was, leading to decreased scour of the river banks and blooms of toxic algae in tributaries. The vast area covered by the TGR has encouraged the spread of *Oncomelania* snails, which carry the parasitic disease schistosomiasis [10]. Another drawback of the TGR arises from its sheer scale—The immense hydrostatic loading created by the water behind the dam may induce seismic activity and the raised water level may result in landslides and erosional processes [11]. The potential risk of dam failure must also be considered because this would affect more than 70 million people living downstream. The dams also impaired the natural habitats of the now-extinct Chinese river dolphin *Lipotes vexillifer* [12].

The second example is provided by the cascade of more than 25 hydraulic projects along the Yellow River, with a total reservoir volume exceeding 55.75 km^3. The Yellow River is the second largest river in China and contributes about 6% of global river sediment load [13]. Surface water evaporation from the upper and middle mainstream reaches of the Yellow River increased by approximately 68% after the construction of these reservoirs, resulting in a corresponding 2.3% decrease in water discharge [14]. The reservoirs in the Yellow River regulate both water discharge and sediment load, and control the amount of sediment deposited in the downstream channel. In the past, especially from the 1970s to the end of the 1990s, low flows in the lower Yellow River led to increased deposition that elevated the riverbed, prolonging the periods during which the river dried up. At Lijin, near the mouth of the river, no discharge was recorded for a total of 226 days in 1997 [15], which hastened the formation of the so-called 'suspended river' [11]. Moreover, the increased deposition aggravated the risk of major floods occurring in a river that has a long history of flood disasters. The cascade of reservoirs has greatly improved the water-sediment situation in the Yellow River: the occurrence of zero flow events has been reduced in recent years and the riverbed is no longer significantly rising. Consequently, most of reservoirs are gradually silting up, even after regular flushing. Furthermore, the barriers created by the dams have resulted in reductions in the numbers of certain indigenous species of semi-migratory fish [16]. It has been reported that about 13 fish species in the Yellow River are already almost extinct, including *Cultrichthys erythropterus*, *Leiocassis crassilabris*, *Oryzias latipes*, and *Mastacembelus aculeatus* [17].

The final and perhaps most contentious example is the dam cascade under construction along the Mekong River. For this case, we consider the river basin as a whole, rather than focusing solely on the upper reach in China. The Mekong River rises in Tibet and flows for 4800 km before eventually discharging into the Gulf of Thailand. The Mekong River Basin occupies a total area of 795,000 km^2 and overlaps parts of China, Myanmar, Thailand, Laos, Cambodia, and Vietnam, and has a mean discharge rate of about 14,500 m^3/s. Since the early 1990s, China has been constructing a cascade of 8 dams with a projected total volume exceeding 40 km^3 along a 750 km reach of the upper Mekong River (known as Lancangjiang in Yunnan province, China). The first four dams at Manwan, Dachaoshan, Jinghong, and Xiaowan were completed in 1993, 2003, 2008, and 2010. Other dams are presently under construction (see Table 1 for more details). The dams provide China with considerable hydropower, which will add up to more than 15,000 MW when all are operational, and should provide the benefits of flood prevention in the wet season and drought alleviation in the dry season through flow regulation. However, field measurements at hydrological stations along the Mekong River have indicated that operation of the Manwan dam may have altered downstream flow and sediment discharge characteristics, causing increased intermittent fluctuations during the dry season and a decline in sediment flux—The latter extending as far as the lowest reach of the Mekong River [18]. More recently, it has been suggested that the historical baseline low-flow levels of the Mekong River will have changed irrevocably by the time that China has completed construction of the upper Mekong cascade of dams [19]. Laos stands to benefit from about US $25 billion in foreign investment and US $2.6 billion in revenue per annum from exported energy [10], but at enormous environmental cost to the Mekong basin. The Xayaburi dam in Laos could render two species of giant catfish, *Pangasianodon gigas* and *Pangasius sanitwongsei*, extinct. There have been warnings that the dams under construction by China and planned by Laos will have dire consequences for the Tonlé Sap (or Great Lake) in Cambodia, which is connected by a tributary to the Mekong River [10]. In Cambodia, the loss of migratory fish species as a result of hydroelectric dams poses a direct threat to the security of food and income. It is estimated that no less than 80% of Cambodia's protein comes from the Tonlé Sap basin, making Cambodia particularly vulnerable to any change in fish stocks in the basin. In the lower Mekong basin, seasonal flooding occurs at an estimated net cost of about US $50 million per annum, and seasonal drought occurs at an estimated cost of about US $10 million per annum solely in terms of loss of rice production [20]. Although completion of the entire cascade of dams will help regulate the usual flood cycle, it should be noted that there will be adverse repercussions for farmers who rely on nutrient-rich sediments being deposited during flood events [18]. With dams proposed along the lower stretches of the Mekong River, this situation is set to become even worse. Vietnam estimates that the Xayaburi dam will reduce nutrient-rich alluvial deposits in the Mekong Delta from 26 to 7 million tonnes per year [21]. This loss of soil nutrients for downstream farmers will result in reduced soil fertility, yields, and income [22]. Some researchers have raised the concern that the cascade of dams could damage the ability of the Mekong River to return to equilibrium, and propose further studies into the impact of the dam cascade on the water–sediment characteristics of the Lower Mekong River. Moreover, the dams may worsen the issue of saline intrusion in Vietnam. Currently, the last 10 to 20 km of the river experience salt intrusion, which impacts farmers in the area. The cumulative effects of another 11 planned hydropower schemes along the lower Mekong River could also be disastrous for smaller migratory fish species, leading perhaps to a more than 40% decrease in fish stocks [23,24].

Table 1. China's 8-dam cascade along the upper Mekong River (Lancang River).

Dam	Catchment Area (km^2)	Mean Flow Discharge (m^3/s)	(Estimated) Completion Year	Maximum Height (m)	Total Reservoir Capacity (million m^3)	Installed Power Capacity (MW)
Gongguoqiao	97,300	985	2012	130	510	750
Xiaowan	113,300	1,210	2011	292	14,560	4,200
Manwan	114,500	1,230	1996	132	920	1,500
Dachaoshan	121,000	1,230	2004	111	933	1,350
Nuozhadu	144,700	1,730	2014	261.5	22,400	5,500
Jinghong	149,100	1,840	2008	108	1,233	1,750
Ganlanba	151,800	1,880	Before 2025	60.5	-	150
Mengsong	160,000	2,020	-	65	-	600

3. Discussion and Perspectives

As a consequence of the negative impacts outlined above, some environmentalists and non-governmental organizations (NGOs) in China have begun to campaign against China's dam construction policy, influenced by the 'anti-dam' movement in the U.S. that began in the 1960s. These campaigners believe that the adverse impact of dams on rivers and riparian communities outweighs the benefits, and propose dam removal as the best way to restore healthy rivers and riverside communities in China. This idea rapidly propagated throughout China at the same time as dam construction accelerated after 2000.

So now China's policy on dams is at a crossroads. Should China follow the economic, flood control, energy, and irrigation imperatives and continue to construct dams? Or should it reject further plans for construction and remove pre-existing dams in order to let the rivers run free, and hopefully return to halcyon days, as viewed from environmental and societal perspectives? It is particularly difficult for stakeholders in China to choose between the options because of the scale of the problems encountered whichever way the decision goes.

There is precedent for the removal of existing dams in developed countries. Hydropower in developed countries is comparatively mature, with those countries having already undergone the period of large-scale hydropower development. Moreover, most of the dams removed in developed countries were dilapidated, had existing problems, and/or were of small capacity. In our opinion, it is both unrealistic and inadvisable to reject arbitrarily all dams in China. Firstly, the natural hydrological conditions in China include a unique monsoon climate, highly non-uniform rainfall distribution, and strong peak river-flow variability. These factors have resulted in frequent drought and flood disasters throughout history. Dams are able to effectively attenuate peak flood discharges during the wet season, while adjusting to the competing water supply and ecological water demands of the middle and lower reaches of a river during the dry season. At the time of writing, dams and reservoirs in China provide flood protection to more than 310 million people and 3.2×10^5 km^2 of arable land. Moreover, grain production from dam-related irrigation areas accounts for ~70% of the total output of grain in China. If all dams were to be removed, the total indigenous grain output in China would be halved, and hundreds of millions of people could then face possible starvation.

Secondly, dams exploit China's abundant renewable hydropower resources, and the electricity produced has helped meet the rapid growth in energy demands associated with economic and societal development in China. The technically exploitable hydropower potential in China is about 2474 TWh/yr, but the actual hydropower generated is about 565.5 TWh/yr [25]. The hydropower development ratio is only 22.86%, far behind that of many developed countries, e.g., Italy (over 90%), France (90%), Japan (84%), USA (80%), and Germany (73%). Moreover, the exploitable hydropower in the western provinces accounts for more than three-quarters of the total exploitable hydropower in China (Figure 1), but the hydropower development ratio there is less than 10%. In order to bridge the gap between energy supply and demand while meeting its economic objectives as a developing country, China must increase its utilization of renewable energy, rather than relying on energy imports.

Thirdly, China has been the world's highest emitter of greenhouse gases since 2007. Global climate change has brought forward a new challenge for modern China. At present, most electricity in China is generated by coal-fired power stations, with only 9% originating from non-fossil-fuel sources in 2010. China has pledged to reduce its carbon intensity (the amount of carbon dioxide produced per unit of economic growth) by 40%–45% of 2005 levels before 2020 [26]. In the case of the TGR, the huge levels of hydropower production can replace 50 million tons of coal equivalent consumption, 0.1 billion tons of CO_2, 2 million tons of SO_2, 10,000 tons of CO, and 370,000 tons of nitrogen oxide, when compared with coal-fired power plants [6]. The 12th Five-Year Plan for 2011–2015, adopted by the National People's Congress of the People's Republic of China in March 2011, states that 15% of energy will be generated from non-fossil-fuel sources by 2020. Considering the return on investment (ROI), development potential, and current development levels of nuclear, solar, and wind energy in China, the 12th Five-Year Plan pointed out clearly that 9% of China's energy will need to come from hydropower by 2020. To meet this target, a massive expansion in hydropower will be needed.

The foregoing hydrological risk mitigation, energy security, and low carbon arguments mean that dam construction remains a necessity during the present stage of development in China. However, the authorities in China are increasingly aware of the painful lessons drawn from the failure in the past to achieve fully sustainable reservoir development, and are taking note of various perspectives including those of the international anti-dam movement as well as the international community of hydropower engineers and scientists. River basin authorities in China should consider making the following adjustments to their policies concerning reservoirs and dams:

(1) Processes for public and stakeholder engagement should be established. The involvement of society at large in decisions about dams is an inevitable consequence of social development, and it is important that the general public be encouraged to participate in the decision-making process. In particular, all relevant information should be assessed openly and objectively.
(2) Eco-environmental monitoring systems should be established for the entire period during which the dam is constructed, in operation, and decommissioned. China has started to construct such monitoring systems, and has already begun to reap the benefits. However, most of the current

monitoring systems only serve reservoirs that are in operation or under construction. Reservoirs that are being decommissioned have so far tended to be ignored.

(3) Eco-environmental impact assessments conducted during studies of reservoir feasibility should be comprehensive and consider all the short- and long-term benefits and drawbacks. These assessments should take into account all competing interests—including positive and negative effects on the people in the upper, middle, and lower reaches of the affected river. This should encompass trans-boundary interests and involve dialogue between affected provinces and nations. Current environmental impact assessments focus heavily on the immediate, and often obvious, outcomes of reservoir construction and may not consider the longer-term effects on the physical and human ecology of an area. More comprehensive assessments would allow longer-term scientific and technical predictions to be coupled with political and social assessments in order to influence decisions over whether reservoirs can be built with minimal environmental impact and in a way that protects the social and environmental rights of affected communities. To date, such considerations have not been granted enough emphasis in the design and construction of dams in China.

(4) Eco-environmental impact and running benefits of existing reservoirs should be evaluated scientifically. Reservoirs provide considerable benefits; however they also unavoidably have severe ecological and societal impacts. Evaluation of existing reservoirs should include integrated input from experts in the fields of sociology, economics, and environmental science, and should also involve public participation. China has started to implement such integrated evaluations for representative reservoirs but, so far, the evaluations have involved only a limited number of experts. This has restricted publicity of the information and awareness of the relevant findings.

(5) Resettlement policies should be strengthened so that they fairly safeguard the legitimate rights, equity, and wellbeing of the people displaced by dams and reservoirs. This involves ensuring that displaced people are involved as stakeholders through consultation, participation, decision-making, and supervisory activities.

China's policy towards dams is at a crossroads. The way forward involves a sensitive compromise between satisfying the economic and societal demands for flood control, water supply, food, and energy security, while also achieving a more balanced approach to river-basin management that is in keeping with Nature as well as Mankind.

Acknowledgments

Funded By CRSRI Open Research Program (CKWV2013214/KY), the National Natural Science Foundation of China (41001166), 948 Project of Ministry of Water Resource (201306), and the State Key Laboratory of Earth Surface Processes and Resource Ecology.

Author Contributions

The research topic was discussed and decided by Chiyuan Miao and Alistair G.L. Borthwick. Honghu Liu and Jigen Liu made important recommendations regarding the paper structure. Alistair G.L. Borthwick also contributed primarily to the discussion on the environmental biodiversity consequences of dams along the Yellow River, the Yangtze River, and the Mekong River. The major part of this manuscript was completed by Chiyuan Miao.

Conflicts of Interest

The authors declare no conflict of interest.

References

1. Berga, L.; Buil, J.M.; Bofill, E.; Cea, J.C.D.; Perez, J.A.G.; Mañueco, G.; Polimon, J.; Soriano, A.; Yagüe, J. *Dams and Reservoirs, Societies and Environment in the 21st Century*; Taylor & Francis Press: London, UK, 2006.
2. Chao, B.F.; Wu, Y.H.; Li, Y.S. Impact of artificial reservoir water impoundment on global sea level. *Science* **2008**, *320*, 212–214.
3. Lehner, B.; Liermann, C.R.; Revenga, C.; Vorosmarty, C.; Fekete, B.; Crouzet, P.; Doll, P.; Endejan, M.; Frenken, K.; Magome, J.; *et al.* High-resolution mapping of the world's reservoirs and dams for sustainable river-flow management. *Front. Ecol. Environ.* **2011**, *9*, 494–502.
4. Ministry of Water Resources the People's Republic of China. *China Water Statistical Yearbook 2011*; China Statistics Press: Beijing, China, 2012; pp. 117–119. (In Chinese)
5. Wang, J.; Sheng, Y.W.; Gleason, C.J.; Wada, Y. Downstream Yangtze River levels impacted by Three Gorges Dam. *Environ. Res. Lett.* **2013**, *8*, doi:10.1088/1748-9326/8/4/044012.
6. China Three Gorges Project Corporation. China Three Gorges Corporation Annual Report 2010. Available online: http://www.ctgpc.com/file/Annual_Report_2010.pdf (accessed on 11 November 2011).
7. Qi, M.M.; Jiang, J.D. Summary of resettlement planning of three gorges project. *Yangtze River* **2013**, *44*, 16–20. (In Chinese)
8. United States Geological Survey. Geographical Overview of the Three Gorges Dam and Reservoir, China. Available online: http://pubs.usgs.gov/of/2008/1241/pdf/OF08-1241_508.pdf (accessed on 11 November 2011)
9. Fu, C.Z.; Wu, J.H.; Chen, J.K.; Qu, Q.H.; Lei, G.C. Freshwater fish biodiversity in the Yangtze River basin of china: Patterns, threats and conservation. *Biodivers. Conserv.* **2003**, *12*, 1649–1685.
10. Stone, R. Mayhem on the Mekong. *Science* **2011**, *333*, 814–818.
11. Li, K.F.; Zhu, C.; Wu, L.; Huang, L.Y. Problems caused by the Three Gorges Dam construction in the Yangtze River basin: A review. *Environ. Rev.* **2013**, *21*, 127–135.
12. Turvey, S.T.; Pitman, R.L.; Taylor, B.L.; Barlow, J.; Akamatsu, T.; Barrett, L.A.; Zhao, X.J.; Reeves, R.R.; Stewart, B.S.; Wang, K.X.; *et al.* First human-caused extinction of a cetacean species? *Biol. Lett.* **2007**, *3*, 537–540.
13. Miao, C.Y.; Ni, J.R.; Borthwick, A.G.L. Recent changes of water discharge and sediment load in the Yellow River basin, china. *Prog. Phys. Geogr.* **2010**, *34*, 541–561.
14. Liu, C.; Zhang, X. Causal analysis on actual water flow reduction in the mainstream of the Yellow River. *Acta Geogr. Sin.* **2004**, *59*, 323–330.

15. Ni, J.R.; Liu, X.Y.; Li, T.H.; Zhao, Y.; Jin, L. Efficiency of sediment transport by flood and its control in the lower Yellow River. *Sci. China Ser. E* **2004**, *47*, 173–185.
16. Guo, Q.Y.; Yang, Z.F. Post-project ecological analysis for the Sanmenxia Dam. *Acta Sci. Circumst.* **2005**, *25*, 580–585.
17. Ru, H.J.; Wang, H.J.; Zhao, W.H.; Shen, Y.Q.; Wang, Y.; Zhang, X.K. Fishes in the mainstream of the Yellow River: Assemblage characteristics and historical changes. *Biodivers. Sci.* **2010**, *18*, 169–174.
18. Lu, X.X.; Siew, R.Y. Water discharge and sediment flux changes over the past decades in the lower Mekong River: Possible impacts of the chinese dams. *Hydrol. Earth Syst. Sci.* **2006**, *10*, 181–195.
19. Wang, J.J.; Lu, X.X.; Kummu, M. Sediment load estimates and variations in the lower Mekong River. *River Res. Appl.* **2011**, *27*, 33–46.
20. Mekong River Commission. State of the basin report 2010. Available online: http://www.mrcmekong.org/assets/Publications/basin-reports/MRC-SOB-report-2010full-report.pdf (accessed on 11 November 2014).
21. Experts Worry about Mekong River Power Plant. Available online: http://en.qdnd.vn/news/experts-worry-about-mekong-river-power-plant/139532.html (accessed on 11 November 2014).
22. Barrington, D.J.; Dobbs, S.; Loden, D.I. Social and environmental justice for communities of the mekong river. *Int. J. Eng. Soc. Justice Peace* **2012**, *1*, 31–49.
23. Mekong River Commission. Strategic Environmental Assessment of Hydropower on the Mekong Mainstream. Available online: http://www.mrcmekong.org/assets/Publications/Consultations/SEA-Hydropower/SEA-FR-summary-13oct.pdf (accessed on 13 January 2015)
24. Grumbine, R.E.; Xu, J.C. Mekong hydropower development. *Science* **2011**, *332*, 178–179.
25. Jia, J.S.; Yuan, Y.L.; Zheng, C.Y.; Ma, Z.L. Dam construction in China: Statistics, progresses and concerned issues. *Water Power* **2010**, *36*, 6–10.
26. Qiu, J. China unveils green targets. *Nature* **2011**, *471*, 149.

Short Global History of Fountains

Petri S. Juuti [1,*], Georgios P. Antoniou [2], Walter Dragoni [3], Fatma El-Gohary [4], Giovanni De Feo [5], Tapio S. Katko [6], Riikka P. Rajala [1], Xiao Yun Zheng [7], Renato Drusiani [8] and Andreas N. Angelakis [9]

[1] International Environmental History Group, University of Tampere, Tampere FI-33014, Finland; E-Mail: riikka.rajala@uta.fi

[2] Department of Architecture Engineering, NTUA, PhDc, Athens 10682, Greece; E-Mail: antonioug@tee.gr

[3] Department of Physics and Geology, Perugia University, Perugia 06123, Italy; E-Mail: walter.dragoni@unipg.it

[4] Water Pollution Research Department, National Research Center, Bohouth Str. Dokki, Cairo 12622, Egypt; E-Mail: fgohary@hotmail.com

[5] Department of Industrial Engineering, University of Salerno, via Ponte don Melillo, Fisciano 1-84084, Italy; E-Mail: g.defeo@unisa.it

[6] Capacity Development in Water and Environmental Services, Tampere University of Technology, Tampere FI-33720, Finland; E-Mail: tapio.katko@tut.fi

[7] Yunnan Academy of Social Sciences, 577 Huan Cheng West Road, Kunming City 650034, China; E-Mail: zhengxy68@163.com

[8] Federutility, Rome 00192, Italy; E-Mail: renato.drusiani@federutility.it

[9] Institute of Iraklion, National Foundation for Agricultural Research (N.AG.RE.F.), Iraklion 71307, Greece; E-Mail: info@a-angelakis.gr

* Author to whom correspondence should be addressed; E-Mail: petri.juuti@uta.fi;

Academic Editor: Enedir Ghisi

Abstract: Water fountains are part of every human settlement, and historical and mythological stories. They are the source from which life-sustaining water was distributed to people until piped systems started providing fresh tap water inside buildings. In many places, people visit fountains to experience the freshness of running water, to prepare for prayers, or to make a wish. Fountains have also provided water for the people of cities under siege, and purified believers as part of holy rites. The *Castalia* shrine in Delphi, Greece, for its part, is a spot where various groups of people come to socialize, which greatly improves the quality of their lives. This paper is a look back through the history of fountains in various parts of the world. Experts from various areas have identified the historic, cultural, and ritualistic aspects of fountains and their findings are summarized. The paper concludes by providing a glimpse into the role of fountains in modern society and their continued influence in our lives today.

Keywords: Chinese civilizations; Egyptians; Etruscans; Minoans; Hellenes; Medieval times; Ottomans; Romans; water distribution; water supply systems

---

1. Prolegomena

Nullusenim fons non sacer.
(There is no spring which is not sacred),
Servius Marius Honoratus,
4th century AD, Rome.

The English word "fountain" derives from the Latin word *Fons*, having the meaning both of natural spring and of artificial construction built for water supply and/or decorative or symbolic purposes. However, this paper sees fountains in a somewhat wider perspective that also covers artifacts such as water kiosks.

Fountains are known to have existed since the early Bronze Age in southeastern Greece and the Indus Valley. They were originally connected to springs or aqueducts to provide drinking water, water for bathing, and water for washing to the residents of palaces, cities, towns and villages. Advanced hydrotechnologies, including fountains, were first developed in Crete, the center of Europe's first advanced civilization, by the Minoans [1]. Later, they were improved and exported to other European and Mediterranean regions by Mycenaeans, Etruscans, Classical and Hellenistic Greeks, Romans, Venetians, and Ottomans [2].

Fountains were also used in several regions as decorations and to celebrate their builders besides being sources of drinking water. Hellenistic and Roman fountains were decorated with bronze or stone masks of animals or heroes. The baroque decorative fountains of Rome in the 17th and 18th centuries marked the arrival of restored Roman aqueducts glorifying the popes who built them [3]. By the end of the 19th century, indoor plumbing became a standard feature which is why many urban fountains—Earlier important as sources of drinking water around the world—Gradually began to be seen as monuments (built e.g., to decorate city parks and squares, honor cultural events and for recreation and entertainment). In addition, water transportation by gravity was replaced by pumping processes. The *Jet d' eau* in Lake Geneva, built in 1951, shoots water 140 m into the air. However, the highest fountain in the world is King Fahd's Fountain in Jeddah, Saudi Arabia, which spouts water 260 m above the Red Sea [4].

The evolution of the major achievements in the field of water fountains with emphasis on the major technologies throughout the centuries is presented and discussed in this study. Valuable insights into ancient water fountain technologies and management including their apparent durability characteristics, adaptability to the environment, and sustainability are also provided. A comparison of water technological developments in several civilizations is also made. These technologies are the underpinnings of the modern achievements in sanitary engineering, and the best proof that "the past is the key to the future". Furthermore, the timeline of the historical development of fountains worldwide over the last 5500 years of humankind's history is established and a chronological order is followed with emphasis on the major civilizations.

It is obvious that the existence of water fountains in a civilization had a lot to do with its location: whether its water supply sources were abundant, and whether they were above or below the level of the cities. The ancient Egyptians had ingenious systems for hoisting water from the Nile for drinking and irrigation. The Minoans, and later the Hellenes, led water down from mountains via aqueducts. Since the source was higher than the outlet, it was easy for the cities to build fountains that spouted or poured water into drinking water basins.

Rapid technological progress in the twentieth century created a disregard for past water technologies that were considered to be far behind the present ones. There were many unresolved problems relating to water distribution networks and, especially, fountains. In the developing world these problems became exacerbated to an unprecedented degree.

Moreover, new problems such as the contamination of surface and groundwater have arisen. Naturally, exacerbation of unresolved problems led societies to revisit the past and reinvestigate successful past achievements. To their surprise, the people that took part in the reinvestigation based on archaeological, historical, and technical evidence were impressed by two things: the similarity of past principles with the present ones and the advanced level of the lavatories engineering and urine separation practices of earlier periods [5].

Our hope is that these small case studies covering various time periods and different parts of the world describe adequately the diversity of water fountains over time.

2. Minoans and Ancient Egyptians

2.1. Water Distribution Systems

Several different types of conduits have been found during archaeological excavations in Minoan Crete, Greece. They were used for water supply as well as removal of storm water and wastewater. Therefore, they are made mainly of stone or terracotta. The terracotta conduits were canals or pipes of rectangular or circular cross-section. The most interesting conduits are the U-shaped and truncated conical terracotta pipes never used before by other civilizations [6]. In the Knossos Palace, water distribution was through a network of terracotta piping located beneath the floors at depths from a few centimeters to 3 m [7]. Minoan terracotta pipes were used for both distribution and collection of water for storage. Consequently, the terracotta pipe had to be designed to perform [8]:

(a) Underground Transportation of Water: This (underground = unseen) system may have been intended for use by a restricted group of people or for a specific activity, which in turn may be indicative of ritual use of water [9].

(b) Transportation of Good Quality Water. The evidence also suggests that the pipes probably fed fountains in open areas of the palaces which were not part of the domestic distribution system, according to several scholars [10].

2.2. Fountains

Despite their aesthetic accomplishments in palatial frescoes and architecture, fine artifacts of pottery, metals, precious and semi-precious stones, the Minoans had to grapple with practical problems related to water and wastewater. Thus, the extensive and elaborate structures they built for water supply included fountains planned and constructed to serve the needs of the growing population of the island.

A remarkable fragment from a fresco composition depicting a fountain of a supposedly Minoan garden was found in the House of Frescoes in Knossos [9]. The fragment has been restored to what appears to have been a *jet d'eau* fountain. Although the actual top of the fresco was not found, the object depicted in the upper part of the field is clearly some type of fountain or *jet d'eau* where the water spout rises from a forked base. Minoan fountains manifest another interesting technology [11]. Most of them are subterranean structures supplied with water directly or from other springs via ducts.

The most typical of these fountains is the *Tykte* fountain, which was discovered in the Guest House (*Caravanserai*) of Knossos (Figure 1a). It is located in the restored "Room of Spring" or "Minoan Fountain". It is a rectangular underground structure with a central cistern-basin (of three steps measuring 2.0 × 1.6 m), which was accessible from the courtyard of the adjacent building with three descending stairs at the end of which on the right was a circular slotted door [7]. The walls of the "Room of the Fountain" were constructed of large limestone blocks and slabs of internally coated gypsum [2,12].

(a) (b)

Figure 1. *Tykte* fountains: (**a**) at *Caravanserai* in Knossos palace and (**b**) at Zakros palace (photo by Andreas N. Angelakis).

The water depth in the cistern could reach 45 cm. The water did not reach that level through a channel, but was supplied by an artesian spring naturally through the pebbles forming the bottom. The plate at the entrance was very worn, perhaps from filling of the vessel with water. The Minoan Fountain was a very sophisticated construction. The conch and double crown square section on the wall opposite the entrance was probably designed to contain a lamp.

Another similar *Tykte* fountain is that of the Zakro palace (Figure 1b). It is located in the southwest corner of the *Hall of* the *Cistern* and has an opening which leads to a small chamber where water was collected and channelled under the floor into a square underground fountain [8,11]. Its dimensions are

3 × 4 m and it dates to the Neopalatial period (*ca.* 1700–1500 BC). For more details, see [8]. Two hand-made tubes about 30 cm long with a narrow opening on the zoomorphic side were found in a building (Figure 2). These artifacts also date back to the Neopalatial period. Floods [13] indicated that the Mediterranean dry-season became more pronounced on Crete during the Late Minoan IB (*ca.* 1500–1450 BC).

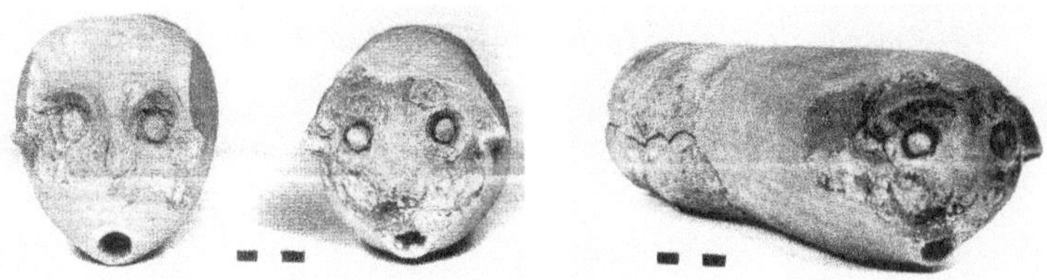

Figure 2. Zoomorphic type waterspouts.

It was perhaps a particularly dry period with more pronounced summer temperatures or less winter precipitation. This position is supported by the increase in the scale of Minoan water-management activities then, especially in the eastern regions of the island. It appears that Minoan Crete responded to the changing climatic conditions sometime during the Middle Bronze Age. At that time, the management of water resources became increasingly important [14]. Technologies such as gutters, wells, dams, ceramic mulches to conserve soil moisture, terraced hillsides, and *Tykte* fountains were widely used.

Minoans, as the later Greek civilizations, avoided regions of fertile soils and high water availability. They usually resided in areas under water scarcity. Thus, it is likely that lack of water triggered progress, particularly in water sciences, among people living in dry are as during the Minoan Era.

2.3. Non-European Ancient Civilizations (from 3000 Onwards)

During the early phases, fountains were used especially for public water supply purposes. The very first evidence of a water supply network emerged when cities of the first ancient civilizations (Egypt, Sumerian and other civilizations in the Middle East, Indus River, and China) were studied. Ancient civilizations of the New World (e.g., Maya, Aztec, and Inca) also developed sophisticated water systems, fountains being an important part of them. According to Hirst [15], the first still remaining fountains originated *in ca.* 3000 BC:

> *"The earliest surviving carved water basin, dating from around 3000 BC, was discovered at the site of Tello, one of the cities of Mesopotamia. At Mari, another of the most important cities, a stone fountain figure dating from around 2000 BC was discovered. The figure can be considered a prototype for the kind of fountains made in gardens for thousands of years thereafter: a female goddess holding a base into which water is piped to cascade forth, symbolizing the source of all life, the ultimate creative force of the garden" [15].*

3. Etruscan Civilization (*ca.* 800 BC–200 AD)

Before delving in to the Roman attitude towards water management, aqueducts and fountains in more detail, we must remind ourselves that since ancient times, Roman hydraulic technology was known to have its roots in the Etruscan civilization. The Etruscan civilization, whose development was influenced by Greece and the Near East, flourished in central Italy from about the eight century BC onwards. Etruscans were masters of land reclamation, aqueduct building and water management in general, *cf.* e.g., [16–18]. The tunnel shown in Figure 3, which still works, was built in order to control the discharge of a small catchment close to the ancient Etruscan city of Veii, a few kilometers from Rome.

Figure 3. The Etruscan *Ponte Sodo,* a tunnel about 75 m long. It is a good example of the Etruscan hydraulic technology; its main purpose appears to have been to control floods and reclaim land close to Veii, the Etruscan city located a few kilometers from Rome (photo by Walter Dragoni).

During its early history, Rome was ruled by Etruscan kings, the last of which was over thrown around the end of 6th century BC. The Etruscan kings transformed Rome into a true city, reclaiming swampy land between the hills and building the first sewers, of which the *Cloaca Maxima* (Greatest Sewer) is the best known. Etruscan rule and the peaceful and bellicose relations between the two ethnic groups left the Romans a patrimony of cultural and religious attitudes and rituals, as well as a sound knowledge of hydraulic technology. By the first century BC, after a few centuries of turbulent relationships and wars, the Etruscans were absorbed by Rome (*cf.* e.g., [19]). After the conquest and the following initial destruction, most Etruscan cities and settlements continued to exist as Roman centers. Hence, over the centuries, most old Etruscan buildings and public structures were substituted by Roman ones. This explains why, at present, no remains of entire Etruscan fountains in good condition exist. Remains of basins and pertinent hydraulic structures and conduits, however, are plentiful, but are mainly part of temples and sanctuaries [20], as in the case of the Etruscan town of Marzabotto, in Northern Italy (Figure 4). The fountain fed by an aqueduct destroyed during excavations has been dated to between *ca.* 500 and 350 BC. These dates are believed to be the (approximate) dates of the foundation and destruction of the Etruscan city by the Gallic tribes.

In spite of the scarcity of remains of Etruscan fountains, some interesting information has been derived from paintings and ceramics. For instance, in the mural painting of a fountain (Tomb of the Bulls, Tarquinia, and Southern Etruria) shown in Figure 5, water flows out of the mouth of two animals, whose positions suggest the presence of pressure pipes. The Etruscans appreciated highly Greek pottery and ceramics. Therefore, Greek cups and vases are often found in Etruscan tombs. In many cases, the ceramics were decorated with scenes where the focus was on one or more fountains. A splendid Hellenic hydria, found in an Etruscan tomb of Vulci, in Southern Etruria is shown in Figure 6. Figure 7 reproduces the scene of another vase, in which a few young women are fetching water from a fountain [21]. It is noteworthy that the drawing in Figure 7 represents a scene which is still common today in many developing countries. Actually, some of the older authors of this paper also remember having seen similar scenes since many Mediterranean villages had no piped water in houses until the second half of the 20th century, and had to rely on a free flowing public fountain. The water jar was also often carried on the head as shown in Figure 7.

There is no doubt that Etruscans had actual fountains similar to those represented on their cups and vessels. An actual faucet of an Etruscan fountain in the shape of a lion's head, quite similar to that of Figures 7 and 8, is shown in Figure 9. It was found in Orvieto at the San Giovenale site and probably dates back to around the end of the 6th century BC [20].

Figure 4. Etruscan Fountain of Marzabotto, northern Italy, *ca.* 500–350 BC (photo by Walter Dragoni).

Figure 5. A painting in the "Tomba dei tori" (Tomb of bulls), in Tarquinia, Southern Etruria (*ca.* 540–530 BC). The painting probably depicts a scene of Greek mythology [22].

Figure 6. Attic vase found in an Etruscan tomb in Vulci, Southern Etruria (Trustees of the British Museum).

Figure 7. Women fetching water from a fountain. Painting reproduced from an engraving in an vase found in an Etruscan tomb [21].

Figure 8. Faucet of an Etruscan fountain found in San Giovenale (Orvieto), reproduced with kind permission of the Museum "Claudio Faina", Orvieto, Italy (photo by Walter Dragoni).

Figure 9. Earthenware well of Early Han Dynasty.

4. The History of Wells and Fountains in Early Chinese Dynasties

In China, water was first distributed to houses through taps in the late 19th century in some major cities. The Chinese generally obtain water for daily use from two sources: dug wells and nearby streams. Wells are the main water supply structure in China and have played a very important role until today throughout the country. In many places, people also use water from springs, which they lead via ditches into residential areas, where they build special structures with a spiritual meaning over the supply point.

Based on evidence from the earliest ruins of a well discovered in He Mu Du in Zhejiang Province in 1974, the history of the well in China dates back about 6000 years. The ruins were those of a well of wooden construction [23]. Later, many other old well sites used in ancient China around that time were discovered. The early civilization of China emerged initially in the mid-basins of the Yellow River and the Yangtze River. The wells dug there allowed settlements to be spread beyond the river banks.

The structure of the well developed over the years. Archaeological findings show that the early wells (around 3000 BC) were built of wood and were usually square. In the Fighting-States period (475–221 BC), the well wall was built of earthenware because it was a popular material then. An earthenware well of the Early Han Dynasty discovered in Yeng County is shown in Figure 9. About 40 wells were discovered in the same ruins each approximately 5 to 7 m deep with a diameter of 1.3 m [24].

During the Han Dynasty (*ca.* 202 BC–220 AD) and subsequent dynasties, wells were commonly built of bricks. That made them firmer and more diversified [25]. However, since around the Tang Dynasty (618–907 AD), the covered structure started gaining popularity in residential areas (compared to wells for irrigational purposes in farming communities). Therefore, people preferred to build a stone structure above their well to make it more firm and beautiful and to serve as a spiritual symbol. A typical such structure (Song Dynasty, *ca.* 960–1279 AD) is shown in Figure 10.

The Dai are an ethnic group in China living mainly in Yunnan. Their villages usually have one or two public wells. The villagers always build a special structure above the well, such as a Buddhist tower, a temple-style house, and a dragon head to beautify it. It makes the well a special place, even holy. Today, tap water is commonly used in these areas, but most villages still following the tradition of using well water for drinking and cooking (Figure 11) ([5], p. 43).

Figure 10. Well structure in Chaozhou City built during the Song Dynasty.

Figure 11. The well of a Dai Village.

5. Greco-Roman European Classical Antiquity

5.1. The Classical and Hellenistic periods (ca.5thc.–1stc. BC)

In ancient Greek cities, shrines and fountains were mainly linked to social and public life in the Agoras and religious beliefs about gods and nymphs living by water. Fountains existed in every ancient Greek city such as Athens, Corinth, and Megara, and in every sanctuary such as Delphi and Epidauros. Some of them supplied water from springs and rivers into the cities since the *ca.* 6th century BC as terminal points of aqueducts.

Greek fountains were made of limestone or marble. Often the water flowing through lead pipes emerged from the mouth of a sculpted mask representing the head of a lion or the muzzle of some other animal. Most Greek fountains flowed by simple gravity, but water pressure and the siphoning principle were also used to make water jet or spout out (as seen on Greek vases). Many images and illustrations of the uses and other aspects of ancient Greek fountains have survived on the vases of that period (Figure 12).

Figure 12. Attic Greek vases: from South Italy, about 480 BC showing Polyxena filling a vessel from a fountain (The Louvre) [26].

The fountains were discovered either in central areas of settlements or the Agoras of Classical and Hellenistic cities, or in sanctuaries or sacred places. In the last mentioned, they were not only for water supply, but also for ritual purposes.

The typical ancient Greek water supply fountain (see Figure 13) consisted of a single- or rarely a double-basin small tank, as in the case of the Enneakrounos fountain, from where water either overflowed through simple stone or ornamented spouts, or was carried out in clay jars and vessels or leather utricles. However, such a tank was in many cases at the front of a larger water reservoir, as in the fountains of "Theagenes" at Megara in Attiki (Figure 14), which was filled by water either from a spring or from an underground water harvesting tunnel network. In very few cases they were supplied with rainwater.

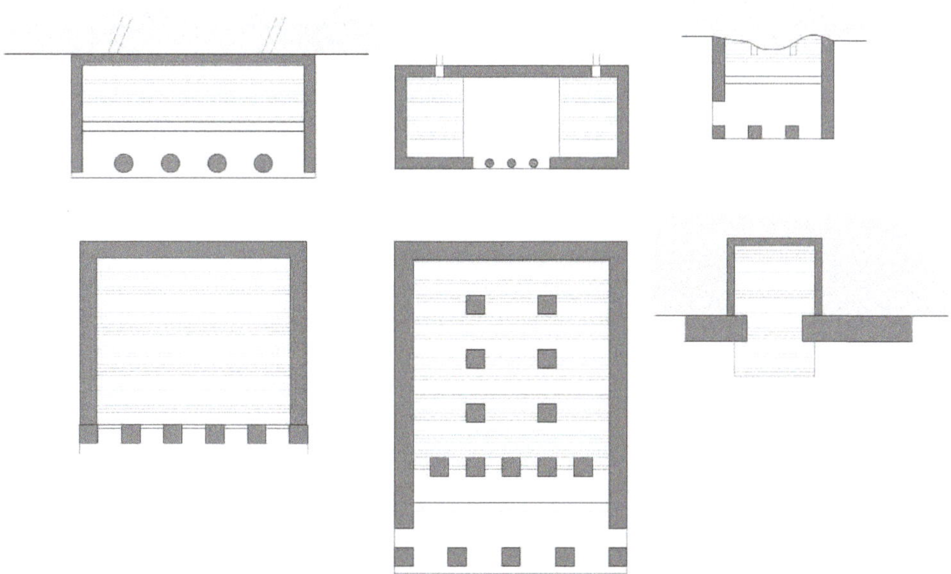

Figure 13. Most common types of ancient Greek fountains by cross-sections, (photo by Georgios P. Antoniou).

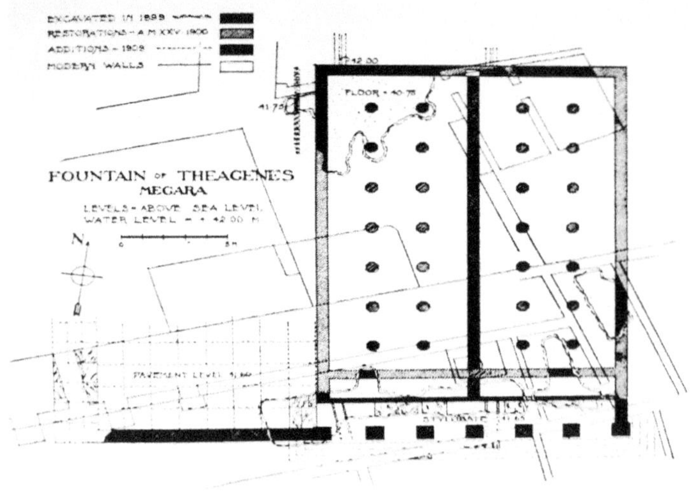

Figure 14. The fountain of "Theagenes" at Megara [27].

Many of the archetypal Classical and Hellenistic fountains were constructed either in a retaining wall such as the Palestra of Sikion and the Asklepeion of Kos, or in the rear wall of a portico such as the south stoa of the Athenian Agora. Quite often, the fountains were attached to the natural rock from which the water was harvested, as in Ialyssos or Corinth where the fountains had been hewn out of the soft local rock. Architecturally, a colonnade or portico usually formed the façade of the fountains (Figure 15). It was usually narrow as at Megara and Ialyssos on Rhodes.

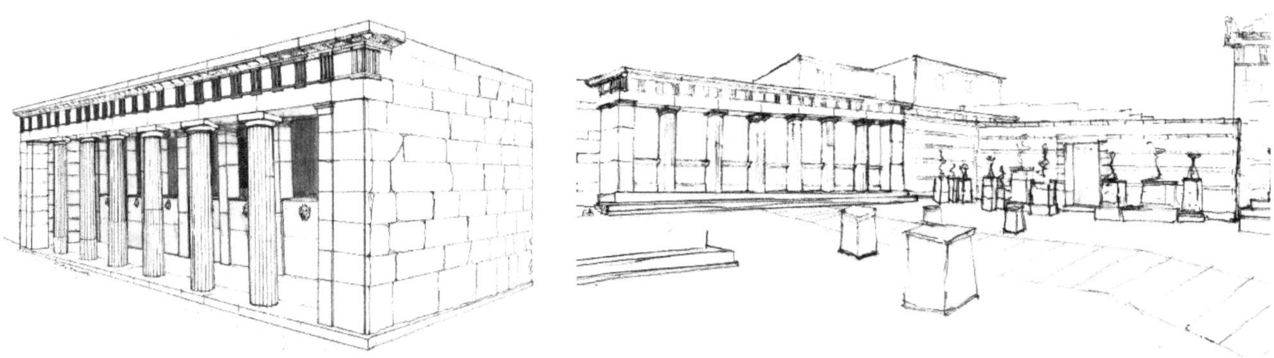

(**a**) (**b**)

Figure 15. Fountains: (**a**) The Doric Fountain at Ialyssos, Rhodes (reconstruction by Georgios P. Antoniou) and (**b**) The Fountain at the Agora of Kameiros, Rhodes (reconstruction by Georgios P. Antoniou).

The other category of ancient Greek fountains consisted of religious ones or those for ritual purposes, as mentioned earlier. Most of them were attached to sanctuaries or sacred places. The *Castalia* fountain was one of them, but public. Many of the ritual fountains were grottos, dug completely or partially in rock. Their structure served the needs of the sanctuary. As mentioned briefly before, there were a few different methods for water acquisition. The simplest one was to use spring water. For instance, the sacred spring *Castalia* in Delphi was supplied by the springs of Maleata's hill and the fountains at the Asklepeion of Kos. Rainwater cisterns and water conveyed by regular aqueducts were not commonly used (Figures 16 and 17).

(a) (b)

Figure 16. (**a**) The archaic spring in Delphi; a type of branch of the famous *Castalia* spring (with permission of Andreas N. Angelakis) and (**b**) Asklepeion Kos; one of the fountains flanking the monumental stairway.

Figure 17. Peirene fountain in ancient Corinth, Greece (construction phase during the Roman Herodes Atticus) (Georgios P. Antoniou).

In contrast, underground water harvesting conduits were the most common method applied since the time of the Peisistratos aqueduct in Athens dating back to the *ca*. 6th century BC. That water supply network collected not only spring water, but also groundwater from the foot of Mt Hymytus, at almost the same time as the well-known qanats [28]. Most of the famous fountains of the Athenian Agora like Enneakrounos were supplied by the aqueduct which is still conveying water to the National Gardens in Athens.

The Doric fountain at Ialyssos, Rhodes was also supplied by an extensive network of water harvesting conduits hewn in the rock to which the fountain was attached. In Corinth, not only the two main fountains of the Agora, Peirene and Sacred, were supplied by an extensive network of water harvesting conduits, but also the Glauke and Lerna, westwards and northwards, respectively. Moreover, the technique not only supplied the fountains attached to a rocky backfill, but also the free standing buildings such as the Fountain of "Theagenes" [29] which received water from similar underground tunnels.

Several small-scale functional constructions facilitated water extraction. They were intended either for filling buckets or flasks or most commonly ceramic jars (hydria). Drainage and other pipes, on the other hand, ensured the removal of surplus water. There were also generally methods and structures that allowed directing a reasonable amount of the water for reuse in lavatories or elsewhere [30]. Moreover, numerous small structures such as intake openings [29], Y-shaped clay pipes [31] and even the tar waterproofing of the "Theagenes" Fountain at Megara [29] are proof of the innovative techniques and construction know-how of the engineers and architects of that era also in this field.

The two well-known springs of antiquity in the Corinthian Agora, Peirene (Figure 17) and the Sacred Spring, were supplied with water by a network of underground water harvesting channels running southwest and partially to the southeast. Another famous spring and fountain of antiquity is the Castalian Spring in the ravine between the Phaedriades at Delphi. It was—and still is—situated at the place where all visitors to Delphi, who came to consult the Delphic Oracle, stopped to purify themselves before meeting the Oracle. At that same spot Apollo killed the monster (Python) according to mythology. Two fountains fed by the spring still exist [32].

5.2. Roman Period (4th c. BC–4th c. AD)

Rome, which according to tradition was founded in 753 BC, inherited and advanced greatly the hydraulic knowledge of the Etruscans. Their technology remained unsurpassed till the end of the 18th century. Since Roman hydraulic technology was closely linked to the Etruscan one for the first centuries after the foundation of Rome, only the period after the construction of the first aqueduct (in 312 BC) will be considered. According to Sextus Julius Frontinus [33,34], the Roman consul who was named *curator aquarum* (guardian of the water) of Rome in 98 AD, Rome had nine aqueducts. Two hundred years later there were eleven, as two more were built in 109 and 226 AD [34]. The aqueducts of Rome, plus a few small local springs, supplied water to hundreds of public fountains and a few dozen monumental fountains and large private and public thermal baths, not counting the water supplied to the Imperial household and owners of private villas. Even in the fourth century, in a deep decline, Rome had 1352 fountains or cisterns [35]. The quantity of water and number of fountains in Rome was so large that the term *Roma regina aquarum* (Rome queen of the waters) has been commonly used since antiquity. The attitude of ancient Rome towards water and fountains left a permanent imprint on Rome (and Italy), which survived the fall of the Roman Empire and probably still exists. Rome is the city with the highest number of monumental and artistic fountains. Today, there are also more than one thousand little fountains along the streets of Rome providing excellent free drinking water to everybody.

"Roman fountains" are such a broad subject that only a few pieces of general information and little known examples can be given here about them. The interested reader can easily find more information on the Internet; for recent printed publications see e.g., [16,35,36].

The Roman Empire extended from the Spanish Atlantic coast to the Rhine River in Germany and the Black Sea, and from the Scottish Lowlands to a few hundred kilometers south of the North African Mediterranean coast [37]. In that large Empire, all Roman cities of any importance had a Forum (the business area), a theatre, an amphitheater, a military barracks, public baths and the necessary aqueducts and fountains to provide water for these structures.

Roman pipes were made of wood, stone, mortar, terracotta and lead; the taps and valves were of high-quality bronze alloys, similar to modern ones. A good example of Roman water supply planning is Pompeii, the city destroyed by the Vesuvius eruption in 79 AD. Pompeii's water was supplied by an aqueduct. So far, over forty of the city's public fountains have been discovered, and it appears that most houses were no more than 50 m from a public fountain. The fountains were supplied by lead pipes under pressure [38,39]. The simplest form of street fountain in Pompeii was generally equipped with an oblong stone basin, typically about 1.5 × 1.8 m^2 and 0.8 m high, into which a spout discharged, and the basin was presumably normally full. The fountains were deliberately designed to overflow in order to clean the streets [16,40–42]. Excellent pictures of Pompeii fountains can be seen in [43].

Fountains were common in the private houses of wealthy families. In Roman cities and private villas, fountains with water jets where not rare. For example, in Pompeii, the House of the Bracelet had a spectacular mosaic fountain with 29 water jets in its triclinium ([44], as reported in [38]). Pliny the Younger describes his villas as having gardens and fountains with jets [45]. The Villa of Emperor Hadrian in Tivoli (Figure 18), had immense large fountains and basins, and fountains with jets [46].

Figure 18. One of the fountains in the Villa of Emperor Adrian (76–138 AD) in Tivoli, close to Rome (photo by Walter Dragoni).

Not far from the city of Pompeii, in the District of Salerno, there is a Roman tunnel in the rock in the village of Sant'Egidio del Monte Albino in the Sarno River basin. The tunnel was constructed in order to supply a public fountain which stands on the grounds of an ancient Roman villa (Villa Helvius). The little known Helvius fountain is quite different from the public fountains in nearby Pompeii. The Helvius fountain was not constructed of matched slabs of limestone or Vesuvian stone. It was built from a single block of white marble. In addition, as shown in Figure 19, it has sculptural decorations on the three exposed sides representing the River Sarno along its path from the spring toward the sea [40,42].

In Roman cities, all buildings planned for frequentation by a large number of people, like theatres and amphitheaters, had many fountains for public service. Hence, they were built according to a rational plan. A good example is the Coliseum in Rome, completed in 80 AD, which could accommodate up to 65,000 spectators for games that could last a few days. That required a sophisticated hydraulic system for latrines and fountains. In addition, indeed, traces of 92 fountains are visible today [47]; those of one are shown in Figure 20.

Figure 19. Helvius fountain in the village of Sant' Egidiodel Monte Albino in the Sarno Valley: (**a**) global view; (**b**) left view; (**c**) front view; and (**d**) right view.

Figure 20. Traces of a public fountain in the Coliseum in Rome which tourists often interpret as a cross. The grooves contained the lead pipes that fed one of the dozens of fountains in the amphitheater (photo by Walter Dragoni).

The previous pages focused on the public fountains of ancient Rome, leaving aside imperial residences and villas of the dominant class, which were numerous and all equipped with monumental fountains. Even some private "middle class" buildings, often used for commercial purposes, had little private fountains located in the inner court. A good example is shown in Figure 21.

Figure 21. Fountain inside a private commercial building, a sort of restaurant, in Ostia, the harbor of Rome at the mouth of the Tiber River (photo by Walter Dragoni).

6. Medieval Times

6.1. The Byzantine Period (ca. 4th c–15th c. AD)

During the early Byzantine period, the structures and technologies of the fountains adhered to Roman tradition and customs. Fewer were constructed and the know-how of the past vanished. The densely built settlements also had less room for elaborate and impressive fountains.

On the other hand, the powerful and wealthy clergy of that period constructed well-designed fountains for churches and, especially, monasteries which survived raids and destruction because of their location (Figures 22 and 23).The typical fountains of that time were the simple arched niches either attached to a building wall or the more rare stand-alone ones where a stone or basin received the water flowing out of a simple or slightly ornamented spout.

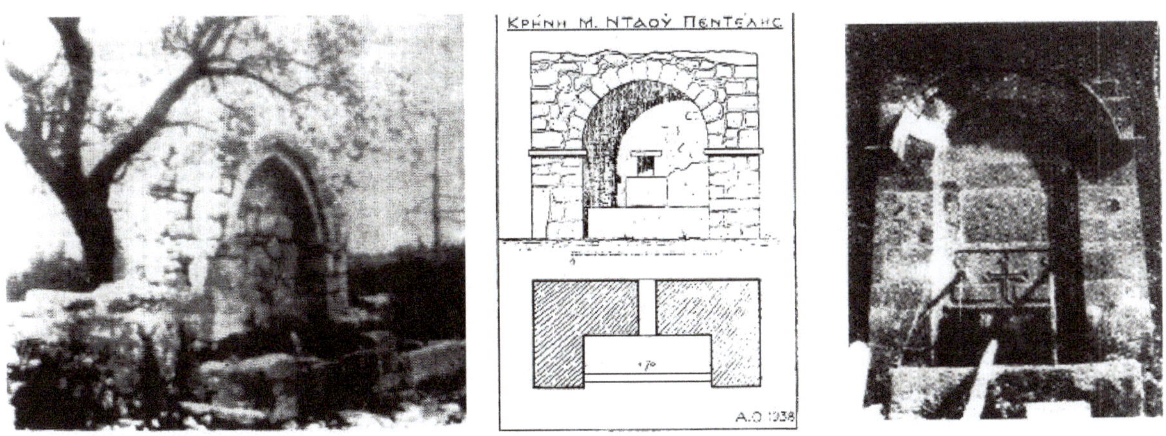

(a) (b) (c)

Figure 22. Typical structures of fountains in monasteries of the Byzantine period: (**a**) *Faneromeni* in Korinth; (**b**) *Daou*, Penteli, Attica; and (**c**) *Filotheou* in Mount Athos.

Figure 23. Typical structures of fountains in the Byzantine period.

Despite the general simplicity, there were also fountains constructed in a more elaborate manner. The stand-alone domed constructions embody the constructional technologies of the arches, domes, colonnades and tie beams of that period. The well elaborated examples also show the typical Byzantine types of masonry. As concerns water supply, it was implemented by the regular methods of the era as mentioned below (Figures 24 and 25):

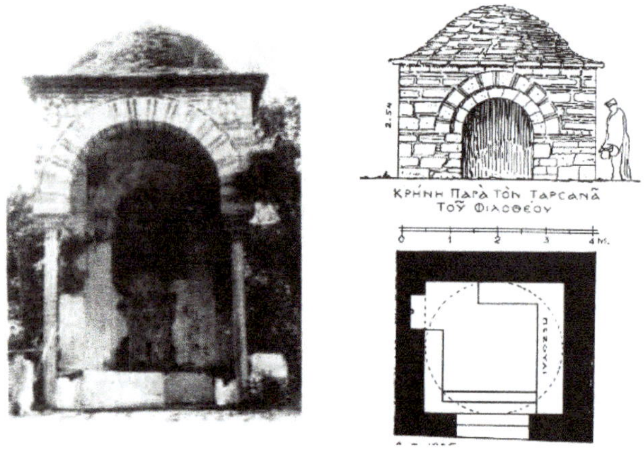

Figure 24. The well elaborated examples show also the typical Byzantine types of masonry.

(a) (b)

Figure 25. (**a**) Meg. Lavras Monastery (*ca.* 1060 AD) and (**b**) Batopedi Monastery.

The most characteristic fountain of the Byzantine period is the "phiale". It had a dome which covered a circular shallow basin at the center, where the holy water was during the various rituals of the monastery. A low parapet separated the covered space from the yard around it. There was no natural flow of water to the basin. Yet, traces of the existence of a "strobilion" which jets water up at the center of the circular basin have been found.

Generally, the water supply for most fountains in Byzantium was provided through springs and pre-existing aqueduct systems. Most underground water harvesting networks had ceased operating or were no longer maintained. Thus, it is likely that the water conveyance systems were not operating well during the last period of the Byzantine Empire. Only in areas conquered by Westerners were the major supply systems functioning a little better.

6.2. Medieval, Renaissance and Baroque Times in Western Europe (ca. 450–1700)

After the fall of the Roman Empire, the great barbaric invasions transformed the empire into a multitude of independent little states, with ever changing boundaries and rulers, in a semi-permanent state of war. During these times, due to the combined effects of natural wear, lack of maintenance and wars, the great aqueducts fell into ruin: a great and complex aqueduct system can only be built and function in a region at peace under the control of one centralized political power with substantial economic resources [36]. With a few notable exceptions, none of these conditions existed in Western Europe from the 5th to the 15th century AD. Consequently, the great Roman fountains, as well as the networks of little city fountains built for public service, were in ruin all over Western Europe until about the 12th century AD. Water supply also generally declined significantly along with the number of the fountains. That is reflected in the scarcity of medieval fountains, which were small, simple, and often made of materials taken from ancient monuments. During the Early Middle Ages, water management was of high quality only in the East Roman Empire (*i.e.*, the Byzantine Empire) and the Islamic kingdoms. In Western Europe, Islamic Spain had an excellent standard of water technology for irrigation, water supply and fountains. In Islamic Spain, old Roman aqueducts were repaired, new ones were constructed, and public and private fountains were built. The description of Islamic water management in Spain would require a separate book. Here, it must suffice to remember that Islamic culture adsorbed and applied Greek and Roman water technology. Under Islamic rule, new inventions were added to the ancient water technology, including the sophisticated pumps of Ibn al-Jazari, who lived *ca.* 11th–12th centuries AD [48]. Today, the fountains built in Spain under Islamic rulers constitute a well-known cultural attraction. They are in general much smaller than the great Roman ones, but they are fascinating for their grace and calm beauty (Figure 26).

In Medieval Western Europe, the technology for water supply and fountains was preserved thanks to the monasteries and abbeys. These were sort of autonomous communities, where water was needed for daily life, and often for irrigation of orchards and gardens. Gardens with fountains were seen as mystic places, a sort of earthly equivalent of Paradise (or the Garden of Eden), by both Christians and Muslims. In abbeys and monasteries, water was not only used for washing and other daily purposes, but also in the religious rite of cleansing hands before services and meals [49]. Wells were not efficient enough so, whenever possible, water was brought into religious communities and delivered through one or more fountains. As they were used simultaneously by many people, they had to have a large circular, polygonal or rectangular basin with multiple taps for discharging water (Figures 27 and 28).

Figure 26. Granada, Alhambra, Fountains of the Lions, *ca.* 11th century AD (photo by Walter Dragoni).

Figure 27. Lavatorium in the Monestir de Santa Maria de Poblet, municipality of Vimbodí and Poblet, Spain. (Photo © Adrian Fletcher, www.paradoxplace.com). [50].

Figure 28. 14th century lavatorium at Gloucester Cathedral (photo by Derek Harper) [51].

Versailles, France (around 1660), Chatsworth in Derbyshire, England (around 1687) and the imperial park of Peterhof, close to St. Petersburg, Russia (around 1710–1720) [35].

Figure 34. Fountain in the Villa d'Este park, in Tivoli, near Rome (photo by Walter Dragoni).

6.3. Fountains and Islam

In Islam, water is considered holy, a gift from God. Islam is a monotheistic religion based upon the Qur'an, and its Mosques have fountains where worshippers can wash themselves before entering the Mosque. The washing act is called "wudu". The Qur'anic mandate for "wudu" is expressed in the sixth ayat of sura 5 (Al-Ma'ida).

No one can refuse to give away surplus water without sinning against Allah and against man. The hadiths say that among the three people Allah will ignore on the Day of Resurrection will be the man who has water in excess of his needs, but refuses to give it to a traveler.

"We made from water every living thing" (Qur'an 21:30).

Water is the primary element that existed even before the heavens and the earth did:

"And it is He who created the heavens and the earth in sixth days, and his Throne was upon the water" (Qur'an 11:7).

The water of rain, rivers, and fountains runs through the pages of the Qur'an to symbolize God's benevolence:

"He sends down saving rain for them when they have lost all hope and spreads abroad His mercy" (Qur'an 25:48).

There are two fundamental precepts that govern the rights to water in the Shari`ah. Shafa, the law of thirst, establishes the universal right for humans to quench their thirst and that of their animals; and shirb, the right of irrigation, gives all users the right to water their crops. From the numerous Quranic references to cooling rivers, fresh rain, and fountains of flavored drinking water in Paradise, we can deduce that water is the essence of the gardens of Paradise. It flows beneath and through them, bringing coolness and greenery, and quenches thirst. The believers will be rewarded for their piety. The water in Paradise is never stagnant; it flows, rushes, unlike the festering waters of Hell. The Qur'an also equates the waters of Paradise with moral uprightness. In the garden, there is no idle talk, and there is a gushing fountain [59].

Islamic Law, the Shari`ah, based on the teachings of the Qur'an and the prophet Mohammed, ensures democratic distribution of water in communities [60].

Ancient Persians already built water channels. These underground channels called Qanats were connected to smaller canals called karizs. These systems were used to lead water for household use and to irrigate farmland and gardens. Later, after the 7th century, gardenswere enclosed by walls to make them resemble paradise. These paradise gardens were laid out in the form of a cross, with the four arms representing the rivers of paradise that divide the world into four parts [61]. In some cases, water spouted from a fountain in the center of the cross representing the spring or fountain described in the Qu'ran as the source of the Paradise rivers [61].

7. Modern Times

7.1. The Ottoman Period

Water is part of the Ottoman religion. Cleansing of the body symbolizes the cleansing of the soul, according to the Koran [62]. Thus, a water fountain was located at the entrance of each mosque. Following the old Moslem tradition, public baths (hammams) and fountains were the major hydraulic works developed in several places during the Ottoman period. At that time (*ca.* 1669–1898 AD), Hammams occupied a key position in the Ottoman culture. They were used as places for meeting, cleaning and bathing, ritual cleansing, *etc.* The hammam is a very old Ottoman institution, and public baths were established all over the Ottoman Empire, including the Island of Crete [62].

Generally speaking, although a large quantity of water was needed for the hammams, no special effort was made to increase the quantity of water coming into major Cretan cities (e.g., Iraklion, Rethymnon, and Chania). The Venetian hydraulic works were just maintained in good condition. Thus, during the Ottoman period, the existing Venetian water cisterns and fountain houses remained in use and new ones were built. Tzombanakis [63] distinguished three different types of fountain houses: (a) large polygonal free-standing, elaborate fountain houses built by mosques (successors of the Christian phiale); (b) fountains usually built inside palaces and mosques; and (c) less elaborate fountain houses built by main roads and in villages.

During the period (*ca.* mid-14th century—1923 AD) ruling pashas often built fountains next to mosques such as the Fountain of Qasim Pasha (1527); the one on Temple Mount, Jerusalem, an ablution and drinking fountain built during the Ottoman reign of Suleiman the Magnificent; and the Fountain of Ahmed III (1728), at the Topkapı Palace, Istanbul. Of course, palaces also had small decorated fountains. They provided cool or hot air, drinking water, and made a calming and soothing sound. One surviving example is the Fountain of Tears (1764) at the Bakhchisarai Palace in Crimea made famous in a poem by Alexander Pushkin [64].

7.2. Evolution of Fountains in Late Chinese Dynasties

After the 1950s, along with the rapid urbanization of China, tap water replaced the well as the source of daily use in most places. Yet, the well has always played an important role in the water history of China.

Spring water is also an important water resource in many places in China. Consequently, spring end structures have also assumed characteristics of fountains in China. The Chinese assign holy and spiritual meanings to springs; they are not merely sources of water. For example, a spring that is called a dragon pool is believed to be the house of the dragon, the God of Water.

Ji Nan City, the provincial capital of Shandong Province, is the city most famous for its springs in China: it has around 700 of them. They are not just water sources, but also special places with complex spiritual meanings. People have built pools around the outlets of these "fountains" and "sub-fountains", to watch and enjoy the gushing of the water. People have also constructed buildings and gardens in memory of famous people to enjoy their "fountains". In summer, watching gushing fountains brings great pleasure to citizens. For example, the history of the BaoTu Qian fountain dates back 2700 years. The fountain structure measuring 30 m long and 20 m wide was built over three spring outlets. In addition, a big garden with many buildings was built near the fountain in memory of saints and famous people, and for common people to enjoy the fountain (Figures 35 and 36) [65]. Even today many spring structures for daily water use are found in the city and the countryside.

Figure 35. The BaoTu Qian fountain.

Figure 36. The fountain of Jinan City.

Another holy spring fountain in Yuxi Prefecture is the Nine-Dragon Fountain. The Nine-Dragon Fountain of Yuxi City, Yunnan Province, in Southwest China is a typical example of Chinese dragon fountain culture. It is located 10 km northwest of downtown Yuxi and comprises nine fountain outlets. The gushing volume of the fountain water is considerable all year round and forms a river called the Nine-Dragon River. Since its utilization started, it has been an important water source for local agriculture which has irrigated tens of thousands hectares of farmland. Since the outlets are surrounded by ominously quiet hills, dense forests, and crystal clear ponds, the scenery is wonderful (Figure 37).

Figure 37. The main outlet of the Nine-Dragon fountain.

At the end of the Ming Dynasty about 300 years ago, people started building many religious and sightseeing buildings near the fountain. An oblong pond called the Large Dragon Pool, about 110 m^2 in area, was built with stones at the outlet of the largest fountain. The main buildings include the "Nine-Dragon Palace", a temple dedicated to the dragon king; the "Five Si Temple" dedicated to the tiger, the monkey, the chicken, the sheep and the rabbit; the "Tower of Three Saints" dedicated to Sakyamuni, the founder of Buddhism, Lao Zi, the founder of Taoism, and Confucius, the founder of Confucianism, and the pavilion where tourists can listen to the sound of the water of the fountain.

Although the Nine-Dragon Fountain is a natural attraction, it also has profound cultural connotations derivingmainly from the relationship between the Chinese culture and water; it is a symbol of Chinese culture. Firstly, because it is linked to the most important natural god of Chinese culture, the Dragon. Chinese people believe that dragons live in fountains and spit out the fountain water, which is why they hold fountains sacred. According to a legend, the water is spit out by nine dragons, hence the name the "Nine Dragon Fountain" [65].

Many modern type fountains have been built in the cities which made them more beautiful and allowed citizens to enjoy the gushing waters. As they became a fashionable feature of urban design, more and more cities wanted them.

7.3. Industrial Times

Water and water fountains are used as cultural allegories. They are often assigned religious and other meanings. For example, in Christian allegory the fountain in "Fountain of Life" is associated with rebirth which happens after baptism. An interesting derivative of the Fountain of Life is the Fountain of Youth, from which one can drink to regain his youth or become immortal. Water fountains are also important for Muslims and Buddhists. Before the introduction of public waterworks, fountains supplied drinking and bathing water to the residents of cities. Yet, artists designed them to be aesthetically appealing. Fountains have always had a calming effect on people as they draw people to relax and socialize around them [5].

During the early history of modern public water systems in the *ca.* 19th and even the 20th century, water from fountains was also used for productive purposes, such as to water animals in Tampere, Finland. In Victoria, Canada, fountains erected near the turn of the 20th century had different levels for various users of water. The fountains provided water not only for travellers and their thirsty horses but also for smaller animals such as dogs and pigs [66].

The water fountain in Trafalgar Square in London, UK, is nowadays a popular destination of sightseeing tours, which resembles the Canaletes Fountain in Plaza Catalunya in Barcelona. Over the years, Trafalgar Square has accumulated additional statues and other works of art. The original Trafalgar Square Fountain erected in 1845 was remolded in 1939, and is actually a set of two fountains. The Trafalgar Square Fountain features mermen and mermaids alongside dolphins, all sculpted in bronze. As pointed out on the city web page: "…this is a popular tourist attraction. Best of all; it is free!" Since its construction, Trafalgar Square has also been a venue of political demonstrations [67].

New types of high-tech fountains, called "water kiosks", have begun to spread in some European countries since the beginning of the twenty-first century [3]. The water they dispense from an aqueduct undergoes certain treatments (active carbon, cooling, addition of carbon dioxide) to improve its organoleptic characteristics such as smell, taste, and temperature. That reduces the use of bottled water and the resulting generation of waste (PET bottles) as well as the CO_2 emissions associated with the transport of the bottles from bottling plants to the consumer. These kiosks have met wide approval in Italy, from the public and municipal authorities alike, to the point that over the last ten years their number has gone from zero to over a thousand, especially in the northern and central parts of the country. A water kiosk in the Turin area is shown in Figure 38 [68].

A water fountain made for fun and aesthetic enjoyment is shown in Figure 39.

Figure 38. Water Kiosk in a park in Turin (with permission of SMAT SpA).

Figure 39. Children enjoying fountains with changing flows and jets in Arvada, Colorado, US (Photo: Dave Hendricks 2008).

8. Discussion

The Minoans of Crete, the center of Europe's first advanced civilization, developed the first advanced hydraulic engineering applications. In the early phases, fountains were used especially for public water supply. The roots of Roman hydraulic technology were in the Etruscan civilization and contemporary mural paintings even suggest the existence of pressure pipes.

In ancient Greek cities, shrines and fountains were associated mainly with social and public life and religious beliefs about gods and nymphs living by water. Most Greek fountains used gravity flow, but water pressure and the siphoning principle were also used to make water jet or spout.

Ancient Rome is known for its aqueducts and public fountains—In the fourth century, during a deep decline, it still had 1352 fountains or cisterns. Fountains were also common in the private houses of wealthy families. The numerous imperial residences and villas were all equipped with monumental fountains. Even some private "middle class" homes had little private fountains in the inner court. More than forty public fountains of Pompeii, the city destroyed by the Vesuvius eruption in 79 AD, have been discovered so far.

In the early 13th century, communes (city-states) in northern and central Italy built new water systems and fountains or repaired old aqueducts. Fountains were increasingly built for public water supply and to adorn cities and placed in the middle of the main piazza (square) between the symbols of religious and political power. Various sources indicate that the people participated actively in the life of the cities.

Hundreds of artistic and monumental fountains were built in Rome after the 15th century including the famous fountains of Piazza Navona, Acqua Paola, and Trevi. The splendid fountains of Villa d'Este in Tivoli (*ca.* 1550) and Villa Aldobrandini in Frascati (*ca.* 1600) gave rise to the European fashion of adorning parkland with great *jeux d'eau*, such as in Versailles, France (*ca.* 1660), Chatsworth in Derbyshire, England (*ca.* 1690) and Peterhof, next to St. Petersburg in Russia (*ca.* 1710–1720).

During the early history of modern public water systems in the 19th and even the 20th century, water from fountains was also used for productive purposes, such as to water animals. The water fountain in Trafalgar Square in London, UK, is nowadays a popular destination of sightseeing tours. More recently new types of high-tech fountains, called "water kiosks", have spread across some European countries to provide safe drinking water. Recently, many fascinating fountains have also been designed for fun and aesthetic enjoyment.

Water fountains have had and continue to have major religious significance such as the "phiale" in the Byzantine period. Gardens with fountains have been mystical places, sort of earthly equivalents of Paradise, for both Christians and Muslims. Islamic mosques have fountains where worshippers can wash themselves before entering. The Chinese assign holy and spiritual meanings to springs called dragon pools.

Fountains have also been symbols of power and wealth which have spread to public buildings and private residences as shown by existing examples. In our built environment, flowing water and fountains seem to be a calming element in many respects.

9. Conclusions

Undoubtedly, a wide variety of fountains have been built over time. On the whole, the emphasis has shifted, or rather expanded, over time along with societal development from just meeting basic needs, such as providing drinking water for people and animals, more towards providing enjoyment and aesthetic values.

The aim of this paper is not to give an exhaustive account of the history of fountains in various parts of the world until today. Instead it presents some characteristic fountains of different civilizations chronologically. Historical fountain technologies may have useful applications even in today's water engineering. Some lessons learned include:

(a) Throughout history, fountains have been an essential part of water supply that provides water resources sustainability for the survival and well-being of humans. Thus, ancient fountains should be considered, not only historical artifacts, but also potential models for sustainable water technologies now and in the future.

(b) Ancient water technologies, including fountains and well- or cisterns-like fountains, were characterized by simplicity, ease of operation, and lack of complex controls, which makes them more sustainable. However, they have lost some of their importance for modern water supply purposes in the developed parts of the world, whereas they are more applicable in the developing parts of the world [8].

In conclusion, globally some 0.8 billion people lack safe water, more than 2.6 billion people do not use improved sanitation and 1.1 billion practice open defecation. Thus, there is a huge need for sustainable and cost-effective water supply and sanitation facilities, such as fountains, particularly in the cities of the developing world [69]. The applicability of select ancient technologies in the contemporary developing world should, therefore, be seriously considered.

Acknowledgments

We thank Academy of Finland AKVA project Water as Social and Cultural Space (No. 263417) for their support. The chapters on Etruria, Rome and the Middle Ages in Western Europe were supported by Provincia di Perugia and Fondazione Cassa di Risparmio di Perugia (project 2010.011.0442, coordinated by W. Dragoni).

Author Contributions

Author Contributions are: Petri S. Juuti: Fountains in general. Georgios P. Antoniou: Fountains in general; Walter Dragoni: Case expert of Italy and Western Europe; Fatma El-Gohary: Fountains in general; Giovanni De Feo: Fountains in general; Tapio S. Katko: Fountains in general; Riikka P. Rajala: Fountains in general; Xiao Yun Zheng: Case expert of China; Renato Drusiani: Fountains in general; Andreas N. Angelakis: Case expert of Greece.

Conflicts of Interest

The authors declare no conflict of interest.

References

1. Mays, L.W.; Koutsoyiannis, D.; Angelakis, A.N. A brief history of urban water supply in antiquity. *Water Sci. Technol. Water Supply* **2007**, *7*, 1–12.
2. Angelakis, A.N.; Spyridakis, S.V. Major Urban Water and Wastewater Systems in Minoan Crete, Hellas. *Water Sci. Technol. Water Supply* **2013**, *13*, 564–573.
3. *Histoire des Jardins*; Ouest, S., Ed.; PREVOT, Philippe: Bordeaux, France, 2006. (In French)
4. SAMIRAD. Saudi Arabia Market Information Resource Directory. Available online: http://www.saudinf.com/index.htm (accessed on 21 March 2015).
5. Hynynen, A.J.; Juuti, P.S.; Katko, T.S. *Water Fountains in the World scape*; Hynynen, A.J., Juuti, P.S., Katko, T.S., Eds.; International Water History Association (IWHA): Delft, The Netherlands, 2012.
6. Angelakis, A.N.; Koutsoyiannis, D.; Papanikolaou, P. On the geometry of the Minoan water conduits. In Proceedings of the 3rd IWA International Symposium on Water and Wastewater Technologies in Ancient Civilizations, Istanbul, Turkey, 22–25 March 2012; pp. 172–177.
7. Evans, S.A. *The Palace of Minos at Knossos: A Comparative Account of the Successive Stages of the Early Cretan civilization as Illustrated by the Discoveries, (Vol. I–IV)*; Macmillan and Co.: London, UK, pp. 1921–1935.
8. Angelakis, A.N. The History of Fountains and Relevant Structures in Crete, Hellas. *J. Glob. Environ. Issues* **2015**, in Press.
9. Angelakis, A.N.; Spyridakis, S.V. The status of water resources in Minoan times: A preliminary study. In *Diachronic Climatic Impacts on Water Resources with Emphasis on Mediterranean Region*; Angelakis, A.N., Issar, A.S., Eds.; Springer-Verlag: Heidelberg, Germany, 1996; Chapter 8, pp. 161–191.
10. Platon, E. Minoan terracotta water spouts. In Proceedings of the 7th International Congress of Cretan Studies, Rethymnon, Greece, 25–31 August 1991; pp. 767–775. (In Greek)
11. Angelakis, A.N.; Dialynas, M.G.; Despotakis, V. Evolution of water supply technologies in Crete, Greecethrough the centuries. In *Evolution of Water Supply throughout Millennia*; Angelakis, A.N., Mays, L.W., Koutsoyiannis, D., Mamassis, N., Eds.; IWA Publishing: London, UK, 2012; Chapter 9, pp. 227–258.
12. Angelakis, A.N.; Kavoulaki, E.; Dialynas, M.G. Sanitation and Stormwater and Wastewater Technologies in MinoanEra. In *Evolution of Sanitation and Wastewater Management through the Centuries*; Angelakis, A.N., Rose, J., Eds.; IWA Publishing: London, UK, 2013; Chapter 1, pp. 1–24.
13. Floods, J. Water Management in Neopalatial Crete and the Development of the Mediterranean climate. Master's Thesis, The University of North Carolina, Greensboro, NC, USA, 2012.
14. Betancourt, P. *Pseira IX: The Archaeological Survey of Pseira Island Part 2*; INSTAP Academic Press: Philadelphia, PA, USA, 2005.
15. Hirst, B.R. *Fountains*; Kingston University: London, UK, 1996.
16. Hodge, A.T. *Roman Aqueducts & Water Supply*, 2nd ed.; Gerald Duckworth: London, UK, 2002.
17. Castellani, V.; Dragoni, W. Opere Arcaiche per il controllo del territorio: Gli emissari sotterranei artificiali dei laghi Albani. In *Gli Etruschi Maestri di Idraulica*; Bergamini, M., Ed.; ELECTA: Venezia, Italy, 1991; pp. 43–60. (In Italian)

18. Bersani, P.; Canalini, A.; Dragoni, W. First results of a study of the Etruscan tunnel and other hydraulic works on the "Ponte Coperto" stream (Cerveteri, Rome, Italy). *Water Sci. Technol.* **2010**, *10*, 561–569
19. Howard Hayes, S. *Etruscan Cities & Rome*; Cornell University Press: Ithaca, NY, USA, 1967; p. 320.
20. Maggiani, A. Le fontane nei santuari d'Etruria. In *Il Fanum Voltumnae e i Santuari Comunitari Dell'Italia Antica*; Fina, G.D., Ed.; Edizioni Quasar: Roma, Italy, 2012; Volume XIX, pp. 265–292. (In Italian)
21. Inghirami, F. *Pitture di Vasi Fittili, Volume 2*; Nabu Press: Fiesole, Italy, 1833. (In Italian)
22. Agguato di Achille a Troilo. Available online: http://it.wikipedia.org/wiki/Tomba_dei_Tori#mediaviewer/File:Etruscan_mural_achilles_Troilus.gif (accessed on 21 March 2015).
23. Jiao, J.J. A 5600-year-old wooden well in Zhejiang Province, China. *Hydrogeol. J.* **2007**, *15*, 1021–1029.
24. The Old Han Well Found in the Worksite. *Yeng Zhao Night Daily*, 30 April 2010.
25. Liu, S. A Exploring on Form of Chinese Old Well. *Agric. Archaeol.* **1991**. (In Chinese)
26. Polyxene Louvre. Available online: http://en.wikipedia.org/wiki/File:Polyxene_Louvre_F366.jpg (accessed on 21 August 2013).
27. Elderkin, G.W. The Fountain of Glauce at Corinth. *Am. J. Archaeol.* **1910**, *14*, 19–50.
28. WaterHistory Home Page. Available online: http://www.waterhistory.org/ (accessed on 21 August 2013).
29. Hellner, N.H. *Κρήνη του Θεαγένους στα Μέγαρα*; Nastou: Athens, Greece, 2009. (In Greek)
30. Antoniou, G.P. Ancient Greek Lavatories: Operation with Reused Water. In *Ancient Water Technologies*; Mays, L.W., Ed.; Springer: Dordrecht, The Netherlands, 2010.
31. Thompson, H.; Wycherley, R.E. *The Athenian Agora, The Agora of Athens: The History*; Shape and Uses of an Ancient City Center: Athens, Greece, 1972; pp. 194–203.
32. Castalian Spring. Available online: http://wikipedia.org/wiki/Castalian_Spring (accessed on 21 March 2015).
33. Frontin. *Les Aqueducs de la Ville de Rome, Translation and Commentary by Pierre Grimal*; Société d'édition Les Belles Lettres: Paris, France, 1944. (In France)
34. Pace, P. *Gli Acquedotti di Roma*; Art Studio S. Eligio: Roma, Italy, 1983; p. 330. (In Italian)
35. Dragoni, W. Rome's Fountains: Beauty and Public Service from Geology, Power and Technology. In *Water Fountains in the Worldscape*; Hynynen, A.J., Juuti, P.S., Katko, T.S., Eds.; IWHA: Delft, The Netherlands, 2012; pp. 18–33.
36. Castellani, V.; Dragoni, W. Opere Idrauliche Ipogee nel Mondo Romano: Origine, Sviluppo e Impatto nel Territorio. *L'UNIVERSO* **1989**, *LXIX*, 100–137. (In Italian)
37. Pounds, N.J.G. *An Historical Geography of Europe 450 B.C.—A, Parte 1330*; Cambridge University Press: Cambridge, UK, 1973; p. 475.
38. Keenan, D. Water supply and wastewater disposal in Pompeii: An Overview. *Anc. Hist.* **2004**, *34*, 149–158.
39. Jansen, G.C.M. Urban Water Transport and Distribution. In *Handbook of Ancient Water Technology*; Wikander, Ö., Ed.; BRILL: Leiden, The Netherlands, 2000; p. 741.

40. De Feo, G.; Laureano, P.; Drusiani, R.; Angelakis, A.N. Water and wastewater management technologies through the centuries. *Water Sci. Technol. Water Supply* **2010**, *10*, 337–349.
41. De Feo, G.; Mays, L.W.; Angelakis, A.N. Water and Wastewater Management Technologies in the Ancient Greek and Roman Civilizations. In *Treatise on Water Science*; Wilderer, P., Ed.; Academic Press: Oxford, UK, 2011; Volume 4, pp. 3–22.
42. De Feo, G.; de Gisi, S.; Malvano, C.; Tortora, A.; del Prete, S.; Maurano, F.; Tropeano, E. The Roman aqueduct and the Helvius' Fountain in Sant'Egidiodel Monte Albino, in Southern Italy: A historical and morphological approach. In Proceedings of the 2nd IWA International Symposium on Water and Wastewater Technologies in Ancient Technologies, Bari, Italy, 28–29 May 2009.
43. Pompeii. Fountain outside VII.1.32 and VII.1.33. Excavated 1846. Available online: http://www.pompeiiinpictures.com/pompeiiinpictures/Fountains/Fountain%2070132.htm (accessed on 21 March 2015).
44. Jashemski, W.F. The Use of Water in Pompeian Gardens. In Proceedings of the 9th History of Water Management and Hydraulic Engineering in the Mediterranean Region, Pompeii, Italy, 1–8 October 1994; pp. 51–57.
45. Prakoso, S. The Laurentine Villa and the Tuscan Villa of Pliny the Younger. *Jurnal Ilmiah Arsitektur UPH* **2006**, *3*, 130–142. (In Bahasa)
46. Mari, Z.; Sgalambro, S. The Antinoeion of Hadrian's Villa: Interpretation and Architectural Reconstruction. *Am. J. Archaeol.* **2007**, *111*, 83–104.
47. Lombardi, L.; Corazza, A. L'acqua e la città in epoca antica. In *"La Geologia di Roma, dal Centro Storico alla Periferia", Part I Memorie Serv. Geol. d'Italia*; S.E.L.C.A: Firenze, Italy, 2008; Volume LXXX, pp. 189–219. (In Italian)
48. Lombardi, L. Camillo Agrippa's Hydraulic Inventions on the Pincian Hill (1574–1578). Available online: http://www3.iath.virginia.edu/waters/Journal5LombardiNew.pdf (accessed on 21 March 2015).
49. St. John Hope, W.H.; Fowler, C.J.T. Recent discoveries in the Cloister of Durham Abbey. *Archaeologia* **1903**, *58*, 437–460.
50. Lavatorium. Available online: http://www.paradoxplace.com/Photo%20Pages/Spain/Navarre_Aragon_Catalonia/Catalonia/Poblet/Images/800/Lavatorium2-Jun06-D9807sAR900.jpg (accessed on 29 April 2015).
51. Lavatorium, Gloucester Cathedral Cloisters. Available online: http://commons.wikimedia.org/wiki/File:Lavatorium,_Gloucester_Cathedral_Cloisters_-_geograph.org.uk_-_876615.jpg#file (accessed on 29 April 2015).
52. Andersen, T.B.; Jensen, P.S.; Skovsgaard, C.S. *The Heavy Plough and the Agricultural Revolutionin Medieval Europe*; Discussion Papers on Business and Economics, No. 6/2013; University of Southern Denmark: Copenhagen, Denmark, 2013; p. 69.
53. Dragoni, W. Some considerations on climatic changes, water resources and water needs in the Italian region south of the 43°N. In *Water, Environment and Society in Times of Climatic Change*; Issar, A., Brown, N., Eds.; Kluwer: Alphen aan den Rijn, The Netherlands, 1998; pp. 241–271.
54. Cavallucci, F. *La Fontana Maggiore di Perugia: Voci e Suggestioni di Una Comunità Medievale*; Quattroemme: San Giovanni, Italy, 1993; p. 237. (In Italian)

55. Soletti, A. *L'acquedotto medievale di Perugia. Quaderno 5—Laboratorio di Disegno Automatico, Università di Perugia*; Soletti, A., Ed.; Istituto di Disegno Architettura Urbanistica: Perugia, Italy, 1992; p. 94. (In Italian)
56. D'Onofrio, C. *Le Fontane di Roma, con Documenti e Disegni inediti*; Staderini Editore: Roma, Italy, 1962; p. 309. (In Italian)
57. Cope, F. *Fontane di Roma*; Rizzoli Libri Illustrati, GruppoSkira: Ginevra, Switzerland, 2004; p. 207. (In Italian)
58. Rinne, K.W. *The Waters of Rome: Aqueducts, Fountains, and the Birth of the Baroque City*; Yale University Press: New Haven, CT, USA, 2011; p. 262.
59. Islamonline Homepage. Available online: www.islamonline.net (accessed on 21 March 2015).
60. Encyclopaedia Britannica Online. Tunis: Encyclopædia Britannica. Available online: http://www.britannica.com/EBchecked/topic/609220/Tunis (accessed on 21 August 2013).
61. Yves-Marie, A.; Christiany, J. *L'Art des Jardins en Europe*; Citadelles & Mazenod: Paris, France, 2006. (In France)
62. Kanetaki, E. *Ottoman Baths in the Greek Territory*; Technical Chamber of Greece: Athens, Greece, 2004. (In Greek).
63. *The Cretan Architecture the 18th and 19th Centuries*; Chryssoula Tzobanaki: Iraklion, Greece, 2005; Volume A1. (In Greek).
64. Fountains of the Islamic World. Available online: http://www.watercharm-cn.com/plus/view.php?aid=165 (accessed on 21 March 2015).
65. Zheng, X.Y. Meeting Place of Chinese Culture and Water: The Case of the Nine-Dragon Fountain of Yuxi City, China. In *Water Fountains in the Worldscape*; IWHA: Delft, The Netherlands, 2012; pp. 36–43.
66. Hynynen, A.J.; Juuti, P.S.; Katko, T.S. *Water Fountains in the Cityscape*; Hynynen, A.J., Juuti, P.S., Katko, T.S., Eds.; American Public Works Association: Kansas City, MO, USA, 2011.
67. Trafalgar Square. Available online: http://www.destination360.com/europe/uk/trafalgar-square.php (accessed on 10 April 2013).
68. Handbook of Water Kiosks from the Site of Federutility. Available online: http://www.federutility.it/openAttachment.aspx?IDFILE=04a49bf6-60a4-4467-b659-fb0d7f4c2038 (accessed on 22 April 2015). (In Italian)
69. Bond, T.; Roma, E.; Foxon, K.M.; Templeton, M.R.; Buckley, C.A. Ancient water and sanitation systems—Applicability for the contemporary urban developing world. *Water Sci. Technol.* **2013**, *67*, 935–941.

11

Comparative Analysis of Spatial Interpolation Methods in the Mediterranean Area: Application to Temperature in Sicily

Annalisa Di Piazza [1,*], Francesco Lo Conti [2], Francesco Viola [2], Emanuele Eccel [3] and Leonardo V. Noto [2]

[1] National Research Council ISSIA—Institute of Intelligent Systems for Automation—UOS Palermo, Via Dante Alighieri 92, Palermo 90141, Italy

[2] Department of Civil Engineering, Environmental, Aerospace, of Materials University of Palermo, University of Palermo, Viale delle Scienze, Building 8, Palermo 90128, Italy; E-Mails: francesco.loconti@unipa.it (F.L.C.); francesco.viola77@unipa.it (F.V.); leonardo.noto@unipa.it (L.V.N.)

[3] Sustainable Agro-ecosystems and Bioresources Department, IASMA Research and Innovation Centre, Fondazione Edmund Mach, Via E. Mach 1, San Michele all'Adige (TN) 38010, Italy; E-Mail: emanuele.eccel@fmach.it

* Author to whom correspondence should be addressed; E-Mail: a.dipiazza@pa.issia.cnr.it;

Academic Editor: Miklas Scholz

Abstract: An exhaustive comparison among different spatial interpolation algorithms was carried out in order to derive annual and monthly air temperature maps for Sicily (Italy). Deterministic, data-driven and geostatistics algorithms were used, in some cases adding the elevation information and other physiographic variables to improve the performance of interpolation techniques and the reconstruction of the air temperature field. The dataset is given by air temperature data coming from 84 stations spread around the island of Sicily. The interpolation algorithms were optimized by using a subset of the available dataset, while the remaining subset was used to validate the results in terms of the accuracy and bias of the estimates. Validation results indicate that univariate methods, which neglect the information from physiographic variables, significantly entail the largest errors, while performances improve when such parameters are taken into account. The best results at the annual scale have been obtained using the the ordinary kriging of residuals from linear regression and from the artificial neural network algorithm, while, at the monthly scale, a Fourier-series algorithm has been used to downscale mean annual temperature to reproduce monthly values in the annual cycle.

Keywords: geostatistics; Sicily; spatial interpolation; temperature

1. Introduction

Climatic data are used for scientific and technical purposes in many different areas of environmental research and resource management [1], hydrology and agriculture [2].

Climate series are often limited in spatial and temporal coverage. The management and maintenance of gauge stations to obtain reliable data can be very expensive and problematic. In order to tackle this problems, it is useful to develop spatial interpolation techniques to assess climatic variables at ungauged sites. In general, spatial interpolation techniques use observations of the same variable made at other sites. The methods used to estimate precipitation and temperature at ungauged sites can be grouped into: (I) weighting methods which, in turn, can be classified into deterministic interpolation methods (inverse distance weighting, non-linear interpolation as spline techniques, *etc.*) and statistic interpolation methods (different varieties of kriging); and (II) data-driven methods (regression, artificial neural networks, *etc.*) [3].

Even though spatial interpolation techniques have been largely analyzed and compared, an optimum spatial interpolation algorithm is difficult to identify. Indeed, the suitability of a spatial interpolator is strongly related to the specific variable considered, the specific context and the spatial scale desired for mapping [4–7]. Moreover, other factors may also influence the choice of the interpolation method and the accuracy of the results. Among these, station density [8], the duration and the nature of the climate variable [9] (e.g., temperature and sea level pressure are continuous in both time and space, whereas precipitation fields are spatially discontinuous on shorter timescales and more continuous on longer timescales), and geographical factors, such as elevation [10], aspect and distance to the coast [11], are probably the most important. The comparison among algorithms based on deterministic and data-driven methods or methods that integrate stochastic and regressive approaches has usually demonstrated the superiority of the latter two approaches, making them a good solution to the spatial interpolation of climatological variables [12–14].

Among data-driven algorithms, the regression methods can be considered as direct and explicit procedures. These methods use ancillary geographical information as independent variables to improve the estimation of the spatial distribution of the considered climate variables. Several studies present GIS-based spatial interpolation algorithms in order to derive tools for the analysis and the characterization of the spatial structure of precipitation for different applications [15]. Claps *et al.* [16] applied regression analysis to annual and monthly average temperatures to characterize the temperature regime in Italy. Data from 738 weather stations were used to estimate the temperature normals. In particular geo-regressive methods were used, taking into account geographic and morphological parameters, and through a stepwise regression analysis, the significant variables were detected. The temperature regime was reproduced with a two-harmonic Fourier series, leading to satisfactory results for large-scale climatic characterization.

Another data-driven algorithm widely used to derive the spatial distribution of climatic data is the artificial neural network (ANN) [17,18]. Among the applications of this approach, Rigol *et al.* [19] used a feed-forward back-propagation network method for the interpolation of geographical data that incorporated both trend and spatial association and applied the method to the daily minimum temperatures in England and Wales. Coulibaly and Evora [20] performed a comparison of six different types of ANN techniques for assessing precipitation and temperature at ungauged sites. The evaluation of the accuracy of the different models, carried out using daily series from 15 validation weather stations, highlighted that multi-layer perceptron (MLP) was the most effective method.

Several geostatistical methods have been used for assessing the performances of the spatial processing of precipitation fields in several works focused on this issue, e.g., [21–23]. As highlighted by Haining *et al.* [24], geostatistical methods can be used to perform the characterization of the spatial structure of the process considered and provide useful insights into the uncertainty associated with the interpolation analysis. Hudson and Wackernagel [25] analyzed the spatial structure of mean temperature using variograms computed for different directions. Since this anisotropic variogram for universal kriging does not fully explain the spatial variation in temperature, the authors exploited the correlation with elevation, adopting the kriging with elevation as external drift, obtaining better results.

Geostatistics were also used together with ANNs. In particular, Demyanov *et al.* [26] proposed a two-step spatial interpolation method named direct neural network residual kriging: the first step is a data-driven approach, which includes estimating the large-scale spatial structure by an ANN, while the second step is the analysis of residuals carried out by using a geostatistical method; final estimates are produced as a sum of ANN estimates and ordinary kriging estimates of residuals. A similar approach was followed by Antonic and Krizan [27], who built empirical models based on neural networks for seven climatic variables using 127 weather stations; the residuals coming from the ANN were spatially interpolated by kriging and used as a model correction.

In order to obtain yearly averaged temperature maps providing complete and continuous information in the territory, several algorithms of spatial interpolation are presented and applied in this study to annual and monthly average temperature data of Sicily (Italy) measured at 84 sites. Considering the relevant role of elevation in estimating temperature, e.g., [28,29], a set of specific data-driven and regressive approaches have been chosen. In these methods, the elevation information, as well as other physiographic information, provided by a digital elevation model (DEM), were added to improve the estimation of missing data. Moreover, simple univariate methods have been considered in order to compare their performances with those of more complex methodologies, evaluating the improvements. These different models were first calibrated using a modeling subset and then compared through a validation procedure. Finally, the best method was used to derive maps of annual and monthly average temperature, aiming at the preparation of a climate atlas for Sicily.

2. Methods and Study Area

The selection of methods has been performed in order to exploit the range of interpolators features and potentialities. Following the considerations reported above, methods have been classified among univariate methods and variable aided methods. Among univariate methods, the inverse distance weighting (IDW) has been selected as the most simple spatial interpolation procedure. The radial basis functions (RBF) provide a general mathematical tool that is able to identify a continuous surface representing the variable behavior. Ordinary kriging (OK) has been selected as the method for the exploitation of the spatial variability, as can be observed by means of specific tools, such as the semivariogram. The evaluation of variable-aided interpolators has been first explored with simple linear regression models. Later, more complex methods have been tested. In particular, the geographically-weighted regression has been applied in order to consider the spatial variability of the regressive relationship, and the ANN has been chosen because it is able to model the non-linear relationship between variables. Finally, residual kriging has been explored across different base variable-aided methods in order to evaluate the effects of further modeling the residual variability by means of other methods. Analyses have been performed using ESRI ArcGIS and MATLAB software.

2.1. Climatic and Geographic Features of the Area

This study concerned the largest island in the Mediterranean Sea, Sicily (Italy), extending over an area of 25,700 km^2. The ancillary information regarding elevation was embedded through the use of a DEM of the region having a resolution of 100 m (Figure 1). The surface of the island is mainly characterized by a hilly morphology (62%) and, for the rest, by a mountainous and flat morphology among which the widest is around Catania, in the eastern part of the island. The overall island average elevation is around 400 m, but the range of variation of elevation is between 0 and 3263 m (the Etna volcano, the highest peak of Sicily).

Sicily is a region with a temperate-mesothermal (Mediterranean) climate, with a dry summer, having an average temperature in the hottest month greater than 22 °C, with a precipitation regime more intense in the coldest season.

Some physiographic parameters were also considered in this study in order to evaluate their relationship with temperature measurements. Following the criteria of Prudhomme and Reed [30], such physiographic and geographic factors, chosen taking into account the temperature pattern over the area, are the minimum absolute distance from the sea, d_{min}, the angle α formed by the minimum distance vector and the south, the distance d_i from the sea in the eight cardinal directions i and the azimuthal angle β_i of the horizon in the eight cardinal directions.

The latter variable is the obstruction factor [31], defined as the angle subtended by the highest topographical barrier in the *i*-th direction, such that $tan(\beta_i) = \Delta H/\Delta X$, with ΔH as the difference of elevation between the barrier and the station, and ΔX as the distance between the two points. These variables have been derived using spatial analysis techniques in a GIS environment, using a vector representation of the Sicilian coastal line and the DEM. These variables were used to produce the three following parameters:

- the distance measure, D_s, that is the geometric mean of the distance from the sea in the eight cardinal directions:

$$D_s = \sqrt[8]{d_1 * d_2 * \ldots * d_8} \tag{1}$$

- the aspect variable, A_s, a combined measure of aspect orientation and sea proximity:

$$A_s = \frac{10\, cos\alpha}{1 + d_{min}} \tag{2}$$

- the concavity index, C, obtained by weighting the azimuthal angle iin the eight directions:

$$C = \sqrt[8]{\prod_{i=1}^{8} 10^{2\, tan\, \beta_i}} \tag{3}$$

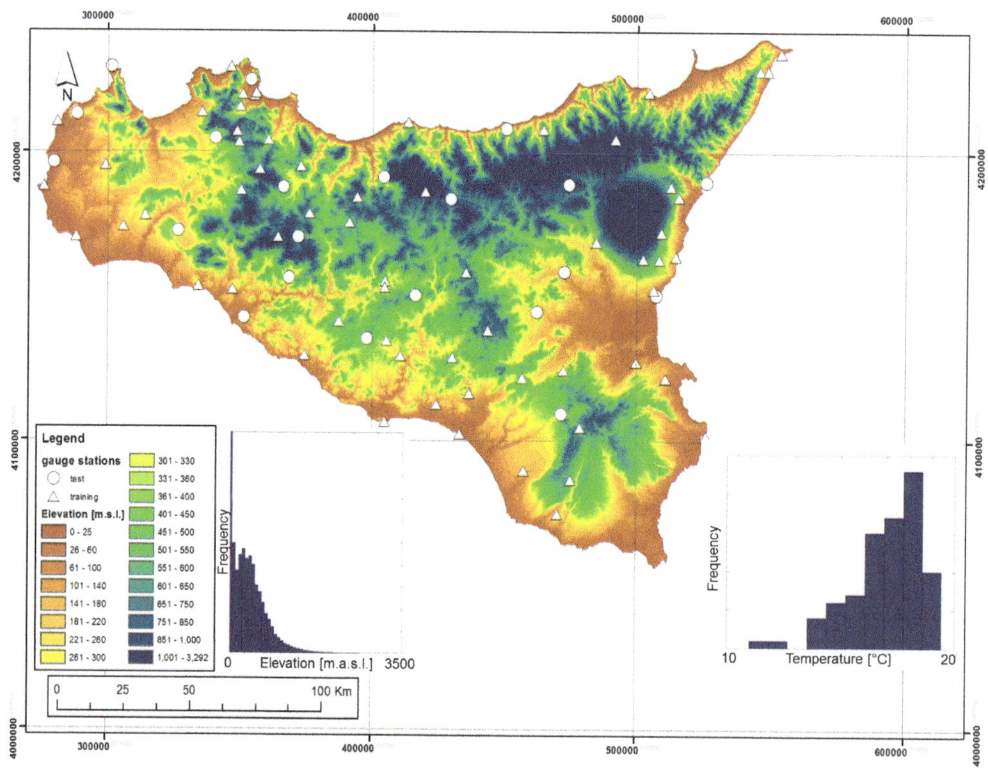

Figure 1. One hundred-meter DEM of Sicily; elevation histograms, based on DEM, and average mean temperature histogram, based on station records, are reported next to the legend and on the bottom-right, respectively.

2.2. Climatic Dataset

The temperature dataset used in this study was provided by OA-ARRA (Osservatorio delle Acque-Agenzia Regionale dei Rifiuti e delle Acque) and comes from 84 weather stations with observations recorded from 1924 to 2006. The entire station dataset was divided into two subsets: (I) the modeling subset, used to calibrate the different interpolation methods, containing about 70% of the stations ($N_t = 59$); and (II) the validation subset, used to validate the calibration results, containing the remaining 30% of the stations ($N_v = 25$). The partition of the dataset into two subsets was done by trying to maintain the distribution of sites, both in the horizontal domain and in the altitudinal distribution.

In order to provide the estimate $\hat{z}$ of the temperature variable z, both annual or monthly, at an ungauged location $\boldsymbol{x}_0$ using temperature data at instrumented sites, denoted with $\{z(\boldsymbol{x}_i), i = 1, 2, ...N\}$, the temperature vector measured at the N sites $\boldsymbol{x}_i$, two different classes of interpolators were used: univariate methods and variable-aided interpolation (VAI) methods. While the former class takes into account only the data and the spatial coordinates $\boldsymbol{x}_i$, the latter also uses supplementary data, such as elevation $q(\boldsymbol{x}_i)$, and other physiographic variables. These methods will be shortly described in the following subsections.

2.3. Univariate Interpolators

The radial basis function [32] represents a large family of exact interpolators similar to those used in geostatistical interpolation (see further), but without the benefit of a prior analysis of spatial correlation. These interpolators, which do not make any assumption on the input data and provide excellent interpolators for a wide range of data, have been used in many disciplines. The simplest variant of this method can be viewed as a weighted linear function of (inverse) distance from grid point to data point, plus a "bias" factor. The temperature at an ungauged site $\boldsymbol{x}_0$ can be estimated by:

$$\hat{z}(\boldsymbol{x}_0) = \sum_{i=1}^{N} \lambda_i \phi(||\boldsymbol{x}_i - \boldsymbol{x}_0||) + \mu \tag{4}$$

where $\phi(\rho)$ is the selected radial basis function, ρ is the radial distance from point $\boldsymbol{x}_0$ to the i-th data point $\boldsymbol{x}_i$ ($\rho = ||\boldsymbol{x}_i - \boldsymbol{x}_0||$) and λ_i are the weights to be estimated together with the bias value μ (or Lagrangian multiplier). These weights are given by the solution of a system of $(n + 1)$ linear equations. The radial basis function used in this work is the thin plate spline, which has the following expression:

$$\phi(\rho) = c^2 \rho^2 ln(c\rho) \tag{5}$$

This interpolation method requires the choice of a minimum number of neighbor points.

The second univariate method taken into account here was the inverse distance weighting (IDW). The temperature is a linear combination of several surrounding observations, while the weights are proportional to the distance between observations and $\boldsymbol{x}_0$:

$$\hat{z}(\boldsymbol{x}_0) = \sum_{i=1}^{N} \lambda_i z(\boldsymbol{x}_i) \tag{6}$$

where the weights λ_i are expressed as a function of distance as follows:

$$\lambda_i = \frac{d_{i0}^{-r}}{\sum_{i=1}^{N} d_{i0}^{-r}} \tag{7}$$

This exact interpolation method requires the choice of the exponent r and of a search radius R or the minimum number of points N.

The third univariate method taken into account, is OK, which belongs to geostatistical interpolators [33,34]. This method uses a particular function, called the semivariogram, instead of simple Euclidean distance functions in order to measure the dissimilarity between observations. The experimental semivariogram is a function of both distance and direction. If anisotropic spatial patterns exist, different semivariograms should be computed along different directions. The semivariogram is used to derive the weights of an unbiased linear interpolator by means of the minimization of the error estimate variance. Thus, the OK can be labeled as a BLUE method (best linear unbiased estimator). The relationship between such weights and the semivariogram results:

$$\overline{\overline{\gamma}} = \overline{\Gamma} \cdot \overline{\lambda} \tag{8}$$

where $\overline{\gamma}$ is the semivariogram vector relative to each sample point and that to which the estimation is referred, $\overline{\overline{\Gamma}}$ is the semivariogram matrix relative to sample points and $\overline{\lambda}$ is the weight vector relative to the point to be estimated. A detailed characterization of this topic can be found, for instance, in Kitanidis [34], Goovaerts [35].

2.4. *Variable-Aided Interpolators*

2.4.1. Linear Models

The most simple method belonging to VAI methods is linear regression (LR). It consists of temperature estimation as a function of the elevation of the instrumented sites $q(\boldsymbol{x}_0)$, in the case of simple linear regression, or as a function of more than one physiographic variable, in the case of multiple linear regression (MLR). The equation used for simple and multiple linear regressions is:

$$\hat{z}(\boldsymbol{x}_0) = f[v_j(\boldsymbol{x}_0)] = \alpha + \sum_{j=1}^{n_v} \beta_j v_j(\boldsymbol{x}_0) \tag{9}$$

where $v_j(\boldsymbol{x}_0)$ are the n_v independent variables for the MLR ($n_v = 1$ simple regression), while the regression coefficients α, β (or β_j), constant over the whole study area, can be estimated, using the least squares method, from the set of temperature and the independent variables data at N ungauged sites $\{z(\boldsymbol{x}_i), v(\boldsymbol{x}_i), i = 1, 2, ..., N\}$.

In order to assess regression coefficients, the standard OLS (ordinary least squares) methodology has been compared with the results provided by the robust regression method (ROB) [36]. Physiographic and geographic parameters used in MLR were the elevation q, the latitude L, the minimum absolute distance from the sea, d_{min}, the angle α formed by the minimum distance vector and the south, the distance d_i from the sea in the eight cardinal directions i, the azimuthal angle β_i of the horizon in the eight cardinal directions and three indices D_s, A_s, C, as previously defined in Equations (1)–(3), respectively.

2.4.2. Geographically Weighted Regression

Geographically Weighted Regression (GWR) [37] considers the variability of coefficients α and β of a regression expression linking temperature to the topographic features allowing these two parameters to change over space using the following relationship:

$$\hat{z}(\boldsymbol{x}_0) = f[q(\boldsymbol{x}_0)] = \alpha(\boldsymbol{x}_0) + \beta(\boldsymbol{x}_0)\, q(\boldsymbol{x}_0) \tag{10}$$

Each coefficient may be thought of as a 3D surface over the geographical study area rather than a single constant value. The estimation of the two coefficients $\alpha(\boldsymbol{x_0})$ and $\beta(\boldsymbol{x_0})$ can be carried out by one of the two following approaches: GWR with fixed spatial kernels and GWR with adaptive spatial kernels. The latter approach, used in this study, is based on a circle of radius *R* drawn around a point to calibrate an ordinary least squares regression model on the basis of the observations that fall within this circle; the coefficients α and β are an estimation of the local coefficients at the point $\boldsymbol{x_0}$. The spatial kernel can be adapted depending on the density of the data [38].

2.4.3. Artificial Neural Networks

ANNs have been extensively applied in the past decade for estimating and forecasting hydrological variables [39,40]. In this work, the MLP method has been applied. An MLP is a feedforward neural network consisting of a number of units (perceptrons or neurons), which are connected by weighted links. The units are organized in layers: an input layer, with a number of neurons defined by the input independent variables, an output layer, with a number of neurons defined by the dependent variable, and, finally, one or more hidden layers. In an MLP, an arbitrary input vector is propagated forward through the network, where the hidden layers of neurons make a linear combination of input signals and convert it through a generally nonlinear function (activation function). Each layer output forms the input to the following layers.

MLP was used like a VAI technique to estimate temperature values at ungauged sites using the elevation information, as well. The implemented MPL consists of three layers: the input layer with three neurons, the output layer with one neuron and one hidden layer.

ANN techniques require a training stage, to learn the dynamics of the causal relationship. In the training phase of an MLP, known input-output pairs are presented to the network, and the weights are updated by following pre-determined learning rules. In our work, the "error backpropagation rule" (EBP) was used.

2.4.4. Use of Residual Spatial Interpolation (Residual Kriging)

The above-mentioned VAI methods (LR, MLR, GWR and ANN) can provide, for each one of the N gauged sites, a residual error defined as the difference between the estimated value $\hat{z_A}(x_i)$ and the observed value $z(x_i)$:

$$r_A(\boldsymbol{x_i}) = \hat{z}_A(\boldsymbol{x_i}) - z(\boldsymbol{x_i}) \qquad with\ A = LR,\ MLR,\ GWR,\ ANN \qquad and\ i = 1,\ 2, \ldots,\ N \quad (11)$$

The spatial correlation of residuals $r_A(\boldsymbol{x_i})$ can be taken into account using a geostatistical algorithm denoted as residual kriging (RK). In this case, the four methods estimate large-scale structures of temperature data, valid for the whole domain, and then, the interpolation of residuals $r_A(\boldsymbol{x_i})$, carried out with the OK method, produces a map of $r_A(\boldsymbol{x_i})$ representing the corrections to apply to each one of the three models. Final estimates at an ungauged site $\boldsymbol{x_0}$ are produced as a sum of the VAI method estimates and ordinary kriging estimates of residuals:

$$\hat{z}_{RK-A}(\boldsymbol{x_0}) = \hat{z}_A(\boldsymbol{x_0}) + \hat{r}_A(\boldsymbol{x_0}) \qquad with\ A = LR,\ MLR,\ GWR,\ ANN \qquad (12)$$

This approach has been extensively used for mapping climatic variables [21,30,41].

2.5. Fitting Monthly Temperature by Fourier Series

Another method, based on the seasonal scaling of annual temperature, was implemented in the case of monthly analysis. This method, described in Claps *et al.* [16], allows one to derive the monthly temperature by means of the estimation of the whole parameterized curve of the temperature regime at each point over the domain. With such an approach, it is possible to describe with a single model the causal relationship between physiographic and geographic predictors and the seasonality of temperature in Mediterranean climates. Starting from estimates of the spatial variability of the mean annual temperatures, the monthly deviations from the annual mean can be analyzed, considering the zero-mean temperature regime $t\left(j\right)$ given by this expression:

$$t\left(j\right) = T\left(j\right) - T_a \tag{13}$$

where j denotes the month of the year, $T(j)$ is the average temperature at month j and T_a is the mean annual temperature. The sequence of 12 monthly temperature deviations $t\left(j\right)$ can be well approximated by means of sinusoidal curves obtained by Fourier series using the following expression:

$$t\left(j\right) = A_0 + \sum_{i=1}^{N} A_i cos\left(\frac{2\pi}{\tau_i}j + \frac{\tau}{\tau_i}\phi_i\right) \tag{14}$$

where A_0 is the mean of $t\left(j\right)$ (here, equal to zero), τ, here equal to 12, is the fundamental period of the cycle, N is the number of the harmonics, A_i is the amplitude, ϕ_i is the phase and τ_i is the period of the i-th harmonic. For the estimation of the Fourier series parameters, the cosine argument in Equation (14) can be decomposed to a polynomial form:

$$t\left(j\right) = A_0 + \sum_{i=1}^{N} \left[b_i * cos\left(\frac{2\pi}{12}j\right) + c_i * sin\left(\frac{2\pi n_i}{12}j\right)\right] \tag{15}$$

where $b_i = A_i cos\left(n_i\phi_i\right)$; $c_i = -A_i cos\left(n_i\phi_i\right)$, and $n_i = \frac{\tau}{\tau_i}$. The amplitude and phase of the i-th harmonic can then be obtained as:

$$A_i = \frac{b_i}{cos\left(n_i\phi_i\right)} \tag{16}$$

$$\phi_i = arctan\left(-\frac{c_i}{b_i}\right) \tag{17}$$

2.6. Evaluation of Different Algorithms

For a general comparison among methods, it was not possible to use the geostatistical cross-validation [33] because of the presence of global interpolation methods, such as LR and GWR. Therefore, the comparison criteria were based on the following different indices used to measure the strength of the statistical relationship between estimated $\hat{z}\left(\boldsymbol{x_i}\right)$ and measured $z\left(\boldsymbol{x_i}\right)$ temperatures in the N_v points of the validation subset.

(1) The RMSE (root mean square error of the prediction), which measures the average square difference between the true temperature and its estimate at the validation points:

$$RMSE = \sqrt{\frac{1}{N_v}\sum_{i=1}^{N_v}\left[\hat{z}\left(\boldsymbol{x_i}\right) - z\left(\boldsymbol{x_i}\right)\right]^2} \tag{18}$$

(2) The MBE (mean bias error) or simply bias:

$$MBE = \frac{1}{N_v}\sum_{i=1}^{N_v}\left[\hat{z}\left(\boldsymbol{x_i}\right) - z\left(\boldsymbol{x_i}\right)\right] \tag{19}$$

(3) The MAE (mean absolute error):

$$MAE = \frac{1}{N_v}\sum_{i=1}^{N_v}\left[\left|\hat{z}\left(\boldsymbol{x_i}\right) - z\left(\boldsymbol{x_i}\right)\right|\right] \tag{20}$$

(4) The CC (linear correlation coefficient):

$$CC = \frac{\sum_{i=1}^{N_v}\left[z\left(\boldsymbol{x_i}\right) - z_m\right]\left[\hat{z}\left(\boldsymbol{x_i}\right) - \hat{z}_m\right]}{\sigma_e\sigma_o N_v} \tag{21}$$

where z_m and $\hat{z}_m$ are the mean of measured and estimated temperature values, respectively, and σ_e and σ_o are the standard deviation of measured and estimated temperature values, respectively.

Another tool used to compare the different methods is the Taylor diagram [42]. This tool is based on a trigonometric similitude between the standard deviation of series (SD), CC and centered pattern root mean square difference (RMSD), and it allows the measuring of estimate performances, in terms of RMSD, given SD and CC values calculated between estimates and measurements for the validation subset.

3. Results

3.1. Univariate Methods

The application of the first univariate method, RBF with a thin plate spline, has been performed by varying the minimum number of points *N*. Table 1 shows the results of the different trials carried out on the validation set to get the best parameter choice for all univariate methods. The best result, in terms of RMSE and CC, is obtained for N = 30, even if the relative MBE value is slightly worse than that obtained for N = 10.

The estimate made by the second method, IDW, depends on the selection of the exponent *r* and the neighborhood search strategy. As can be seen in Table 1, several attempts were done to get the optimal combination of these settings. These trials were carried out by fixing the exponent r and changing the minimum number of points *N* and the search radius *R*. The best performance was achieved for $r = 2$ and $R = 40$ km, both in terms of RMSE and bias. Instead, the best result, in terms of CC, is obtained for $r = 2$ and $N = 5$. The bias of RBF and IDW is negative, pointing out that these univariate methods underestimate the annual temperature at the validation sites.

Table 1. Comparison of goodness of interpolation for the univariate methods applied to mean annual temperature.

Methods				RMSE [°C]	MBE [°C]	MAE [°C]	CC
RBF			$N = 10$	1.3400	−0.1961	1.0211	0.6883
			$N = 12$	1.3556	−0.2205	1.0218	0.6964
			$N = 15$	1.3223	−0.2355	0.9902	0.7146
			$N = 20$	1.3139	−0.2296	0.9752	0.7157
			$N = 25$	1.3023	−0.2231	0.9655	0.7203
			$N = 30$	1.2883	−0.2117	0.9527	0.7250
			$N = 35$	1.2878	−0.2117	0.9528	0.7250
			$N = 40$	1.2878	−0.2117	0.9528	0.7250
IDW	Number of points	$r = 2$	$N = 3$	1.5518	−0.4396	1.1974	0.5490
			$N = 5$	1.5154	−0.3744	1.1578	0.5526
			$N = 10$	1.5440	−0.3173	1.1936	0.4990
			$N = 20$	1.5388	−0.2515	1.2020	0.4840
		$r = 3$	$N = 3$	1.5510	−0.4311	1.1905	0.5497
			$N = 5$	1.5603	−0.3926	1.1893	0.5205
			$N = 10$	1.5557	−0.3582	1.1855	0.5114
			$N = 20$	1.5557	−0.3419	1.1858	0.5055
		$r = 5$	$N = 3$	1.6190	−0.5010	1.2637	0.5293
			$N = 5$	1.6188	−0.4891	1.2680	0.5212
			$N = 10$	1.6161	−0.4828	1.2667	0.5202
			$N = 20$	1.6160	−0.4816	1.2669	0.5196
	Radius search	$r = 2$	$R = 20$ Km	1.7914	−0.5241	1.3502	0.4114
			$R = 40$ Km	1.5046	−0.3018	1.1520	0.5348
			$R = 60$ Km	1.5546	−0.2492	1.2088	0.4737
			$R = 100$ Km	1.5425	−0.1678	1.2127	0.4639
		$r = 3$	$R = 20$ Km	1.7874	−0.5299	1.3585	0.4204
			$R = 40$ Km	1.5331	−0.3893	1.1610	0.5452
			$R = 60$ Km	1.5538	−0.3586	1.1778	0.5158
			$R = 100$ Km	1.5501	−0.3272	1.1822	0.5056
		$r = 5$	$R = 20$ Km	1.7843	−0.5399	1.3718	0.4286
			$R = 40$ Km	1.6096	−0.4891	1.2580	0.5299
			$R = 60$ Km	1.6142	−0.4818	1.2638	0.5221
			$R = 100$ Km	1.6153	−0.4805	1.2662	0.5197
OK				1.4990	−0.2453	1.1050	0.5343

For the OK method, the possible presence of anisotropic patterns was explored, despite the limited number of available stations. Indeed, the experimental semivariogram is a function of both distance and direction, so it can account for anisotropic patterns revealed by the data. In particular, it was observed that the sill varies along different directions, determining the phenomenon known as zonal anisotropy [35]. The zonal direction, *i.e.*, the direction of maximum variability, was found at 20 degrees measured clockwise from the north (Figure 2).

The theoretical zonal anisotropic semivariogram was operationally modeled by summing an isotropic semivariogram, given by the first term, to a geometrical anisotropic semivariogram with a small anisotropy ratio (according to the procedure suggested for the GSTAT software; see Pebesma and Wesseling [43]). With regard to the average annual temperature data, the theoretical semivariogram, calibrated with a manual procedure, assumes the following form:

$$\gamma = 1.3\,Sph(25000) + 1.7\,Gau(2.5 \times 10^4,\, 110,\, 0.0001) \tag{22}$$

In particular, 1.25 °C^2 is the partial sill of the theoretical semivariogram at the principal direction (called c_1, in GSTAT nomenclature); 1.75 °C^2 is the value that was summed to c_1 to obtain partial sill in the zonal direction (called c_2); $25,000$ m is the maximum range of the variogram model (called a_1), 1.25×10^4 m is the maximum range amplified (called a_2); 110 degrees is the angle for the principal direction (called p); 0.0001 is the range ratio (*i.e.*, the ratio between the minor range and the maximum range, called s). It has been decided not to consider the nugget effect, because information about possible sensor errors was not available. Furthermore, data have been preprocessed and possible data anomalies checked. In order to model possible high variability in the short range, the spherical model (Sph), almost linear at its origin, has been selected. The additive model used for the anisotropic formulation is the Gaussian model (Gau), varying very smoothly at the origin and allowing for a gradual development with respect to the principal direction semivariogram.

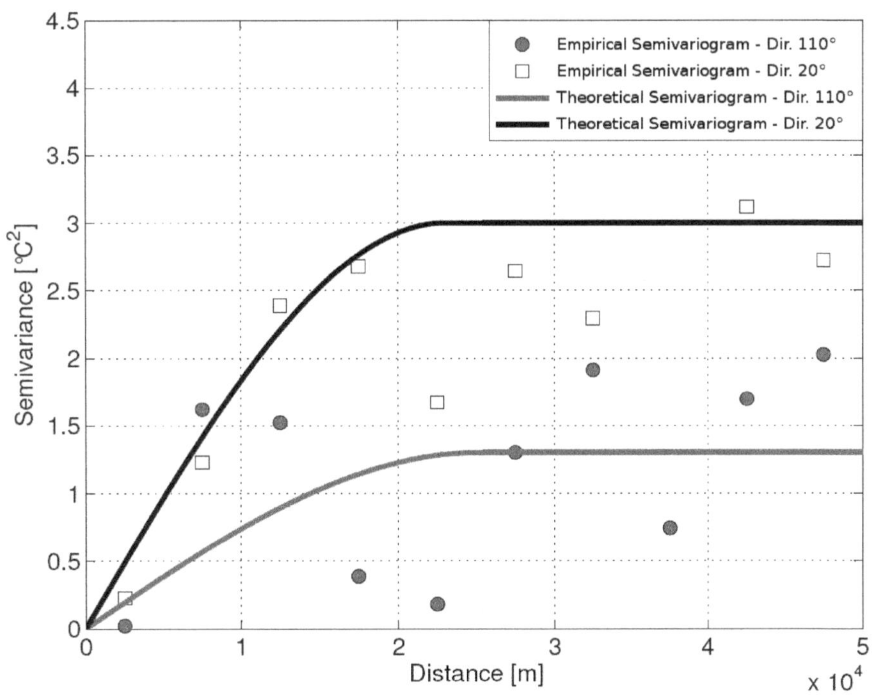

Figure 2. Empirical and theoretical semivariograms of mean annual temperature in the zonal direction (20°) and in the principal direction (110°).

Performance indices (RMSE, MBE and CC) of OK are comparable to those obtained with the best IDW trial, but worse than those obtained with RBF, which can be definitely considered the best method, among the univariate, in terms of accuracy, bias and correlation for annual temperature field.

3.2. Variable Aided Methods

The results of the application of VAI algorithms for the spatial interpolation are reported in Table 2. The ordinary least squares (OLS) and the robust (ROB) regression methods were used for regression coefficient estimation. The first of these provides the best results in terms of both MBE and RMSE, but the differences in terms of accuracy and unbiasedness between the two methods are negligible, suggesting the absence of outliers in the temperature dataset.

Table 2. Comparison of indices of interpolation goodness of the variable-aided (VAI) interpolation methods applied to mean annual temperature.

Methods			RMSE [°C]	MBE [°C]	MAE [°C]	CC
LR	OLS	z = f(q)	0.4886	–0.0014	0.376	0.9598
		log(z) = f(q)	0.5369	–0.0039	0.4196	0.9445
		sqrt(z) = f(q)	0.5113	–0.0028	0.3954	0.9535
	ROB	z = f(q)	0.4909	–0.0098	0.3784	0.9598
		log(z) = f(q)	0.5423	–0.0058	0.4218	0.9452
		sqrt(z) = f(q)	0.5165	–0.0012	0.4	0.9537
MLR	SW	$Ts = 18.68 - 0.0057\ q - 8 \times 10^{-6}\ Ds$	0.501	0.0159	0.3617	0.9572
GWR		$N = 5$	0.7363	0.1176	0.5366	0.9113
		$N = 10$	0.5376	–0.0064	0.4081	0.9515
		$N = 15$	0.5585	–0.0026	0.4509	0.9469
		$N = 20$	0.4996	0.028	0.3939	0.9588
		$N = 25$	0.5131	0.0298	0.4132	0.9562
		$N = 30$	0.5176	0.0762	0.4179	0.9592
		$N = 35$	0.5068	0.0688	0.4034	0.9595
ANN		MLP 3 - 3 - 1	0.4464	–0.0153	0.3478	0.9663
		MLP 3 - 4 - 1	0.4193	–0.0158	0.3161	0.9703
		MLP 3 - 6 - 1	0.4849	–0.0186	0.3455	0.9607
		MLP 3 - 8 - 1	0.51	–0.005	0.3689	0.9556
		MLP 3 - 10 - 1	0.5833	–0.0816	0.4592	0.9435
		MLP 3 - 15 - 1	0.6304	–0.127	0.4984	0.9348
RK		GWR: $N = 20$	0.3592	–0.0582	0.274	0.9692
		LR-OLS; z = f(q)	0.33	–0.0124	0.1098	0.9797
		ANN-MLP 3 - 4 - 1	0.3655	0.015	0.2505	0.9779
		ML	0.347	0.0171	0.2677	0.9676

The second VAI method (GWR) was applied adopting a searching strategy based on the number of closest points to be considered in the estimate. The best performance, in terms of accuracy and correlation, was obtained for N greater than 20, while, in terms of bias, for N equal to 15. In general, the best results of GWR are comparable to those of LR.

The third VAI method is the multiple linear regression (MLR), which estimates temperature by exploiting its dependence on geographic and physiographic features. Once the indices listed in Section 2.4 were derived, a stepwise regression procedure was applied to the training set to build the model that links the mean annual temperature to the morphological parameters. This involves the generation of multiple regression models of increasing complexity, according to the number of independent variables considered. These models have been compared and evaluated using the coefficient of determination R^2, the RMSE and the test on coefficients significance of *T stat*, given by the ratio between R^2 and the standard error. The best results in terms of RMSE and R^2 were obtained by the use of q (elevation), L and D_s as independent variables, but the value of *T stat* for L suggests neglecting the latter variable. Finally, the best model for mean annual temperature is only based on elevation q (m a.s.l.) and the geometric mean of the distance from the sea in the eight cardinal directions D_s (m), as follows:

$$\hat{T}_a = 18.68 - 0.0057\,q + 8 \times 10^{-6} D_s \tag{23}$$

The performance indices provided by this method are satisfactory, both in terms of RMSE (lower than for GWR and LR) and for MBE. Moreover, the CC value, equal to 0.96, confirms an excellent agreement between estimates and observations.

3.2.1. Artificial Neural Networks

The results of the application of ANNs are reported together with those of VAI methods in Table 2. The ANN was trained by the scaled conjugate gradient algorithm, and the initial weights were randomly extracted from a standardized normal distribution. The learning rate decreased exponentially from 0.01 to 0.0001. Six network architectures, having different numbers of neurons in the hidden layer, were compared, and on the basis of RMSE and MBE performances, the best resulted in being a network with one hidden layer, having four neurons with a hyperbolic tangent activation function (MLP 3-4-1). In terms of RMSE, the ANN performances overcame those of LR, MLR and GWR, while the MBE value was slightly worse than that provided by MLP 3-8-1. Moreover, the correlation coefficient between estimated and measured data for the validation set was equal to 0.97, confirming an excellent agreement within the validation dataset. This confirms the ANN as a suitable algorithm to estimate nonlinear processes, such as temperature, as well as other hydroclimatic variables.

3.2.2. Residual Kriging

The correlation coefficient between measured data and residuals provided by LR, equal to −0.44, highlights that not all of the spatial data correlations can be explained by LR methods, and a similar observation can be done for the MLR and MLP with four neurons in the hidden layer, whose correlation coefficient between measured data and residuals is equal to −0.4 and −0.35, respectively. This highlights that the residuals obtained by VAI methods still show a spatial correlation that can be exploited to improve the variable estimation. Starting from these considerations, residual kriging (RK) was used as a further VAI method by applying OK to the residuals derived from LR, MLR, GWR and ANN (MLP 3-4-1). In this case, an isotropic behavior in the residuals coming from the four different methods was observed when deriving the experimental semivariograms. As previously explained, the estimated values of residuals were added to the LR, MLR, GWR and ANN, respectively. In terms of accuracy and correlation, the performances of these methods are better than the univariate method and the other VAI methods (Table 2), while the bias is comparable to that of the other VAI methods. The best method, in terms of RMSE, MBE and CC, appears to be the RK-LR (with OLS).

3.3. Monthly Analysis

The interpolation methods with the best performance, described for the case of average annual temperature data, have been also applied to mean monthly temperature. Particular attention was paid to the estimation of the semivariograms of both the average monthly data (*i.e.*, OK) and the residuals coming from the use of four VAI and ANN methods (*i.e.*, LR, MLR, GWR and MLP 3-4-1). In this regard, the presence of a zonal anisotropy was observed in mean monthly data for all months with the exception of the summer months (JJA). Moreover, for the mean monthly temperature, another method has been taken into account: the Fourier series with the estimation of parameters by an MLR method. In particular, the first attempt to reconstruct the zero-mean temperature regime was made with a one-harmonic (N = 1) Fourier series (F1H), with $\tau_1 = 12$ months. The parameters A_1 and ϕ_1 of the Fourier equation were then correlated with the geographical and physiographic descriptors (q (m a.s.l.), L (hexadecimal degrees), D_s (m), A_s (m^{-1}), C), defined before, using the stepwise procedure to check the presence of any statistically significant relationship. Results of this multi-regressive analysis point out that the most efficient models found for amplitude A_1 and phase ϕ_1 are:

$$A_1 = -1.1 \times 10^{-3}\, q + 0.3024\, L + 1.16 \times 10^{-4}\, D_s - 18.69 \tag{24}$$

$$\phi_1 = -5.88 \times 10^{-5}\, q + 7.59 \times 10^{-7}\, D_s - 0.8385 \tag{25}$$

Despite fair model results, significant errors occur in correspondence with the pairs of consecutive months with either the lowest or highest temperature (December and January, and July and August, respectively). In order to improve the temperature cycle reproduction and to avoid this problem, a second Fourier wave (F2H) with $\tau_2 = 6$ months was used. In this case, least-squares regressions of the four parameters for all stations produced estimates of A_1 and ϕ_1 almost identical to those obtained considering only one harmonic. The analysis of the second harmonic parameters showed that A_2 and ϕ_2 can be considered constant in space, with an average value of 0.12 and 2.03, respectively.

Finally, monthly mean temperatures can be reconstructed using the following combined model:

$$\hat{T}\,(j) = T_a + t\,(j) = T_a + A_1 cos\left(\frac{2\pi}{12}j + \phi_1\right) + A_2 cos\left(\frac{4\pi}{12}j + 2\phi_2\right) \tag{26}$$

where T_a is the mean annual temperature, obtained with one of the above-mentioned methods. In this application, T_a is the estimated mean annual temperature obtained by RK-ANN (MLP 3-4-1).

In Table 3, an overview of the results is shown for monthly mean temperature. The best method is given for each month together with the corresponding value of the relative statistical indices. For RMSE values, the best results were always obtained with a VAI method. In particular, in seven of the 12 months, the lowest value of RMSE was achieved using the two-step VAI method, *i.e.*, the Fourier series with the estimation of parameters by an MLR approach (F2H). For the other five months, the lowest value of RMSE was given by two-step VAI methods (RK-OLS, RK-MLP, RK-GWR). Differently from what was obtained at the annual scale, the best results in terms of unbiasedness were achieved with the VAI methods (direct application or two-step). Therefore, on the basis of these considerations, the F2H was chosen as the interpolator to be used to create the mean monthly air temperature maps.

Table 3. Results of monthly analysis; each box reports the interpolation model leading to the best result for each index, and the corresponding value of the index.

Month	RMSE [°C]	MBE [°C]	MAE [°C]	CC
January	F2H 0.39	RK-OLS 0.005	RK-GWR 0.29	RK-GWR 0.99
February	RK-OLS 0.19	RK-GWR –0.004	RK-GWR 0.25	F2H 0.99
March	F2H 0.27	RK-GWR 0.000	RK-GWR 0.25	F2H 0.99
April	RK-MLP 0.41	MLP 0.002	RK-OLS 0.27	F2H 0.99
May	RK-OLS 0.26	RK-OLS 0.027	RK-OLS 0.28	RK-OLS 0.96
June	F1H 0.31	MLP 0.027	F2H 0.26	F2H 0.99
July	RK-GWR 0.44	MLP –0.002	RK-OLS 0.32	F2H 0.99
August	RK-OLS 0.46	LR 0.008	RK-OLS 0.33	RK-OLS 0.94
September	FH2 0.19	RK-GWR –0.023	F2H 0.16	RK-OLS 0.96
October	F2H 0.44	RK-OLS 0.019	RK-OLS 0.27	F2H 0.99
November	F2H 0.37	RK-OLS –0.006	RK-MLP 0.28	RK-MLP 0.98
December	F2H 0.32	RK-OLS –0.004	F2H 0.25	F2H 0.99

4. Discussion and Application of the Best Interpolation Method

The performances of the best methods at the annual scale, respectively in terms of RMSE and MBE, for both univariate and VAI methods are shown in Figure 3. RBF with $N = 30$ can be considered the best univariate method. IDW skill has the worst performances, yielding higher RMSE and MBE than other univariate methods, while OK has intermediate skills between IDW and RBF. However, all univariate methods are very biased, providing a systematic underestimation of temperature. In order to improve the performance, it is necessary to switch from univariate methods to the VAI ones, by introducing the elevation information, in the case of LR, GWR and ANN methods, or by introducing elevation and physiographic information, in the case of ML. Indeed, from a comparison of the performance indices in Tables 1 and 2, it can be observed that with the VAI methods, RMSE and MBE subside dramatically. In particular, while the LR is characterized by the lowest bias, but relatively high RMSE, methods, such as GWR, ANN and MLR, provide good results, with acceptable values of RMSE and with MBEs substantially higher than those obtained with LR. The best results, in terms of RMSE, are achieved by RK and, in particular, RK-LR, which gives the lowest RMSE and the highest CC. In general, the application of OK to the residuals resulting from any VAI method remarkably decreases the RMSE of the underlying VAI method, as can be observed also in Figure 3a. However, taking into account the values of the two indices, RMSE and MBE, at the same time, one can note that a simple method such as LR (ROB), characterized by an RMSE value 36% higher than for RK -LR, has an improvement, in terms of MBE, of about 90% if compared to the MBE value of RK-LR. The values of CC are comparable.

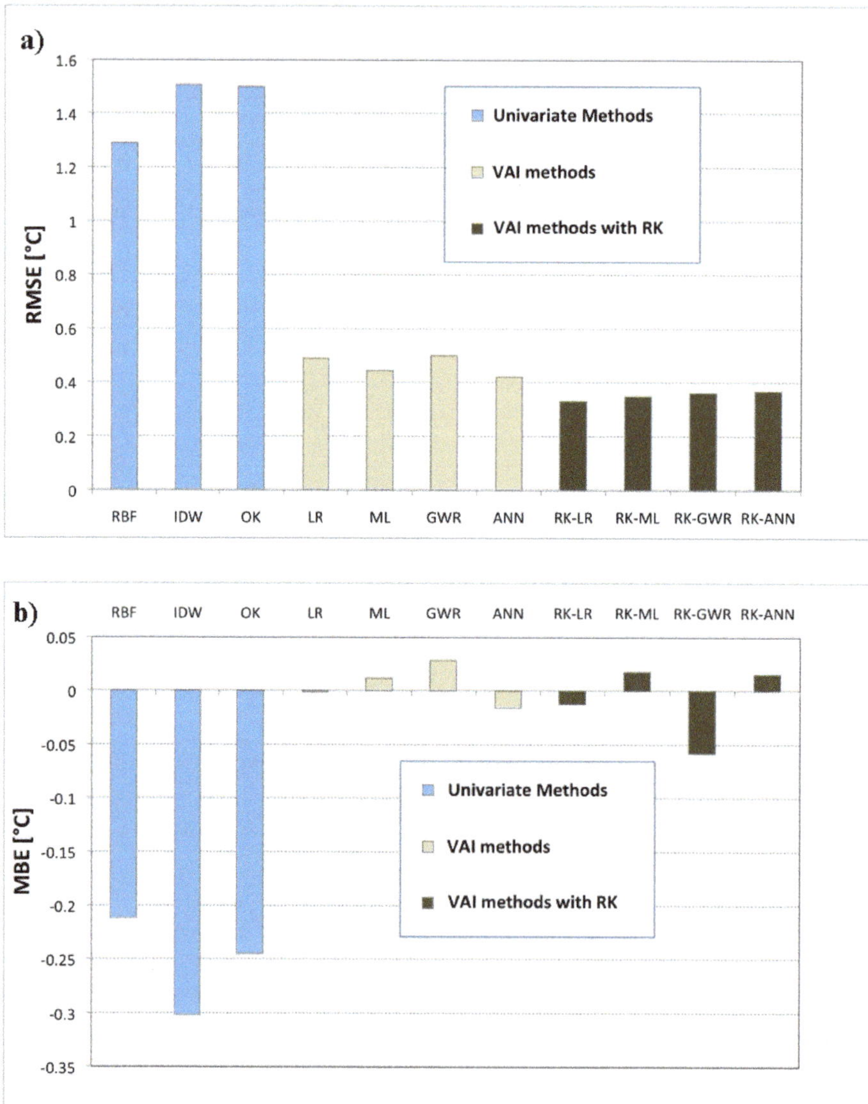

Figure 3. (**a**) RMSE of the best univariate and VAI interpolation methods; (**b**) mean bias error (MBE) of the best univariate and VAI interpolation methods.

On the basis of the previous considerations, both LR and RK-LR (with OLS) could be chosen as the best method and used for spatial interpolation of the mean annual temperature. The mean annual temperature map obtained by RK-LR is shown in Figure 4.

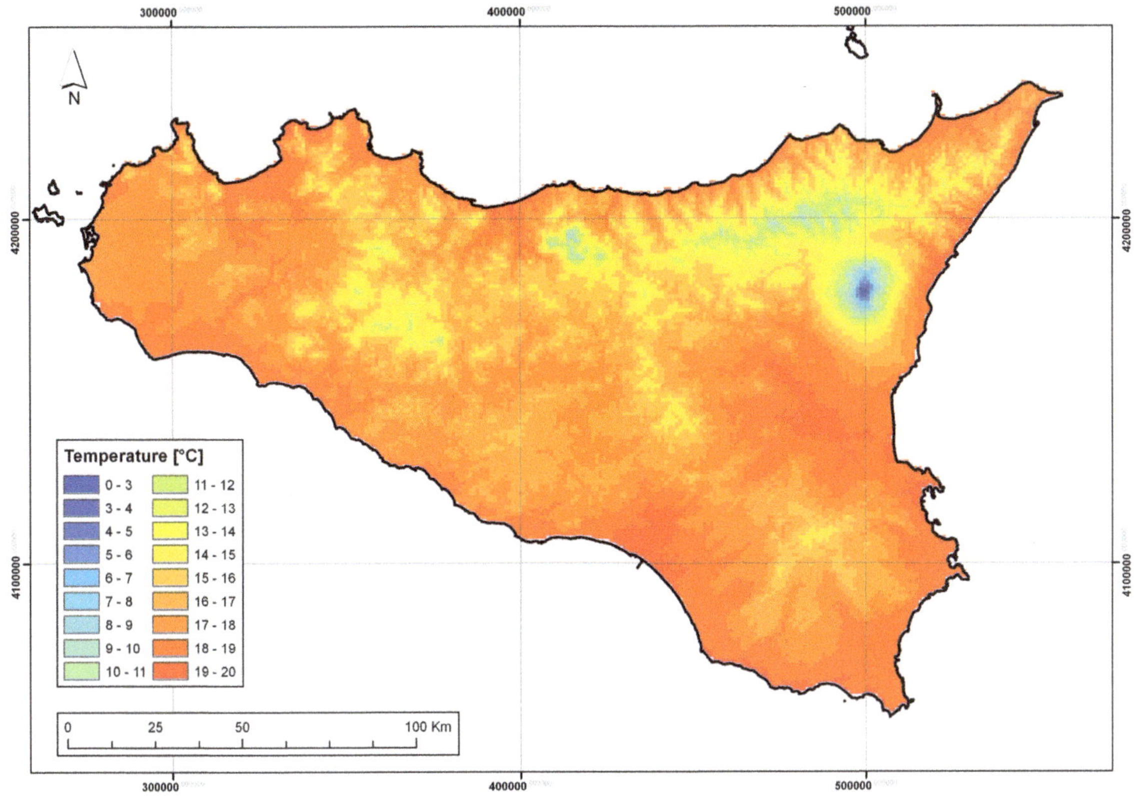

Figure 4. Mean annual temperature interpolated using residual kriging with linear regression (RK-LR) *z-q* method.

Regarding the analysis at the monthly scale, the spatial variation of the cyclical temperature pattern was found to be linearly dependent on the distance from the sea and latitude, in addition to elevation. The parameters of the two-harmonic Fourier series reproducing the temperature regime have been related to the above-mentioned physiographic predictors through a linear multivariate model, without the need to build 12 different models, one for each month. The results of the regression analysis clearly show that the amplitude and phase of the temperature cycle are affected by elevation and distance from the sea and also by latitude (only amplitude). Different spatial dependencies had been found by Claps *et al.* [16], who applied the procedure over the entire Italian territory, pointing out another advantage of this method: it allows one to easily assess the influence of the various parameters at different spatial scales.

Retrieved standard deviation (SD) and correlation coefficient (CC) values were used as polar coordinates on the Taylor diagram (Figure 5), which displays the relationships of these values with the centered pattern error component, therefore not considering the bias component. OK and IDW have very similar performances, while RBF shows better performances in terms of centered root-mean-square difference (RMSD). VAI methods are clustered, with definitely better skill than the univariate methods thanks to the higher CC and SD values, at least in two of the three univariate methods, closer to those of observations. Standard deviation values for the two groups seem related to their mean CC; moreover, it seems that a minimum RMSD level related to the corresponding mean CC is reached by both groups. This effect could be due to the calibration operations performed for each interpolation method by means of parameter selection.

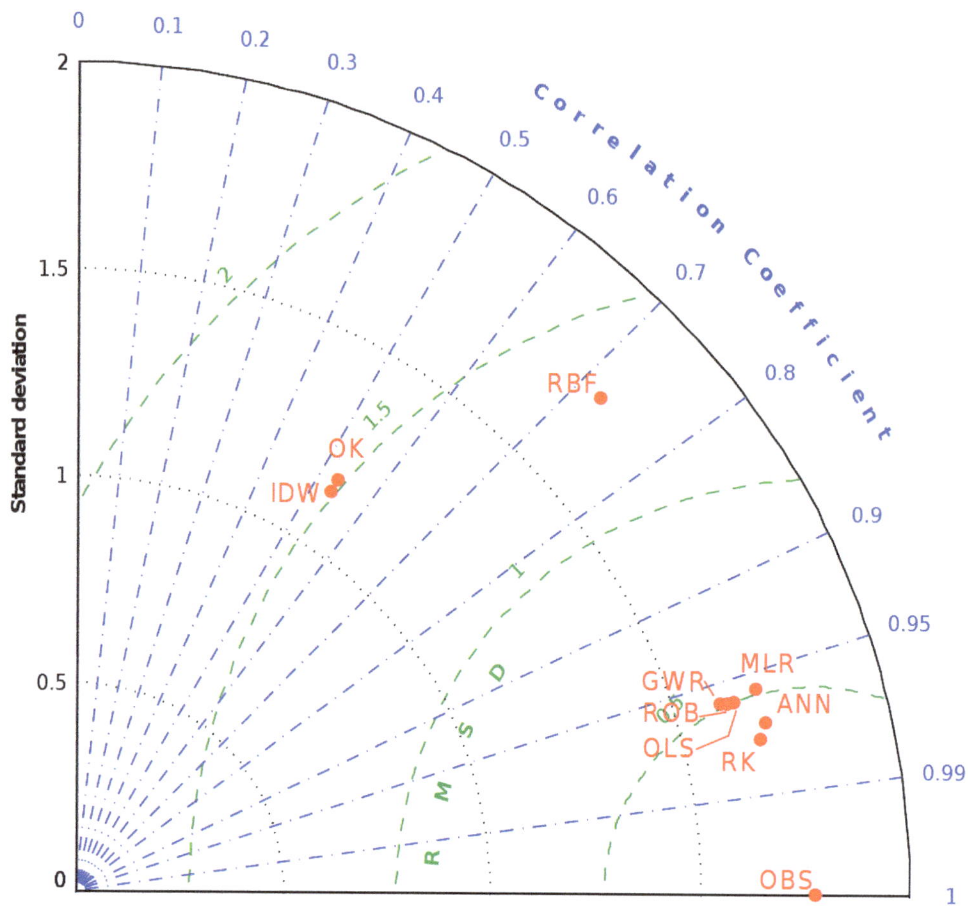

Figure 5. Taylor diagram for mean annual temperature interpolation models. The label OBS is used to represent the reference data (*i.e.*, observations) to which estimator performances are referred.

5. Conclusions

The availability of climatic maps, a key issue for agricultural, hydrological and environmental management, has increased in the last few decades thanks to the spread of new and efficient spatial interpolation techniques and GIS tools. Due to the relative abundance of methods, many algorithms are currently applied, and investigations continue, aiming at the definition of the "best" method for each single climatic variable. This study calibrated and compared nine different methods for spatial interpolation of temperature at the annual and monthly scale over Sicily, a Mediterranean island and a region with complex topography.

Different families of methods were considered: univariate, multivariate, artificial neural networks, a model to describe the cyclical nature of monthly temperature and a two-step application of methods. This comparison highlighted that among the univariate methods, the best performance, in terms of accuracy and unbiasedness, was obtained with RBF, performing better than the geostatistical method OK. Indeed, although RBF is a simple method that does not take into account any external variable, it yielded quite low RMSE (1.28 °C) and bias (−0.21 °C).

Several methods were implemented that take into account elevation and other geographic and physiographic parameters, from the simpler deterministic methods (*i.e.*, linear simple or multiple regression and geographically-weighted regression) to more sophisticated ones. In particular, a reduction of 74.4% of the RMSE is possible using one of the best variable-aided interpolation methods instead of the best univariate one. The introduction of geographic and physiographic variables improves the interpolation performances, highlighting the importance of this information for modeling temperature in a complex morphology region, such as Sicily. On the whole, the best results, for mean annual temperature were achieved by both the application of a geostatistical method to the result of a deterministic one (the RK based on LR) and by the use of a simple method, such as LR.

At the monthly scale, the best results in terms of accuracy and unbiasedness have been obtained by the F2H application, which also takes into account the contribution of a second six-month harmonic. These results open the way for the implementation of a very efficient two-step interpolation technique for monthly temperatures resulting from the interpolation at the annual scale, carried out with linear regression and ordinary kriging of residuals, and the eventual monthly downscaling of annual results by the Fourier series method.

Finally, the results lead to the elaboration of reliable temperature maps, which can significantly contribute to the proper application of hydrological modeling, e.g., for water management, for the monitoring and management of the hydrogeological risk and for the climatological characterization of the territory.

Acknowledgments

This paper has been partially founded by the MIUR project "Messaggeri della conoscenza", with ID 177, and CUP B73D13000470001, titled "Ecohydrology of Mediterranean ecosystems", and by the Flagship Project RITMARE (Italian Research for the Sea) coordinated by the Italian National Research Council and funded by the Italian Ministry of Education, University and Research within the National Research Program 2011–2013.

Conflicts of Interest

The authors declare no conflict of interest.

References

1. Krajewski, W.; Kruger, A.; Smith, J.; Lawrence, R.; Gunyon, C.; Goska, R.; Seo, B.; Domaszczynski, P.; Baeck, M.; Ramamurthy, M.; *et al.* Towards better utilization of NEXRAD data in hydrology: An overview of Hydro-NEXRAD. *J. Hydroinform.* **2011**, *13*, 255–266.
2. Liuzzo, L.; Noto, L.; Vivoni, E.; la Loggia, G. Basin-scale water resources assessment in Oklahoma under synthetic climate change scenarios using a fully distributed hydrologic model. *J. Hydrol. Eng.* **2010**, *15*, 107–122.
3. Burrough, P.; McDonnell, R.; McDonnell, R. *Principles of Geographical Information Systems*; Oxford University Press: Oxford, UK, 1998; Volume 333.

4. Vicente Serrano, S.M.; Sánchez, S.; Cuadrat, J.M. Comparative analysis of interpolation methods in the middle Ebro Valley (Spain): Application to annual precipitation and temperature. *Clim. Res.* **2003**, *24*, 161–180.
5. Hofstra, N.; Haylock, M.; New, M.; Jones, P.; Frei, C. Comparison of six methods for the interpolation of daily, European climate data. *J. Geophys. Res. Atmos.* **2008**, *113*, doi:10.1029/2008JD010100.
6. Serbin, S.P.; Kucharik, C.J. Spatiotemporal mapping of temperature and precipitation for the development of a multidecadal climatic dataset for Wisconsin. *J. Appl. Meteorol. Climatol.* **2009**, *48*, 742–757.
7. Li, J.; Heap, A.D. A review of comparative studies of spatial interpolation methods in environmental sciences: Performance and impact factors. *Ecol. Inform.* **2011**, *6*, 228–241.
8. Willmott, C.; Robeson, S.; Feddema, J. Estimating continental and terrestrial precipitation averages from rain-gauge networks. *Int. J. Climatol.* **1994**, *14*, 403–414.
9. Li, J.; Heap, A.D. Spatial interpolation methods applied in the environmental sciences: A review. *Environ. Model. Softw.* **2014**, *53*, 173–189.
10. Price, D.; McKenney, D.; Nalder, I.; Hutchinson, M.; Kesteven, J. A comparison of two statistical methods for spatial interpolation of Canadian monthly mean climate data. *Agric. For. Meteorol.* **2000**, *101*, 81–94.
11. Daly, C.; Gibson, W.; Taylor, G.; Johnson, G.; Pasteris, P. A knowledge-based approach to the statistical mapping of climate. *Clim. Res.* **2002**, *22*, 99–113.
12. Taskinen, A.; Sirviö, H.; Vehviläinen, B. Interpolation of daily temperature in Finland. *Nord. Hydrol.* **2003**, *34*, 413–426.
13. Teegavarapu, R.S.V. Estimation of missing precipitation records integrating surface interpolation techniques and spatio-temporal association rules. *J. Hydroinform.* **2009**, *11*, 133–146.
14. Tobin, C.; Nicotina, L.; Parlange, M.B.; Berne, A.; Rinaldo, A. Improved interpolation of meteorological forcings for hydrologic applications in a Swiss Alpine region. *J. Hydrol.* **2011**, *401*, 77–89.
15. Naoum, S.; Tsanis, I.K. A multiple linear regression GIS module using spatial variables to model orographic rainfall. *J. Hydroinform.* **2004**, *6*, 39–56.
16. Claps, P.; Giordano, P.; Laguardia, G. Spatial distribution of the average air temperatures in Italy: Quantitative analysis. *J. Hydrol. Eng.* **2008**, *13*, 242–249.
17. Arnone, E.; Francipane, A.; Noto, L.V.; Scarbaci, A.; la Loggia, G. Strategies investigation in using artificial neural network for landslide susceptibility mapping: Application to a Sicilian catchment. *J. Hydroinform.* **2014**, *16*, 502–515.
18. Hettiarachchi, P.; Hall, M.; Minns, A. The extrapolation of artificial neural networks for the modeling of rainfall-runoff relationships. *J. Hydroinform.* **2005**, *7*, 291–296.
19. Rigol, J.; Jarvis, C.; Stuart, N. Artificial neural networks as a tool for spatial interpolation. *Int. J. Geogr. Inf. Sci.* **2001**, *15*, 323–343.
20. Coulibaly, P.; Evora, N. Comparison of neural network methods for infilling missing daily weather records. *J. Hydrol.* **2007**, *34*, 27–41.

21. Goovaerts, P. Using elevation to aid the geostatistical mapping of rainfall erosivity. *Catena* **1999**, *34*, 227–242.
22. Di Piazza, A.; lo Conti, F.; Noto, L.; Viola, F.; la Loggia, G. Comparative analysis of different techniques for spatial interpolation of rainfall data to create a serially complete monthly time series of precipitation for Sicily, Italy. *Int. J. Appl. Earth Obs. Geoinf.* **2011**, *13*, 396–408.
23. Karavitis, C.A.; Alexandris, S.; Tsesmelis, D.E.; Athanasopoulos, G. Application of the standardized precipitation index (SPI) in Greece. *Water* **2011**, *3*, 787–805.
24. Haining, R.P.; Kerry, R.; Oliver, M.A. Geography, spatial data analysis, and geostatistics: An overview. *Geogr. Anal.* **2010**, *42*, 7–31.
25. Hudson, G.; Wackernagel, H. Mapping temperature using kriging with external drift: Theory and an example from Scotland. *Int. J. Climatol.* **1994**, *14*, 77–91.
26. Demyanov, V.; Kanevsky, M.; Chernov, S.; Savelieva, E.; Timonin, V. Neural network residual kriging application for climatic data. *J. Geogr. Inf. Decis. Anal.* **1998**, *2*, 215–232.
27. Antonic, O.; Krizan, J. Spatio-temporal interpolation of climatic variables over large region of complex terrain using neural networks. *Ecol. Model.* **2001**, *138*, 255–263.
28. Willmott, C.; Matsuura, K. Smart interpolation of annually averaged air temperature in the United States. *J. Appl. Meteorol.* **1995**, *34*, 2577–2586.
29. Wu, T.; Li, Y. Spatial interpolation of temperature in the United States using residual kriging. *Appl. Geogr.* **2013**, *44*, 112–120.
30. Prudhomme, C.; Reed, D. Mapping extreme rainfall in a mountainous region using geostatistical techniques: A case study in Scotland. *Int. J. Climatol.* **1999**, *19*, 1337–1356.
31. Faulkner, D.; Prudhomme, C. Mapping an index of extreme rainfall across the UK. *Hydrol. Earth Syst. Sci.* **1998**, *2*, 183–194.
32. Bishop, C.M. *Neural Networks for Pattern Recognition*; Clarendon Press: Oxford, UK, 1995.
33. Isaaks, E.; Srivastava, R. *An Introduction to Applied Geostatistics*; Oxford University Press: Oxford, UK, 1990.
34. Kitanidis, P. *Introduction to Geostatistics: Applications in Hydrogeology*; Cambridge University Press: Cambridge, UK, 1997.
35. Goovaerts, P. *Geostatistics for Natural Resources Evaluation*; Oxford University Press: Oxford, UK, 1997.
36. Andrews, D.F. A Robust Method for Multiple Linear Regression. *Technometrics* **1974**, *16*, 523–531.
37. Brunsdon, C.; McClatchey, J.; Unwin, D. Spatial variations in the average rainfall-altitude relationship in Great Britain: An approach using geographically weighted regression. *Int. J. Climatol.* **2001**, *21*, 455–466.
38. Fotheringham, S.; Brunsdon, C.; Charlton, M. *Geographically Weighted Regression: The Analysis of Spatially Varying Relationships*; Wiley: Chichester, UK, 2002.
39. Govindaraju, R., Rao, A. (Eds.) *Artificial Neural Networks in Hydrology*; Kluwer Academic Publishers: Dordrecht, The Netherlands, 2000.

40. Bahrami, J.; Kavianpour, M.R.; Abdi, M.S.; Telvari, A.; Abbaspour, K.; Rouzkhash, B. A comparison between artificial neural network method and nonlinear regression method to estimate the missing hydrometric data. *J. Hydroinform.* **2011**, *13*, 245–254.
41. Goovaerts, P. Geostatistical approaches for incorporating elevation into the spatial interpolation of rainfall. *J. Hydrol.* **2000**, *228*, 113–129.
42. Taylor, K.E. Summarizing multiple aspects of model performance in a single diagram. *J. Geophys. Res.* **2001**, *106*, 7183–7192.
43. Pebesma, E.; Wesseling, C. Gstat: A program for geostatistical modeling, prediction and simulation. *Comput. Geosci.* **1998**, *24*, 17–31.

12

Towards a Comprehensive Valuation of Water Management Projects When Data Availability Is Incomplete—The Use of Benefit Transfer Techniques

Michael Ahlheim [1,*], Oliver Frör [2,†], Jing Luo [3,†], Sonna Pelz [1,†] and Tong Jiang [4,5]

[1] Institute of Economics, University of Hohenheim, Stuttgart 70593, Germany; E-Mail: s.pelz@uni-hohenheim.de

[2] Institute for Environmental Sciences, University of Koblenz-Landau, Landau 76829, Germany; E-Mail: froer@uni-landau.de

[3] Research Center for China's Borderland History and Geography, Chinese Academy of Social Sciences, Beijing 100732, China; E-Mail: ljing25@163.com

[4] National Climate Center, China Meteorological Administration, Beijing 100081, China; E-Mail: jiangtong@cma.gov.cn

[5] School of Geography and Remote Sensing/Collaborative Innovation Center on Forecast and Evaluation of Meteorological Disasters, Nanjing University of Information Science and Technology, Nanjing 210044, China

† These authors contributed equally to this work.

* Author to whom correspondence should be addressed; E-Mail: ahlheim@uni-hohenheim.de;

Academic Editor: Markus Disse

Abstract: In this paper we deal with the problem of missing data in environmental cost-benefit analysis. If government pursues the goal of maximizing social welfare, this implies that public funds should be allocated to those uses where they generate the highest net social benefit. This criterion makes it necessary to conduct cost-benefit analyses for public projects. While the assessment of project costs is typically rather straightforward, a comprehensive assessment of the project benefits is more complicated because one has to consider that also people living far away from the project site might benefit from that project. Neglecting these so-called passive use benefits would lead to a systematic undervaluation of environmental projects, thereby reducing their chances of being realized.

A comprehensive cost-benefit analysis would, therefore, require benefit assessment studies in all areas where passive use values might occur. Obviously, this would be impossible. In this paper we show how the assessment of the social benefits from environmental projects can be enhanced even with an imperfect database by using benefit transfer techniques. This is also illustrated empirically using an example from Northwest China.

Keywords: sustainable water management; contingent valuation method; benefit transfer; passive use values

---

1. Introduction

In this paper we deal with the problem of missing data for benefit assessment in the context of environmental cost-benefit analysis. If government pursues the goal of maximizing the overall welfare of society, which it is expected to do, this implies that scarce public funds should be allocated to those uses where they generate the highest (marginal) net social benefit. In the U.S., these principles were already laid down by Ronald Reagan in his Executive Order 12291: "... *(b) Regulatory action shall not be undertaken unless the potential benefits to society for the regulation outweigh the potential costs to society; (c) Regulatory objectives shall be chosen to maximize the net benefits to society; ... (e) Agencies shall set regulatory priorities with the aim of maximizing the aggregate net benefits to society* ..." [1]. These guidelines of public decision making were confirmed by Bill Clinton in 1993 in his Presidential Executive Order 12866: "*In deciding whether and how to regulate, agencies should assess all costs and benefits of available regulatory alternatives, including the alternative of not regulating. ... Further, in choosing among alternative regulatory approaches, agencies should select those approaches that maximize net benefits* ..." [2].

This allocation decision criterion makes it necessary to carry out cost-benefit analyses for all public projects of major importance, especially in an environmental context where the accruing benefits are often not easy to quantify. The assessment of the implementation costs of public projects is typically rather straightforward since they mainly consist of the costs of inputs that are traded in markets, so that they can be valued on the basis of market prices. This also holds true for the calculation of foregone profits, which often play an important role in the context of land use changes for environmental preservation. Things are different when it comes to appraising the social benefits accruing from an environmental project because for many environmental "*goods*" like biodiversity, clean air, or a beautiful landscape, no market prices are available. Therefore, for their assessment, special techniques like, e.g., the Contingent Valuation Method (CVM) or Attribute-Based Choice Methods like Choice Experiments must be employed [3–6].

Another complication of environmental valuation arises from the fact that, in many cases, the beneficiaries of an environmental project, for instance a reforestation project, are not located in one single place only (like the project site) but are distributed over many different places. The reason for this phenomenon is that environmental improvements often also have a symbolic importance, which is perceived and appreciated also by people living far away from the project site. This importance is independent of a direct utilization of the environmental goods in question. The protection of e.g., the rain forests in a country, may also be appreciated by people who never visit such a forest. Though they have never been there and do not intend to go there in the future personally, they experience an increase in utility when they read in the papers that measures have been taken to protect some forest areas against economic exploitation or that they have been reforested. Other people might feel happy when they learn that water quality in a formerly polluted river has been improved such that people now can go fishing and swimming in that river in spite of the fact that they personally will never go there. Adamowicz *et al.* [6] refer to this kind of value as passive use value: "*Passive use value is the economic value arising from a change in environmental quality (or any other situational change) that is not reflected in any observable behavior. That is, changes in recreational demand, purchasing behavior, or other observable action do not necessarily take place.*" It is clear that passive use values accrue not only to people on site but, at least potentially, to people living all over a country, and, in the case of rain forests or the preservation of certain keystone species even all over the world [7]. For a cost-benefit comparison as an allocation decision criterion the Total Economic Value (TEV) of an environmental improvement has to be compared to the costs of the project that caused that improvement. The TEV comprises (active) use values, which are connected with observable utilization activities on the one hand and passive use values on the other (*cf.*, e.g., [6]). Neglecting passive use values, therefore, leads to a systematic undervaluation of environmental improvements, thereby reducing the chances of the respective environmental projects being realized if their implementation depends on a cost-benefit comparison.

A comprehensive cost-benefit analysis of an environmental project would therefore require conducting nationwide (in some cases even global) benefit assessment studies while the costs of the project could be measured locally, *i.e.*, on the project site. Obviously, such a comprehensive benefit assessment would be difficult if not impossible under organizational as well as financial aspects. Under real-world conditions, cost-benefit practitioners must do with incomplete data sets collected locally at some pilot sites and then try to infer the total benefits of environmental projects from this imperfect database. One possibility to deal with this problem of missing data is the application of so-called benefit transfer techniques in combination with Contingent Valuation Method (CVM) studies. Benefit transfer requires conducting a primary CVM study in a more or less characteristic study site. Then other regions are identified which are not too different under socio-economic aspects from the study site and where people are also expected to receive benefits from the environmental project in question. In a next step, the CVM-results from the study site are extrapolated to these additional sites via value transfer or value function transfer as described below. This procedure allows approximation of at least roughly also the benefits accruing to people not living in the study site, though no additional CVM study is conducted with them. Since it can be expected that people living in the region where an environmental project (like a reforestation project) is implemented will obtain active use values as well as passive use values (like e.g., existence values or bequest values *etc.*) and since these values cannot be separated properly in a CVM study, two different CVM surveys will have to be conducted for a comprehensive assessment of the overall benefits accruing from such an environmental project. One CVM study must be conducted at the project site itself to assess the active and passive use benefits accruing to the people living there. For the ensuing benefit transfer study then a primary CVM survey must be conducted at a so-called "*study site*" which is characteristic of sites where people perceive only the passive use values of the environmental project in question. Then the results of this study can be extrapolated via benefit transfer as explained below to all other sites where passive use values are expected but not assessed explicitly through additional CVM surveys. Since this paper focuses on the problem of dealing with missing data in benefit appraisal studies, we will neglect the cost assessment aspects of environmental cost-benefit analysis here.

The working of the benefit transfer technique is also demonstrated empirically in this paper using an example where the social benefits accruing from a water management project in the Tarim Basin in Xinjiang Autonomous Region in Northwest China are assessed not only in the project area but also in five megacities of China where, in only one of these cities (Beijing), a CVM survey has been conducted. Based on this example we discuss the advantages and disadvantages of the benefit transfer method as a tool for the assessment of the total value of water projects which generate a high degree of passive use values accruing in various different places which are widely scattered geographically, so that it is not feasible to conduct separate valuation studies in each of these sites in order to assess the total value of the water project comprehensively.

The paper is structured as follows. Section 2 reviews the two environmental valuation techniques used in the present study–contingent valuation and benefit transfer. Section 3 of this paper introduces the environmental site in question and the water management project to be valued. Section 4 reports results. This opens with a consideration of the data gathered through a contingent valuation survey in Beijing in 2013. The section continues with an exemplary application of benefit transfer and ends with a rough approximation of the megacity-wide social value of a more sustainable water management in the Tarim Basin.

2. Methods

2.1. The Contingent Valuation Method

Welfare economic theory postulates that an increase in society's welfare is a function of the utility changes of all members making up a society. Individual utility changes, in turn, can be measured in monetary terms by the Hicksian Compensating Variation which can be interpreted as an individual's willingness to pay (WTP) to obtain a desired good or willingness to accept (WTA) compensation to forego enjoying the good [8]. Applying this concept to the environmental context leads to the well-established practice of assessing individuals' WTP for the realization of a particular environmental project in a first step and extrapolating individual WTP values in a second step to obtain the social value of this project. This would be done ideally by summing up the WTP of all households affected by the environmental project in question [9]. In practice, of course, it is not possible to elicit the individual WTP of all households concerned. Instead, the average WTP of a representative sample of the affected population is assessed and then multiplied by the number of all households. If an environmental change leads to decreases in the utility of people (e.g., after the occurrence of environmental damages) the Hicksian Compensating Variation equals the minimum amount of money they would have to be given in order to be compensated in utility terms for that loss. This amount is usually termed their willingness to accept compensation (WTA).

The WTP for environmental goods, which are typically not bought and sold in markets, is empirically not observable. Due to the absence of market prices and demand functions researchers have to rely on specific methods to obtain WTP values. The present study makes use of the Contingent Valuation Method (*cf.*, [10,11]). Contingent valuation is a survey-based technique; it simulates a hypothetical market for a particular environmental good and assesses the value of this good by directly asking respondents to state their WTP for its provision. An analogous procedure for the elicitation of people's WTA by directly asking them would be problematic because of the obvious incentive for respondents to exaggerate the money amounts they claim for compensation. For this reason, literature generally advises against using WTA measures in environmental valuation studies [12]. Ideally, the sample of a CVM survey is representative of all households affected by the environmental project in question. If this is the case, the sample's average WTP reflects the average WTP of all those who would benefit from the environmental project. Hence, multiplication of the sample's average WTP by the number of all beneficiaries yields the social value of the considered project.

CVM surveys are a suitable way for assessing the total economic value (TEV) of environmental projects. As mentioned above, the TEV is commonly decomposed into so-called use values and passive use values (s. [6] or [3]), which are sometimes also called nonuse values (*cf.*, [13]). Use values refer to the benefits accruing to the users of an environmental good (e.g., spending time at a waterside). Passive use values encompass all kinds of benefits which people can enjoy without actively utilizing an environmental good (*cf.*, [6]). Revealed preference techniques, like the travel cost method or hedonic pricing, exclusively focus on the users of environmental goods (the visitors of parks, the owners of houses located at immediate distance of a natural resource, *etc.*) and are unsuitable for the assessment of passive use values [9]. Therefore, valuation methods that assess people's WTP directly, like the CVM, are preferable over revealed preference methods whenever an environmental good generates substantial passive use values [3].

Water management projects of national importance may have, in addition to their positive effects on local people's livelihood, considerable passive use values. This argument also holds true for the water management project considered in our empirical example below (*cf.*, Section 3.2). One peculiarity of passive use values is that not only those living in immediate proximity to the project site would benefit from the accruing environmental improvements, but also people living far away may experience an increase in their wellbeing. Since it is expected that people all over China might at least psychologically benefit from a more sustainable water management in the Tarim Basin as considered in our example below, a comprehensive CVM survey should assess the preferences of all people living in China. Since such a direct benefit assessment covering the whole of China is not feasible, a less ambitious but realistic strategy consists of conducting a CVM survey in a "*typical*" study site, e.g., in one of China's megacities, and then extrapolating the WTP results gained there to all other Chinese megacities by means of the benefit transfer methodology. This strategy is illustrated in our empirical example in Sections 3 and 4 below.

2.2. Benefit Transfer

The term "*benefit transfer*" describes the extrapolation of the benefits accruing from some environmental project assessed at one or several "*study sites*" to some other target site or "*policy site*" where no primary benefit assessment study has been conducted [14]. In the context of the present study, this means that we transfer the passive use value of the Tarim water management project assessed in Beijing (the "*study site*") to other Chinese megacities (the "*policy sites*"). Benefit transfer is frequently used in practice. This holds especially true for the U.S., where agencies rarely conduct original environmental valuation studies but mostly rely on benefit transfer [15]. In this section, different variants of benefit transfer shall be introduced and the criticism towards the methodology will be briefly addressed. Comprehensive overviews on the benefit transfer methodology have been provided elsewhere (*cf.*, e.g., [16,17], Chapter 17).

There are different kinds of benefit transfer variants discussed in the literature, namely value transfer and benefit function transfer (*cf.*, [18]). Value transfer is the simplest variant; it consists of transferring a WTP point estimate for a particular environmental good identified in the literature to a policy site for which the value of the same (or a similar) environmental good is needed. Hence, value transfer relies on the assumption that the population at the policy site P has the same average WTP as the population at the study site S. Put formally,

$$WTP_S = WTP_P \tag{1}$$

As can be inferred from its name, the second variant of benefit transfer involves the transfer of the value function or WTP function estimated econometrically at the study site to the policy site. For example, in a simple and practical form, WTP might be perceived as a linear function of socio-economic (e.g., income) and demographic characteristics of the population (e.g., age) at the study site, *i.e.*,

$$WTP = \beta_0 + \beta_1(income) + \beta_2(age) \tag{2}$$

Assuming that the environmental good is provided in equal quantity and quality at both sites, Equation (2) allows the analyst to tailor the WTP per household at the policy site. This is done by substituting site-specific values for income and age into the WTP function, while maintaining the regression coefficients, β_1 and β_2, and the constant term β_0 assessed at the study site. Classical as well as Bayesian statistical approaches can be used to combine data from the study site and from the policy site (*cf.*, [19]). There is little agreement among researchers about which variant of benefit transfer—Simple value transfer or function transfer—Provides the most accurate results. [14] show that simple value transfers yield more accurate results than function transfers when the policy site context resembles the study site context in terms of its population and the characteristics of the environmental good. However, function transfer is preferable when the study site and the policy site are dissimilar. Likewise, there is no consensus about whether classical statistical methods or Bayesian approaches are more appropriate when transferring a value function from one site to another [19].

When it comes to function transfer, analysts have two possibilities at hand for obtaining a value function for a particular environmental good: adopting a function which has been reported in a single study or estimating a regression equation based on the results of multiple environmental valuation studies assembled by meta-analysis. Meta-analysis typically encompasses the following steps (*cf.*, [18]): first, researchers collect all available environmental valuation studies on the environmental good in question (e.g., a particular ecosystem, national parks, *etc.*); second, they create a dataset by coding these studies with respect to WTP per unit, attributes of the study site like the quantity of the environmental good available or the availability of substitute goods (e.g., proximity to other similar ecosystems or national parks), the methodological details of the studies (e.g., stated or revealed preference methods, survey mode, sampling approach, *etc.*) and the population statistics (e.g., income levels, age, gender, *etc.*) reported in the primary studies. Based on such a data set, the WTP per unit can be analysed as a function of the set of available variables by means of regression analysis. Doing so, the analyst obtains a meta-analysis equation that provides the basis of the benefit transfer exercise.

It is beyond all question that meta-analysis is more time-intensive and, hence, more costly than the simple variant of benefit function transfer. Given this drawback, the question arises whether ([18], p. 83) and other proponents of meta-regression analysis are right when claiming that this approach is "one of the biggest advances in benefit transfer".

Boyle *et al.* [15] conclude that there is substantial evidence that pooled data approaches are superior to simple function transfer. This is because meta-analysis-based benefit transfer overcomes one of the biggest drawbacks of the standard function transfer; it relaxes the assumption that the statistical relationship between WTP and the explanatory variables is the same between the policy site and a single study site (ibid.). However, meta-analysis-based benefit transfer requires a sufficiently large number of primary studies on similar environmental goods with well-reported and statistically robust results as well as a researcher's ability to identify these studies. The frequently praised Environmental Values Reference Inventory (EVRI), which is a database where the results of multiple environmental valuation studies have been assembled with the overall objective of facilitating benefit transfer, for example, is only open for citizens of member countries. Since neither Germany nor China belongs to the group of member countries, the EVRI database was not much help in the context of the present study. Furthermore, estimating WTP functions based on pooled data is also prone to a lot of errors. In many practical applications, researchers assembled environmental valuation studies which employed different, and in many cases, incompatible valuation approaches. For example, WTP functions are regularly estimated based on the results of studies that employed travel cost models (TCM) on the one hand and studies that made use of contingent valuation studies on the other. This means that valuation techniques assessing only use values (TCM) and techniques assessing use and passive use values (CVM) were mixed together. It is also common practice to include WTP and willingness to accept compensation (WTA) simultaneously as welfare measures into the same dataset from which the value function is then be estimated.

In view of enhancing the soundness of the methodology, some guidelines for conducting benefit transfer have been formulated. The proponents of benefit transfer still cite the three "ideal criteria" proposed by [20] in Water Resources Research and suggest following these criteria as closely as possible. These criteria require that (1) the environmental good valued at the study site and the policy site are identical; (2) that the populations affected by changes in the quality or quantity of the environmental good have the same characteristics in both sites; and (3) that the assignment of property rights at both sites implies the same welfare measure, *i.e.*, WTP or WTA. However, even if these guidelines are strictly followed by all practitioners, the validity of environmental values assessed with the benefit transfer method can still be questioned. Maybe the greatest shortcoming of benefit transfer is that individual preferences of those living at the policy site are completely neglected [21]. It is assumed that the members of the target population have the same preferences as the participants of the primary study. This assumption is, of course, highly questionable, even in the ideal case that both populations resemble each other in terms of their demographic and cultural characteristics.

One decade ago, renowned environmental valuation researchers such as [11], highlighted that benefit transfer "*cannot, at the moment, be relied upon to produce valuation estimates which are statistically indistinguishable from 'true' values*". With the development of meta-analysis and benefit transfer databases such as EVRI, the field of benefit transfer has, certainly, advanced. In spite of this, most researchers still conclude that benefit transfer "*will never take the place of a carefully conducted primary study*" [15]. Nevertheless, in cases where the benefits from a public project accrue in many different places which are geographically widely scattered so that it is not possible to assess all these benefits by separate CVM studies, it seems to be better to approximate the overall benefits by using benefit transfer than to ignore all benefits accruing in other places than the study site. This will be demonstrated in the following sections at the example of a water management project in the Tarim Basin.

3. Water Management in the Tarim Basin

3.1. Water Shortage in the Tarim Basin

The Tarim River, Central Asia's longest inland river, is located in a desert region in Northwest China. This river gives the Tarim Basin its name, which covers the southern part of Xinjiang Uyghur Autonomous Region. The inhabitants of the Tarim Basin largely depend on the supply of water by the Tarim River and its tributaries. Any kind of agricultural and industrial activity in this extremely dry region depends on surface and groundwater [22]. The extension of agricultural and industrial production, accompanied by an impressive population growth, has contributed to the economic development of the region. However, the natural environment has been deteriorating due to unsustainable water management. The extensive water use in the upper reaches of the Tarim River prevents a permanent water supply at the lower reaches. As a consequence, the unique dryland ecosystems of the Tarim Basin are threatened [23]. Due to the deterioration of the riparian poplar forests and grasslands, the inhabitants of the Tarim Basin are increasingly exposed to sandstorms, dust, and rising temperatures. Furthermore, there is a loss in biodiversity and the progressing desertification renders the formerly beautiful landscape into a lifeless desert. In view of the warnings of several climate experts (*cf.*, e.g., [24]), a more sustainable water management is urgently needed. The local impact of global warming may cause a complete desiccation of the Tarim River. Under this scenario the southern part of Xinjiang would become uninhabitable and many rare plant and animal species would become extinct.

The Chinese Government has already undertaken considerable effort to restore the natural ecosystems at the lower reaches of the Tarim River. [25] estimates that public investment into water allocation projects amounts to 11 billion Chinese RMB (approximately 1.5 billion EUR). However, even though there is increasing awareness of the water-related environmental issues among scientists and policy makers, the integration of natural ecosystems into water management policy remains difficult. One reason why policy makers still tend to neglect environmental parameters is the lack of information about the value of a more sustainable water management in the Tarim Basin for Chinese society. Until now, little research has been conducted on the effect of a more sustainable water management in the Tarim Basin on society's wellbeing. The existing studies are hardly helpful for determining the overall benefit of this project. Krutilla [26], for example, used market-price based techniques to estimate unit values of different ecosystems in the lower reaches of the Tarim. However, these values do not reflect the multiple passive use values of the natural ecosystems. Loomis *et al.* [27] estimated the value of China's forest ecosystem services. Similar to the previously cited study, this assessment may underestimate the actual value of forest, because market-price-based methods dominated this assessment. Interestingly, however, the value of Xinjiang's forest ecosystem services was the lowest in the whole of China in this study. Loomis *et al.* [28] conducted a CVM survey to assess the social value of poplar forests in the Tarim Basin. The authors assessed only the preferences of Han Chinese farmers working in the State Farms along the Tarim River. Hence, the WTP estimate does neither reflect the benefits accruing to the permanent residents of the Tarim Basin nor to people living in other parts of the country.

In the study included here, we demonstrate the possibilities to reach a more comprehensive valuation of a sustainable water management scheme in the Tarim basin by also taking into account the passive use values created by this project for people living far away from the project site (the exact description of the project scenario can be found in Table A1). As a demonstration, we first assess the passive use benefits accruing to people living in Beijing as an example of a Chinese megacity (our "*study site*") and then extrapolate these results to the citizens of Shanghai (our "*policy site*") to show how benefit transfer works in practice. In the next step, we generalize our results using benefit transfer to China's "*big five*" megacities, *i.e.*, Beijing, Shanghai, Shenzhen, Guangzhou and Tianjin.

3.2. Social Benefits of an Improved Water Management Scheme

In the present study the following value components of a more sustainable water management regime in the Tarim Basin can be identified. Due to its positive impact on the natural vegetation along the Tarim River, the water management project in question is expected to have considerable use value for the local populations. For example, the riparian forests along the Tarim River protect the Basin's oasis cities from sandstorm and dust. Hence, if these ecosystems are restored, the local households and firms would be less exposed to these two kinds of environmental hazards. Since a more sustainable water management scheme is particularly urgent in view of the wellbeing of future generations, the project has important bequest values. In view of an increasing number of Chinese households which have the chance and budget to travel, many people may wish to preserve the possibility to visit the Tarim Basin one day in the future; hence, this so-called option value further adds to the passive use values of the water management project. As first theorized by [29], people may gain satisfaction from the pure knowledge about the existence of an environmental good. In line with this supposition, it is expected that the natural ecosystems of the Tarim Basin also have a considerable existence value.

Except for the project's direct use value, which could only be enjoyed by the inhabitants of the Tarim Basin, all other value components are likely to contribute also to the wellbeing of people living in other parts of China. Therefore, not accounting for its passive use values would lead to a dramatic underestimation of the water management project's overall social benefit. However, conducting an environmental valuation study that is representative of all Chinese people is organizationally and financially unfeasible. Therefore, we conducted two CVM surveys at two exemplary sites. We explored the preferences of local people during workshops in different cities in Xinjiang. Then, we interviewed approximately 2500 residents in Beijing. The present study focusses on the Beijing survey and the extrapolation of its results to other megacities in China; the results of our Xinjiang survey have been reported elsewhere [30,31]) and are not part of this analysis due to our focus on the passive use values of the Tarim water management scheme.

3.3. A CVM Survey in Beijing

We gathered our data through personal interviews that were based on a standardized questionnaire (*cf.*, Table 1). We opted for this comparatively time- and cost-intensive interview mode because of the complexity of the survey topic and in order to minimize selection bias. In Beijing, many people, especially the elderly and the poor, are not acquainted with the internet and, as it turned out during the survey preparation phase, some of them are not even able to read. Hence, personal interviews appeared to be the best option for gathering a representative sample and maximizing data quality. During these interviews we informed the respondents about the current situation in the Tarim Basin, thereby focusing on water shortage and its environmental impacts. Then we introduced the "*Tarim Environmental Preservation Plan*", a hypothetical water management project that would lead to restoration and conservation of the natural ecosystems along the Tarim River (*cf.*, Table 2). Subsequently, we informed the respondents that the new water management project could not be financed out of existing funds alone. Therefore, the central government would have to increase taxes, implying a higher cost of living for households all over China. We assessed respondents' WTP using the dichotomous choice elicitation question format (*cf.*, [32]) with bids from 10 to 200 RMB. Respondents were told that the government would implement the "*Tarim Environmental Preservation Plan*" only if the majority of survey participants voted in favor of the project. The exact wording of the payment scenario and the referendum question is displayed in the appendix. Using a quota-based intercept survey mode, we obtained a sample of 2438 observations. Respondents' characteristics closely reflect the socio-economic and demographic characteristics of the general population of urban Beijing (Table 3 summarizes the sample characteristics).

Table 1. Description of the contingent valuation method (CVM) questionnaire.

Part	Description
(A) Warm up questions	Previous knowledge about the Tarim River and environmental problems affecting people's livelihood in the Tarim Basin
(B) Description of the status quo	Description of the key environmental features of the Tarim River and the riparian ecosystems; the environmental problems caused by the extensive use of water resources in the past; and some related questions
(C) Description of the water management project	Depiction of the "*Tarim Environmental Preservation Plan*" (*cf.*, Table 2 and the project scenario in Table A1)
(D) Value assessment	Payment scenario, followed by a dichotomous choice question to assess respondents' WTP for a more sustainable water management project in the Tarim Basin (*cf.*, Table A1)
(E) Follow-up questions	Collection of relevant information regarding respondents' - motivation to support or reject the water management project - socio-economic characteristics - attitudes towards several aspects of life and respondents' personality

Table 2. Description of the environmental change resulting from a more sustainable water management.

Scenario	Description
Status quo situation	Extensive water use in the upper and middle reaches of the Tarim River has led to a destruction of the natural environment. This development has, among other things, the following consequences: - more sandstorms - more dust days - extinction of native plant and animal species - progressive desertification of the landscape
Situation subsequent to an improved water management scheme	Expected impacts of a science-based water management: - secured water supply in the lower reaches - restoration of riparian forests and grasslands - less sandstorms and dust - survival of native animals and plants - prevention of desertification and hence conservation of the area for future generations

Table 3. Characteristics of the population at the study and policy site.

Site Characteristics	Beijing (Study Site)	Shanghai (Policy Site)	Difference between Sites
Number of households	4.006 Mio.	6.969 Mio.	−2.963
Consumer Price Index [1]	105.6	105.2	0.4
Mean distance from Tarim Basin			
- Linear distance (km)	3039	3825	−786
- Flight (hours)	6 h 50	6 h 15	0 h 35
	Beijing (N = 2438)	**Shanghai (N = 791)**	**Difference between Samples**
Respondent Characteristics	**Mean (std. dev.)**	**Mean (std. dev.)**	***p*-value from *t*-test of Equal Means in the Study and Policy Site**
Monthly disposable household income (in 1000 RMB)	8.485 (7.749)	10.512 (7.594)	0.000
Male	0.504 (0.506)	0.529 (0.499)	0.023
Age	40.02 (15.408)	40.08 (17.0)	0.534
Education [2]	4.318 (1.318)	4.306 (1.209)	0.666
Household size	2.872 (2.872)	3.095 (1.145)	0.000

Notes: [1] Figures published in the Statistical Yearbooks 2012 of Beijing and Shanghai [34,38]; [2] A respondent's highest education level was assessed by means of seven categories. 4 means "*senior high*" and 5 "*college graduate*".

4. Results and Analysis

4.1. Primary Study: Beijing Residents' WTP for a Sustainable Water Management Project in Northwest China

To assess the sample's average WTP we estimate a simple univariate probit model, thereby regressing the binary variable, derived from a respondent's answer to the referendum question ("*yes*" or "*no*"), on the bid variable. The software package STATA13 provides the following output (in rounded figures):

$$Prob(yes) = \beta_0 + \beta_1(bid) = 0.642 - 0.006(bid) \tag{3}$$

Given the negative sign of the parameter estimate, the likelihood of agreement, *i.e.*, that a respondent's answer to the referendum question is "*yes*", decreases as the bid level increases. This result is, of course, what economic theory suggests and confirms the plausibility of the collected data. The estimated regression coefficient $\hat{\beta}_1$ and the constant term $\hat{\beta}_0$ build the basis for calculating the sample's average WTP. As shown by [33], average WTP equals the negative ratio of the constant term and the parameter estimate, *i.e.*,

$$WTP = -\frac{\beta_0}{\beta_1} = 107 \tag{4}$$

Thus, in urban Beijing, households are, on average, willing to contribute approximately 1.3% of their disposable monthly income to the water management project in the Tarim Basin. According to the population census, there are 4.006 million households in urban Beijing at the time of the survey [34]. Hence, the social benefits accruing to Beijing residents per month from the "*Tarim Environmental Preservation Plan*" amount to 428.642 million RMB. Naturally, this figure has only limited practical relevance. This monthly WTP of all Beijing residents would have to be extrapolated in terms of the project duration, discounted and adjusted for inflation in order to attain the net present value this project has for all Beijing residents taken together.

To learn about the main determinants of WTP as well as the plausibility of this estimate, respondents' answers to the referendum questions have been explored by means of regression analysis. The results of two probit models are summarized in Table 4. Table A1 in the appendix offers a detailed variable description. In model 1, merely the bid and the income variable are included as control variables. Like in the model without control variables (*cf.*, Equation (3)), the magnitude of the bid negatively affects the likelihood of agreeing with the referendum question. Furthermore, as income increases, the probability of answering "*yes*" to the elicitation question also increases. In other words, respondents with higher incomes have a higher WTP, which is, of course, plausible. The marginal effect of the income variable (d*y*/d*x*) indicates that a 1000 RMB increase in a respondent's income would increase the likelihood of agreeing with the referendum question by 0.4%. Hence, there is an observable, but moderate, income effect.

Table 4. Probit regression models displaying determinants of supporting the water management program.

Dependent Variable: WTP	Model 1			Model 2		
	Coef.	s.e.	dy/dx	Coef.	s.e.	dy/dx
CONSTANT	0.559 ***	0.053		0.238	0.093	
BID	−0.006 ***	0.000	−0.002	−0.006 ***	0.000	−0.002
INCOME	0.010 ***	0.004	0.004	0.006 *	0.004	0.002
MALE				0.128 **	0.053	0.046
EDUCATION				0.038 *	0.020	0.013
HH_SIZE				0.043 **	0.020	0.016
Observations	2400			2400		
Log likelihood	−1531			−1525		
Pseudo R^2	0.075			0.079		

Notes: *** means significant at 1%-level; ** 5% level; *10% level.

In model 2, several demographic variables are added as controls. Based on this extended model conclusions regarding the distribution of the benefits of a more sustainable water management in the Tarim Basin can be drawn. Gender, education and the number of people living in a respondent's household all have a significant effect on WTP. Men have a higher WTP than women; as the education level increases so increases WTP; and respondents with larger households have a higher WTP than others, ceteris paribus. The positive relationship between education and WTP has been reported in other environmental valuation studies, including some CVM surveys in China [35]. Hence, we conclude that a more sustainable water management in the Tarim Basin would be particularly beneficial for people with a relatively high education level. Literature provides mixed results regarding the effects of gender and household size on WTP. A positive relationship between the latter variable and WTP is plausible, because the number of those who would enjoy the environmental improvement to be valued is greater in larger households than in smaller households. However, many CVM studies did not discover such an effect or even reported negative effects [36], which might be due to the fact that budgets are tighter in households with many children. We explain the fact that men have higher WTP than women in our sample by the everyday observation that Chinese men tend to be more generous than Chinese women. In many Chinese families, women manage the household budget and seem to be tighter in spending it than their husbands. Taken together, the results of the regression analysis are plausible and robust, which is an indicator of the validity of the CVM data collected in Beijing.

4.2. Secondary Study: Using Benefit Transfer to Assess Shanghai Residents' WTP

As explained above, policy makers could use the WTP value measured in Beijing to approximate the benefits of the Tarim water management project accruing to people in other parts of the country. The present study is particularly suitable for an application of the benefit transfer methodology. This is because the environmental good which has been valued in Beijing is identical to the environmental good to be valued in other Chinese megacities–we wish to assess the benefits from one and the same water management project and we expect that the benefits accruing from this project, which are mainly passive use values, are perceived similarly in all Chinese megacities. Furthermore, the population characteristics of at least some Chinese megacities resemble each other. For example, the residents of

Shanghai and Beijing have approximately the same average education level and age structure (*cf.*, Table 3). Furthermore, when it comes to the benefit assessment of utility increasing public projects, the appropriate welfare measure is always a WTP measure, in other words, property rights in the context of this water management project are the same for all Chinese megacities. Hence, the three "*ideal criteria*" for benefit transfer according to [20] are broadly fulfilled in the present study.

However, as soon as we wish to apply a more sophisticated form of benefit transfer, we require comprehensive population statistics for all megacities of interest. In China such data is, however, difficult to access. This holds especially true for non-Chinese researchers. Even though the statistical yearbooks of most Chinese cities are nowadays available in English, the published data is in many cases not precise enough to serve as an input for a benefit transfer application. For example, in the Beijing Statistical Yearbook a reader finds statistics about people's education levels in the municipality of Beijing, which encompasses also economically underdeveloped rural areas, but no separate figures regarding the education level of the urban population.

In view of this lack of official data, we selected the city of Shanghai as an exemplary policy site. Since the authors conducted a representative CVM survey also in Shanghai in 2013 (for a different environmental project) we collected also comprehensive population data for Shanghai in this context (*cf.*, [37]). A particular advantage is that the CVM questionnaire used in Shanghai closely resembles the one used in Beijing. Household characteristics such as income and household size were assessed by means of identical questions and answer scales. Table 3 summarizes the characteristics of the policy site (Shanghai) and compares them to the characteristics of the study site (Beijing) of our present benefit transfer analysis here.

The data provides information on the characteristics of 791 residents living in Shanghai's 12 districts and counties. This area counted 6.969 million households at the time of the data collection [38]. Beijing and Shanghai differ in terms of the monthly disposable household income of an average household, gender ratio and household size. These differences are obvious when comparing the data of our two surveys displayed in Table 3; the official figures also reflect these dissimilarities. Households in Shanghai have, on average, higher incomes, more members and there are relatively more men living in Shanghai than in Beijing. Since all of these three characteristics have a statistically significant effect on WTP (*cf.*, Table 4), these differences should be taken into account when transferring the WTP value assessed in Beijing to Shanghai. In addition to demographic differences, it is to be noted that the linear distance to the Tarim Basin is longer from Shanghai than from Beijing. However, flight time, for example from Shanghai or Beijing to the city of Aksu, an oasis city located at the upper reaches of the Tarim River, is approximately the same (*cf.*, Table 3). Anyway, while it is plausible that the geographical distance from the project site and therewith the time needed to get there has an effect on the use value of a project it should not influence its passive use value.

To assess the benefit of the Tarim water management project accruing to Shanghai citizens we have two possibilities at choice: transferring the point estimate of WTP assessed in Beijing to Shanghai; or transferring the value function or WTP function estimated for the Beijing sample to Shanghai, thereby replacing the mean values measured at the study site by the mean values of the Shanghai sample. The more sophisticated variant of function transfer, which is the meta-regression approach, is not suitable in the case of the present study, because hardly any similar environmental valuation studies have been conducted so far.

The results of the value transfer exercise and the function transfers are displayed in Table 5. If we simply transfer the point estimate of 107 RMB per household and month, we obtain a social value of approximately 746 million RMB for Shanghai. This figure is much higher than the one that we have computed for Beijing, which is due to the greater number of households living in urban Shanghai. The mean WTP estimate for Shanghai becomes slightly higher if we apply a value function transfer. Table 5 shows the outcome of two kinds of function transfer. Bishop *et al.* [14] recommended employing only "*theory-driven*" value functions in benefit transfer exercises. This means that the econometric model underlying a value function transfer should only encompass variables which economic theory postulates to be related to WTP. Such economic variables are, for example, the distance to the site where a particular environmental improvement takes place, the magnitude of the environmental improvement and an individual's income. Since the present study focuses on the same water management project for both sites and because all respondents live at a similar distance from the improvement site (and this distance does not matter anyway for passive use values), we can confine the arguments of the value function to the latter variable in such a theory-driven function, *i.e.*, on income. Hence, we first estimated a model which encompasses merely the bid level and a respondent's income as control variables (*cf.*, model 1 in Table 4). Second, we used the more fully specified model, which includes the bid, income, gender and education as control variables (*cf.*, model 2 in Table 4) to tailor a value function for Shanghai. Bishop *et al.* [14] advise against using models with "*ad-hoc variables*", *i.e.*, with any demographic variable in addition to income. This is because the relationship between several demographic variables and WTP is not consistent across studies. For instance, unlike in the Beijing CVM survey, we observed neither an effect of education nor of gender when determining the key determinants of Shanghai residents' WTP for the protection of rare species in Southwest China (*cf.*, [37]). Hence, if we transfer a value function from one site to another, this goes along with the strong assumption that all statistical effects discovered in the primary site would also be replicated in the policy site–this assumption is reasonable for the income effect, which most CVM studies report, but less realistic for any other respondent characteristic.

Table 5. Willingness to pay in RMB of Beijing and Shanghai residents (benefit transfer).

Transfer Mode	Beijing (Study Site)		Shanghai (Policy Site)	
	Mean WTP (95% c.i.) [1]	**Social Value**	**Mean WTP (95% c.i.)**	**Social Value**
Value transfer	107 (98; 116)	431 Mio.	107 (98; 116)	746 Mio.
Function transfer (theory-driven model)	107 (98; 116)	429 Mio.	111 (101; 120)	774 Mio.
Function transfer (model with ad-hoc variables)	107 (98; 116)	429 Mio.	111 (102; 121)	774 Mio.

Note: [1] Mean estimates and confidence intervals have been calculated with the delta method.

Nevertheless, the results of both variants of function transfer lead to very similar value estimates since the population values of the variables gender, education and household size are very similar for Beijing and Shanghai. Irrespective of whether we use a theory-driven model with income as the only explanatory variable or the more fully specified model, we obtain a mean WTP of 111 RMB per month and household for Shanghai. This figure is slightly higher than the point estimate derived from the

Beijing sample. This difference is a result of the difference in household income levels at both sites; since, in Shanghai, households dispose of approximately 2000 RMB per month more than an average Beijing household and because of the positive relationship between income and WTP, mean WTP in Shanghai is higher than in Beijing. Since the price level of consumer goods is approximately the same in both cities, we conclude that Shanghai residents show a higher appreciation of the water management project than Beijing residents.

Hence, depending on the variant of benefit transfer technique we apply, we obtain different WTP estimates. While the difference between mean WTP per household obtained from the two approaches is not statistically significant, the two social value estimates differ substantially due to the difference in population size. Unfortunately, we lack information in order to assess whether 107 RMB or 111 RMB comes closer to a Shanghai resident's "*true*" WTP for the Tarim water management project. Furthermore, we cannot even conclude that at least one of the two figures is valid. In order to assess the validity of the transferred values, we would have to assess Shanghai residents' WTP for improved environmental conditions in the Tarim Basin in a separate CVM study.

4.3. Approximating the Megacity-Wide Social Value

Finally, even though we lack sufficiently comprehensive data for this purpose, we want to present at least a rough approximation of the overall passive use values accruing from a more sustainable water management to people living in Chinese megacities far away from the Tarim Basin. Of course, including the WTP of megacity residents in a cost-benefit analysis of the Tarim water management project leads to a better approximation of the overall social benefit generated by such a project but it still underestimates the overall benefit since also people in smaller cities might appreciate that project. Including them also in a cost-benefit analysis using benefit transfer would be problematic at least when using the value transfer approach because we cannot expect the average WTP of people in smaller (and poorer) cities to be the same as in Beijing and Shanghai. In our demonstration example here, we confine ourselves to the largest Chinese megacities, namely Beijing, Shanghai, Guangzhou, Tianjin and Shenzhen.

As displayed in Table 6, the households in these five megacities have quite different levels of income at their disposal. Due to these differences in one of the key determinants of WTP, function transfer appears more appropriate than simple value transfer. Using the value function estimated based on the primary CVM study in Beijing (*cf.*, Model 1 in Table 4) WTP in Guangzhou amounts to 109 RMB per household; we obtain an average WTP estimate of 108 RMB for Tianjin and 112 RMB for Shenzhen (*cf.*, Table 6). Multiplying these numbers with the total number of households gives the city-specific social value estimates. We arrive at an overall social value of 2.4 billion RMB per month by summing up these estimates for all five megacities. Needless to say, this figure dramatically exceeds the figure we would have obtained when only accounting for the households directly affected by the water management project in question, *i.e.*, households in the Tarim Basin.

It is to be noted, however, that this estimate is full of uncertainty due to the above mentioned methodological weaknesses of the benefit transfer approach as such and also because of several assumptions we had to make in order to calculate the megacity-wide social value in the present study. First of all, it turned out to be difficult to find income data for Guangzhou, Tianjin and Shenzhen. With exception of the city of Shenzhen, none of these cities publish their statistical yearbooks online.

For this reason, we had to rely on other available sources that are less complete and not strictly comparable to the data published by the municipal bureaus of statistics. For example, we lack information regarding average household size for Guangzhou and Tianjin and found only information on disposable per capita income. Assuming that households in Guangzhou and Tianjin are approximately as small as households in Beijing and Shanghai, monthly disposable household income was approximated by multiplying monthly per capita income by three. According to the Statistical Yearbook, an average household in Shenzhen counts 3.67 members. Since this figure is substantially higher than the (assumed) household size in the remaining cities, average monthly disposable household income is also the highest in Table 6. It could be the case, however, that households in Guangzhou and Tianjin consist of more than the assumed three persons. If this holds true, we systematically underestimate the average household income and hence average WTP in these two cities.

Table 6. Approximation of the overall social value.

City Characteristics and Welfare Measures	Beijing	Shanghai	Guangzhou [1]	Tianjin [2]	Shenzhen [3]	Total
Monthly disposable household income (in 1000 RMB)	8.485	10.512	9.513	8.164	11.101	--
Monthly WTP per household (in RMB)	107	111	109	108	112	--
Number of households	4.006 Mio.	6.969 Mio.	4.280 Mio.	3.841 Mio.	2.872 Mio.	21.968
Social value (in RMB)	429 Mio.	774 Mio.	467 Mio.	415 Mio.	322 Mio.	2407 Mio.

Notes: [1] Values approximated based on [39]; [2] Values approximated based on [40]; [3] Values reported in [41].

The validity of our social value assessment is further threatened by some degree of uncertainty inherent to the number of households in the five cities of interest. While the Statistical Yearbooks for Beijing and Shanghai provide detailed information regarding the number of urban households, we lack this essential information for Guangzhou, Tianjin and Shenzhen. The figures reported in Table 6 are likely to overestimate the actual number of urban households because we only dispose of data about the permanent residents, including urban and rural households, in these cities. These considerations show how problematic the use of the benefit transfer method is in practice.

Summing up, the benefit transfer exercise reported in this section indicates that the passive use value of a more sustainable water management in the Tarim Basin seems to be very high as compared to its use value since this project seems to be appreciated also by people living far away from the project site. For the assessment of this appreciation in terms of WTP the benefit transfer technique offers a rather cost-saving possibility to enhance a cost-benefit analysis conducted on site only. Nevertheless the assumptions to be made in order to apply this approach in practice are rather heroic as has been shown in this section. Therefore, one should be very careful when interpreting the results of benefit transfer studies.

5. Conclusions

In this paper we deal with the problem of missing data in the context of environmental cost-benefit analysis. Since environmental projects of greater ecological importance (like our example of fighting desertification in Northwest China) typically create benefits not only for the people living on the project site but potentially for all people living in a country a benefit assessment study carried out at the project site only would lead to a systematic underestimation of the total social benefits accruing from such a project. Since it is not possible from a financial as well as an organizational perspective to conduct benefit assessment studies like CVM surveys all over China, we suggest "*approximating*" the total social value of environmental projects by using the benefit transfer technique. We suggest carrying out one CVM survey at the project site where use values, which accrue to people actively utilizing the environment, as well as passive use values, which are not connected to any (observable) utilization activities, accrue to the people there. Then different classes of other possible study sites (e.g., rural areas and villages, small cities, middle-sized cities, megacities), which are far away from the project site, should be defined and, for each category, one representative study site (e.g., one megacity) should be selected. In these remote study sites, people will typically obtain only passive use benefits from the environmental improvement realized at the project site. At each of the selected representative study sites, a CVM survey can be conducted and its results can be extrapolated to the whole class of sites it represents by making use of the benefit transfer technique. Of course, the overall cost of these assessment studies will increase as the number of representative study sites and with it the number of CVM surveys to be conducted increases. In our empirical example here, we confined our analysis to the class of five Chinese megacities. As a representative study site we chose Beijing and then we showed how the WTP results from a CVM survey conducted there can be extrapolated to the other megacities. Such a confinement benefit transfer allows the enhancement of the quality and comprehensiveness of benefit assessment in the environmental sector without boosting the valuation costs.

Our empirical study clearly demonstrates the shortcomings of the benefit transfer methodology and has, of course, its own limitations. First, the primary study underlying the benefit transfer exercise has several weak points, including the limited sample size, covering only 0.06% of all households in Beijing, the non-random sampling approach and the well-known drawbacks of personal interviews (social desirability bias, interviewer effects, *etc.*). Second, we had to adopt one of the central and at the same most criticized assumptions underlying the benefit transfer approach, namely the assumption on equal preferences in the primary population and the target population. It may be the case, however, that in other Chinese megacities WTP is driven by other factors than those identified in Beijing or that a substantially larger share of people is completely indifferent about the environmental impacts of a more sustainable water management in the Tarim Basin. Third, in the exemplary benefit transfer exercise presented in Section 4.2, we only calculated monthly values. For the conduct of a cost-benefit analysis these values would have to be extrapolated over the project duration, which would involve additional assumptions regarding the appropriate discount rate, expected inflation, *etc.*, implying even more uncertainty about the overall social benefit of the water management project. Fourth, we lack information to verify whether our benefit transfer exercise was "successful". Further research on the benefits of a more sustainable water management in the Tarim Basin accruing to the inhabitants of other Chinese megacities would be needed in order to test the validity of the WTP estimates provided in Section 4.2. Finally, we did not carry out a complete cost-benefit analysis but concentrated on the benefit assessment. Unlike the social benefits, the social costs of water management projects can be calculated relatively easily based on market prices, labor and material costs. However, such data is sometimes difficult to access, especially in countries like China with strict data protection rules.

Acknowledgments

This research is part of the Sino-German joint research project SuMaRiO. SuMaRiO is funded by the German Ministry of Education and Research (Award number 01 LL0918F) and also supported by the National Basic Research Program of China (973 Program, 2012CB955903). We would like to express our thanks to Dai Tang-Ping, professor of sociology at Minzu University of China, whose students conducted the CVM interviews in Beijing.

Author Contributions

Michael Ahlheim and Oliver Frör jointly led the SuMaRiO subproject "*Contingent Valuation of a more sustainable oasis management in the Tarim Basin*". Sonna Pelz was responsible for the data collection and data analysis. Jiang Tong and Luo Jing provided important advice on the questionnaire design, sampling frame and sampling mode. Both supported and facilitated the fieldwork in China. All authors contributed to the preparation of the manuscript.

Appendix

A.1. Project and Payment Scenario

Project Scenario:

> "*Scientists have developed a program with the overarching goal to improve the living conditions in the area along the Tarim River for man and nature. This program is called the Tarim Environmental Preservation Plan and implies a science-based water management that ensures that more and more water arrives in the lower reaches of the Tarim River, so that the riparian forests and grasslands can recover there. Once the river and its natural environment has fully recovered, the area will be less exposed to sandstorms and dust; typical animals and plants will survive; also, the living conditions of future generations will improve.*"

Payment Scenario plus Referendum Question:

> "*In order to get the Tarim Environmental Preservation Plan financed, Central Government needs to transfer more money to the Tarim area. In order to finance these transfer payments, government would have to increase taxes if TEPP was realized. This would lead to rising monthly expenditures for households. Economists estimate that the proposed program would increase an average Beijing household's monthly expenditures by approximately (10/25/50/75/100/200) RMB. We would like to find out whether Beijing citizens support the implementation of the Tarim Environmental Preservation Plan although it implies an increase in their monthly expenditures. Considering that your monthly household expenditures would increase by approximately [BID] RMB through the program, would you personally be willing to support it?*"

A.2. Variable Description

Table A1. Description of variables used in the regression models.

Variable	Description	Mean	Std. Dev.	Min.	Max.
WTP	*"Considering that your monthly household expenditures would increase by approximately (BID) RMB through the program would you personally be willing to support it?"* (1 = yes, 0 = no)	0.540	0.498	0	1
BID	Bid amount	89.440	68.270	10	200
INCOME	Monthly disposable household income in 1000 RMB	8.498	7.755	0.5	50
MALE	Gender of the respondent (1 = male, 0 = female)	0.506	0.500	0	1
AGE	Age of the respondent	40.227	15.420	18	84
EDUCATION	Level of education of the respondent (1 = did not graduate from primary school, 7 = master degree or higher)	4.320	1.342	1	7
HH_SIZE	Number of people living in the respondent's household	2.871	1.400	1	9

Conflicts of Interest

The authors declare no conflict of interest.

References

1. Executive Order 12291. *Federal Register*; National Archives: College Park, MD, USA, 1981; Volume 46.
2. Executive Order 12866. *Federal Register*; National Archives: College Park, MD, USA, 1993; Volume 58.
3. Loomis, J.; Kent, P.; Strange, L.; Fausch, K.; Covich, A. Measuring the total economic value of restoring ecosystem services in an impaired river basin: Results from a contingent valuation survey. *Ecol. Econ.* **2000**, *33*, 103–117.
4. Freeman, A.M.; Herriges, J.A.; Kling, C.L. *The Measurement of Environmental and Resource Values: Theory and Methods*; Resources for the Future: New York, NY, USA, 2014.
5. Hanley, N.; Mourato, S.; Wright, R.E. Choice modelling approaches: A superior alternative for environmental valuation? *J. Econ. Surv.* **2001**, *15*, 435–462.
6. Adamowicz, W.; Boxall, P.; Williams, W.; Louviere, J. Stated preference approaches for measuring passive use values: Choice experiments and contingent valuation. *Am. J. Agric. Econ.* **1998**, *80*, 64–75.
7. Mills, L.S.; Soule, M.E.; Doak, D.F. The Keystone-Species Concept in Ecology and Conservation. *BioScience* **1993**, *43*, 219–224.
8. Ahlheim, M. *Handbook of Utility Theory*; Barberà, S., Ed.; Kluwer: Dordrecht, The Netherlands, 1998; pp. 447–568.
9. Ahlheim, M.; Frör, O. Valuing the non-market production of agriculture. *Agrarwirtschaft* **2003**, *52*, 356–369.

10. Carson, R.T.; Hanemann, W.M. *Handbook of Environmental Economics, Volume 2: Valuing Environmental Changes*; Maler, K.-G., Vincent, J.R., Eds.; Elsevier, North-Holland: Amsterdam, The Netherlands, 2005; pp. 821–936.
11. Bateman, I.; Carson, R.T.; Day, B.; Hanemann, M.; Hanley, N.; Hett, T.; Jones-Lee, M.; Loomes, G.; Mourato, S.; Özdemiroglu, E.; *et al. Economic Valuation with Stated Preference Techniques: A Manual*; Elgar: Cheltenham, UK, 2002.
12. Arrow, K.; Solow, R. *Report of the NOAA Panel on Contingent Valuation*; Resources of the Future: Washington, DC, USA, 1993.
13. Nunes, P.A.L.D. *The Contingent Valuation of Natural Parks: Assessing the Warmglow Propensity Factor*; Elgar: Cheltenham, UK, 2002.
14. Bateman, I.J.; Brouwer, R.; Ferrini, S.; Schaafsma, M.; Barton, D.N.; Dubgaard, A.; Hasler, B.; Hime, S.; Liekens, I.; Navrud, S. Making benefit transfers work: Deriving and testing principles for value transfers for similar and dissimilar sites using a case study of the non-market benefits of water quality improvements across Europe. *Environ. Resour. Econ.* **2011**, *50*, 365–387.
15. Richardson, L.; Loomis, J.; Kroeger, T.; Casey, F. The role of benefit transfer in ecosystem service valuation. *Ecol. Econ.* **2015**, *115*, 51–58.
16. Wilson, M.A.; Hoehn, J.P. Valuing environmental goods and services using benefit transfer: The state-of-the art and science. *Ecol. Econ.* **2006**, *60*, 335–342.
17. Pearce, D.; Atkinson, G.; Mourato, S. *Cost-benefit Analysis and the Environment: Recent Developments*; OECD: Paris, France, 2006.
18. Loomis, J.; Richardson, L.; Kroeger, T.; Casey, F. Valuing ecosystem services using benefit transfer: Separating credible and incredible approaches. In *Valuing Ecosystem Services. Methodological Issues and Case Studies*; Ninan, K.N., Ed.; Edward Elgar Publishing Limited: Cheltenham, UK, 2014; pp. 78–89.
19. Johnston, R.J.; Rosenberger, R.S. Methods, Trends and Controversies in Contemporary Benefit Transfer. *J. Econ. Surv.* **2010**, *24*, 479–510.
20. Boyle, K.J.; Bergstrom, J.C. Benefit transfer studies: Myths, pragmatism, and idealism. *Water Resour. Res.* **1992**, *28*, 657–663.
21. Ahlheim, M.; Lehr, U. Nutzentransfer: Das Sparmodell der Umweltbewertung. *Perspekt. Wirtsch.* **2003**, *3*, 85–104.
22. Thevs, N. Water Scarcity and Allocation in the Tarim Basin: Decision Structures and Adaptations on the Local Level. *J. Curr. Chin. Aff.* **2011**, *3*, 113–137.
23. Zhang, J.; Wu, G.; Wang, Q.; Li, X. Restoring environmental flows and improving riparian ecosystem of Tarim River. *J. Arid Land* **2010**, *2*, 43–50.
24. Chen, Y.; Xu, C.; Chen, Y.; Liu, Y.; Li, W. Progress, Challenges and Prospects of Eco-Hydrological Studies in the Tarim River Basin of Xinjiang, China. *Environ. Manag.* **2013**, *51*, 138–153.
25. Lu, Z.; Zhao, L.; Dai, J. A study of water resource management in the Tarim Basin, Xinjiang. *Int. J. Environ. Stud.* **2010**, *67*, 245–255.
26. Huang, X.; Chen, Y.; Ma, J.; Chen, Y. Study on change in value of ecosystem service function of Tarim River. *Acta Ecol. Sin.* **2010**, *30*, 67–75.

27. Niu, X.; Wang, B.; Liu, S.; Liu, C.; Wei, W.; Kauppi, P.E. Economical assessment of forest ecosystem services in China: Characteristics and implications. *Ecol. Complex.* **2012**, *11*, 1–11.
28. Xu, C.Z.; Li, Q.; Wang, W.X. Contingent Valuation Method Assessment of the Ecological Benefits of Populus Euphratica Forest Based on Consumer Perspective-Taking the Typical Area in Tarim River Basin as an Example. *Adv. Mater. Res.* **2014**, *962–965*, 703–709.
29. Krutilla, J. Conservation reconsidered. *Am. Econ. Rev.* **1967**, *56*, 777–786.
30. Ahlheim, M.; Frör, O.; Luo, J.; Pelz, S.; Jiang, T.; Yiliminuer. The Social Value of Environmental Improvements in the Tarim Basin-toward a comprehensive assessment in a heterogeneous setting. *Environ. Nat. Resour. Res.* **2015**, *5*, doi:10.5539/enrr.v5n2p49.
31. Rumbaur, C.; Thevs, N.; Disse, M.; Ahlheim, M.; Brieden, A.; Cyffka, B.; Duethmann, D.; Feike, T.; Frör, O.; Gärtner, P.; *et al.* Sustainable management of river oases along the Tarim River (SuMaRiO) in Northwest China under conditions of climate change. *Earth Syst. Dynam.* **2015**, *6*, 83–107.
32. Bishop, R.C.; Heberlein, T.A. Measuring Values of Extra-Market Goods: Are Indirect Measures Biased? *Am. J. Agric. Econ.* **1979**, *61*, 916–930.
33. Hanemann, W.M. Welfare evaluations in contingent valuation experiments with discrete response data: Reply. *Am. J. Agric. Econ.* **1989**, 71, 1057–1061.
34. Beijing Municipal Bureau of Statistics. *2012 Beijing Statistical Yearbook*; China Statistics Press: Beijing, China, 2013.
35. Ahlheim, M.; Börger, T.; Frör, O. Replacing rubber plantations by rain forest in Southwest China—Who would gain and how much? *Environ. Monit. Assess.* **2015**, *187*, 1–20.
36. Ahlheim, M.; Schneider, F. Considering household size in contingent valuation studies. *Environ. Econ.* **2013**, *4*, 112–123.
37. Ahlheim, M.; Frör, O.; Langenberger, G.; Pelz, S. Chinese urbanities and the preservation of rare species in remote parts of the country: The example of eaglewood. *Environ. Econ.* **2014**, *5*, 32–43.
38. Shanghai Municipal Statistics Bureau. *2012 Shanghai Statistical Yearbook*; China Statistics Press: Shanghai, China, 2013.
39. *Gzstats 2013*; Guangzhou Statistical Manual: Guangzhou, China, 2013.
40. 2013 Tianjin Economic and Social Development Statistics Bulletin (Economic and Social Statistic for Tianjin, 2013). Available online: http://www.ce.cn/ (accessed on 14 May 2015).
41. Shenzhen Municipal Bureau of Statistics. *2012 Shenzhen Statistical Yearbook*; China Statistics Press: Shenzhen, China, 2013.

13

CO_2 Reduction Potential of Water Saving in Vietnam

Takayuki Otani [1], Kanako Toyosada [2] and Yasutoshi Shimizu [3,*]

[1] ESG Promotion Department, TOTO Ltd., 2-1-1, Nakashima, Kokurakita-ku, Kitakyushu City 802-8601, Japan; E-Mail: takayuki.otani@jp.toto.com

[2] Department of Environmental Science, International College of Arts and Science, Fukuoka Women's University, 1-1-1, Kasumigaoka, Higashi-ku, Fykuoka City 813-8529, Japan; E-Mail: kanako.toyosada@fwu.ac.jp

[3] Water and Lifestyle Research Laboratory, 31-3, Munakata, Fukuoka 811-4141, Japan

* Author to whom correspondence should be addressed; E-Mail: yshimizu@wlrlab.com;

Academic Editor: Enedir Ghisi

Abstract: In a previous study, we showed that widespread adoption of water-saving equipment had the potential to reduce CO_2 emissions by 1% in Japan. The usage of already diffused equipment was used as an evaluation baseline. This was an evaluation model of developed countries. In order to evaluate the potential benefits of water-saving in developing countries, it is necessary to set the baseline, as cities in developing countries are expected to have the necessary infrastructure in place in the near future. In this paper, the potential for reducing CO_2 emissions by water saving in Vietnam was evaluated. Based on the development of water infrastructure, and envisioning a society in which the latest high-efficiency flush toilet bowls and showers installed in Hanoi and Ho Chi Minh City are used in all Vietnamese houses as a near future baseline, we evaluated the potential reduction when a water-saving project is implemented. Under these conditions, an 8.8% reduction in CO_2 emissions in Vietnam would be achieved by the widespread adoption of water-saving equipment. Following the recognition of the large environmental contribution potential of water saving, a water-saving project has been planned for implementation in Vietnam in the near future.

Keywords: global warming; CO_2 reduction; water; saving water; Vietnam

1. Introduction

In developing countries, it is expected that rapid economic expansion will result in the mass consumption of energy and resources. The transition of developing countries' growth from mass-consumption to sustainable green growth is an urgent issue facing the world. With respect to energy issues, the framework of a smart city that enables utilization of renewable energies is gaining popularity as a solution for sustainable development, not only in developed countries, but also in developing countries.

Regarding water issues, urbanization and the construction of water infrastructures in developing countries will cause global water consumption to increase 1.8-fold, and act as a brake on the sustainable development of the Earth [1]. Table 1 shows the rate of sustainable access to basic sanitation and domestic water consumption per capita, and it can be seen that the rate of sustainable access to basic sanitation, which means the development of water infrastructure, significantly increases domestic water consumption [2]. Such an increase in domestic water use causes a further increase in energy consumption as the waterworks and sewer systems are operated. The population of urban areas is growing quickly, such as in Vietnam and Indonesia, has reached a global majority. Control of water demand at the time of the construction of waterworks and sewer systems in these areas is important not only for the preservation of water resources, but also for energy security.

Table 1. Relationship between rate of continuous use of sanitary facility and domestic water consumption.

Rate of Sustainable Access to Basic Sanitation	Population in 2007 (million)	Typical Countries	Domestic Water Usage (L/capita/day)
100%	685.3	Japan USA Canada	442
Less than 100% More than 30%	4061.2	Indonesia Vietnam Mexico	132
Less than 30%	1471.0	India Cambodia Nepal	114

It should be noted that the authors have revealed in previous studies that the use of water is closely related to energy consumption, and that water saving contributes to energy saving and a 1% reduction in CO_2 emission in Japan [3]. As a result, water saving is now employed as one of the energy security political measures the Japanese government has adopted against global warming. In addition, water saving is incorporated into a project to support the shift to a low carbon Asia conducted by the Japanese government [4]. This low carbon goal by the Japanese government for developing countries is being pursued through a bilateral offset credit scheme called the Joint Crediting Mechanism (formerly the Bilateral Offset Credit Mechanism). This scheme reduces CO_2 emissions by developing countries using funds and technology provided by Japan, and treats that reduction as part of Japan's global environmental contribution. The details of the scheme were described in a previous paper [5]. This research was performed as a part of a Joint Crediting Mechanism feasibility study project conducted by the Japanese government that evaluated CO_2 reduction potential through the introduction of water-saving showerheads and toilet bowls in Vietnam. In the crediting scheme, the credit amount is evaluated as the difference between emissions following implementation of a low-carbon project and a near-future baseline. A conceptual diagram is shown in Figure 1. The near-future baseline (the Business as Usual or BaU line) is a forecast that assumes that the current new high-efficiency equipment in Vietnam is commonly used in all Vietnamese houses. The CO_2 reduction potential in Vietnam was evaluated in the same manner of the crediting scheme.

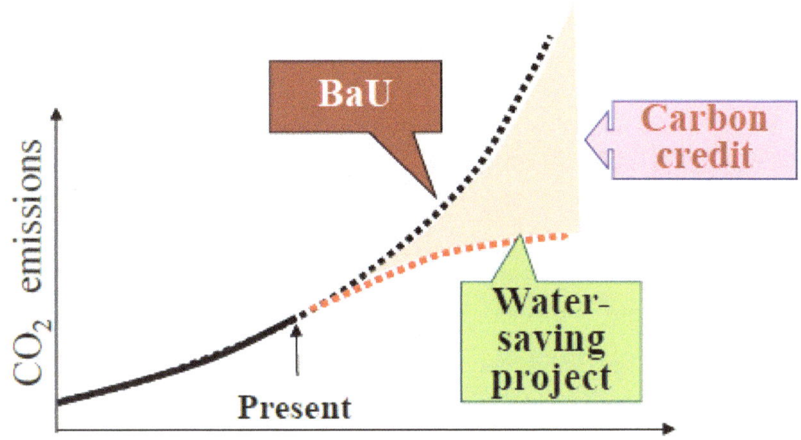

Figure 1. Conceptual diagram of water-saving project evaluation in developing countries. BaU: Business as usual.

2. Survey Method

To determine the BaU line, the data for Hanoi and Ho Chi Minh Cities were extracted as a model of the city to be generalized in Vietnam in the near future.

The evaluation formulae are shown in Equations (1) and (2). RI_i is the reduced CO_2 emission amount. ef_i is the CO_2 emission factor of water. Δq_i is the reduced water amount. pb_i, pp_i and n_i are the equipment performances of BaU and the project, and the equipment usage number, respectively.

$$RI_i = \sum(ef_i \times \Delta q_i) \quad (1)$$

$$\Delta q_i = (pb_i - pp_i) \times n_i \quad (2)$$

In this survey, with the cooperation of the Vietnamese Ministry of Construction and the Energy Conservation Center, treated water volume and energy consumption data were collected by visiting waterworks and sewage facilities in Hanoi and Ho Chi Minh Cities, and we calculated the energy consumption rates and CO_2 emission factors of the facilities in the same manner, as shown in a previous paper [6].

The equipment usage number: n_i was ascertained by a questionnaire survey of 100 residents in each of the cities of Hanoi ($n = 383$) and Ho Chi Minh Cities ($n = 387$), with 10 residents of each group of 100 residents being interviewed. At this time, the performance of their water use equipment was ascertained by conducting measurements at each of the 10 residences. Photographs of existing toilet bowls and showerheads and three-dimensional measurement data of toilet tanks, attached to the 100 questionnaires, were also used to determine performance. Data logging flow meters (logging interval of 0.5 s) were installed in shower flow lines at 5 of the 10 residencies of interviewed survey participants to measure showering time. In addition, wattmeters were installed to measure the monthly energy consumptions of water pumps. The energy consumption rates of pumps were calculated by the energy consumption data collected and the values of the water meters of waterworks companies. To ascertain the shower flow rate, all interviewed residents' showerheads and hose sets were brought back to Japan and measured to determine the optimal flow rate [7]. Furthermore, stores selling toilet bowls and showers were visited, and the performance of equipment on the market and the behavior of users with respect to purchasing criteria were researched to estimate equipment performance in the near future: pb_i.

3. Results and Discussion

3.1. Current Status and Problems of Water Infrastructure in Vietnam

Urbanization has advanced rapidly in Vietnam, and four of its cities have populations of more than one million. In those cities, although waterworks are developing, sewage systems are less developed [8]. Overall, infrastructure development is not keeping pace with the expansion of cities, and there are many regions where a 24-h water supply does not exist. Also, due to the high rate of water leakage, the pressure in waterworks is set to a low value. Therefore, as shown in Figure 2, it was found that residents were installing water tanks to secure a ready supply of domestic water.

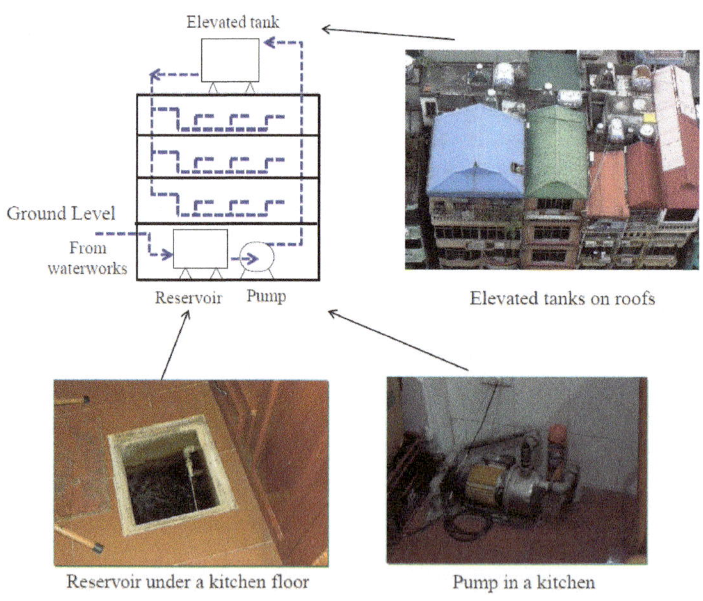

Figure 2. Example of water supply system in a Vietnamese house.

The energy required for waterworks and sewer systems in Vietnam was analyzed by setting boundaries equivalent to the analysis of the system in Japan [3,6]. It was evaluated as energy consumption per unit of water, and calculated by dividing the total sum of energy consumption in all water supply and sewage facilities in Hanoi and Ho Chi Minh Cities by the total amount of treated water (see Table 2). Rapid sand filtering is employed in waterworks, and the activated sludge method is employed in the sewage systems. The system configurations are not much different from those in Japan, and the energy consumption rate of water is within the range of the rate distribution of all facilities in Japan and, therefore, the calculated result was determined as appropriate.

Table 2. Energy consumption rate of water in Vietnam.

Process	Energy Consumption Rate of Water (MJ/m^3)
Waterworks system	1.44
Sewer system	1.05
Water supply system in a house	1.99

The energy consumption rate of water derived from water pumps in detached houses was also calculated (see Table 2). The annual electricity consumption per capita in Vietnam is around 872 kw·h (one-eighth that of Japan) [9], and energy consumption by water pumps accounts for about 3% of this figure.

3.2. Characteristics of Residents and Current and Future Situation Regarding Water Use Equipment

To study the potential of environmental load reduction by introducing water-saving toilets and showers, the performance of equipment used and the behavior of equipment use were surveyed in two cities: Ho Chi Minh City, where there are only dry and rainy seasons; and Hanoi City, where there are four seasons. There are about 21 million people in Vietnam, and 99% of them live in detached houses, only 230,000 (4% of residents in city areas) people live in collective housing, such as apartments in the inner city area [10]. Accordingly, detached houses were selected as the research target. Figures 3 and 4 show an example of water-related equipment in one of the houses we visited. Many residences in Hanoi and Ho Chi Minh City are like the one shown in the photograph: the building structure has a narrow frontage, great depth, and three floors. All surveyed residences had a toilet and shower room on each floor. According to the interview results, most Vietnamese families in large cities have two wage earners, and grandparents often live in the house to support in child-raising. As each generation uses one floor, it is typical for them to install a toilet/shower room on each floor. The average household size is 4.1 persons/household in Ho Chi Minh City, and considering that young, single people are coming into the large city, it appears that the average household size is large [11].

Figure 3. Outside view of the residence visited in Ho Chi Minh City, Vietnam.

Figure 4. Photograph of a washroom in the residence.

All the toilets surveyed were Western-style, flushing toilets. All the residences had showers. Tables 3 and 4 show the average performance of the toilets and the result of elimination behavior. The details for the calculation were shown in the previous paper [12]. The results of preliminary survey and modeling of Japan and China are included. Although the daily number of times of using the toilet was similar in Japan, China, and Vietnam, which is 7–8 times per day, there was a slight difference in the number of times using and flushing the toilet at home. The number of times the toilet was used in the house was extracted from the 24-h data, using the house stay rate data in Figure 5. Vietnamese family members stay home longer than family members do in Japan, as many company employees go home for lunch, and schools have a double system of morning/afternoon (see Figure 5), and this is reflected in household toilet usage. Regarding toilet equipment performance, the amount of water in a flush is much less than the average toilet in Japan. It is thought that this is because many Vietnamese households use a bidet-type water nozzle for post-excretion cleaning purposes instead of using toilet paper. Thus, the waste volume in a toilet and the required flushing water in Vietnam are smaller than those of Japan. People do not dispose of toilet paper in toilets in China, either.

Table 3. Comparison of toileting behavior in Vietnam, Japan and China (number/capita/day).

Toileting Behavior		Vietnam	Japan [13]	China [14]
Number of eliminations	Excrement	1.1	1.1	1.3
	Urine	6.1	5.4	6.7
	Total	7.2	6.5	8.0
Number of toilet flushes at home	Full flush	2.1	2.1	1.8
	Half flush	4.8	2.6	4.3
	Total	6.9	4.7	6.1

Table 4. Typical toilet bowl used in Vietnam, Japan and China.

Typical Toilet Bowl	Equipment Commonly Used: Vietnam	Equipment Commonly Used: Japan	Equipment Commonly Used: China	Water-Saving Equipment
Appearance			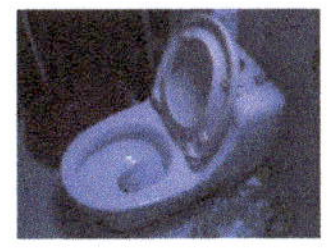	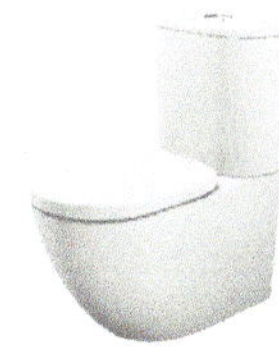
Performance [L/flush]				
Full flush	**6.0**	**13.0** (One mode)	**5.0**	**3.8–4.8**
Half flush	**5.0**		**3.5**	**3.0–3.6**

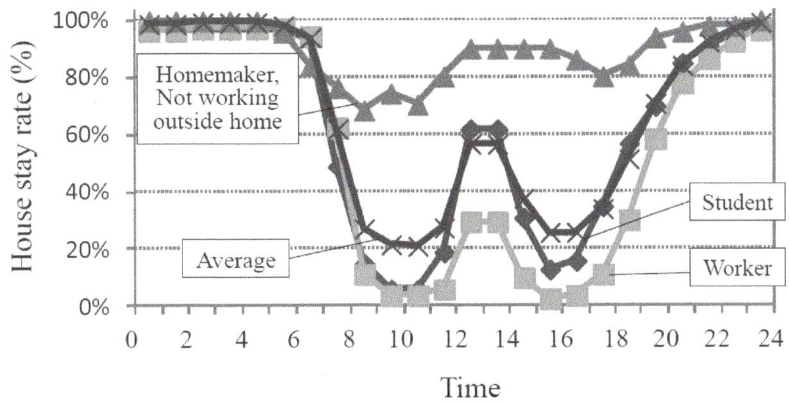

Figure 5. Time spent at home in Vietnam by occupation (weekdays).

Table 5 shows the flow rates of the showerhead being used, and Table 6 shows the number of times and length of time using the shower. Showering behavior was affected by latitude and thus the number of showers taken in the tropical city of Ho Chi Minh City was higher than in Hanoi. The water heater ownership rate was 90% in Hanoi and 68% in Ho Chi Minh City, based on surveyed residents and it was learned that taking a hot shower has become fairly popular for many families. The most common type of water heater varies between the north and south. Instantaneous electric water heaters were common in Ho Chi Minh City, while hot-water storage electric heaters were common in Hanoi City. The heaters are shown in Table 7. According to distributors, storage heaters are selected in Hanoi because the tap water temperature is low in Hanoi City during winter and thus instantaneous heaters do not have the capability to warm water to a suitable temperature.

Table 5. Typical showerhead used in Vietnam, Japan and China.

Typical Showerhead	Equipment Commonly Used: Vietnam	Equipment Commonly Used: Japan	Equipment Commonly Used: China	Water-Saving Equipment
Appearance	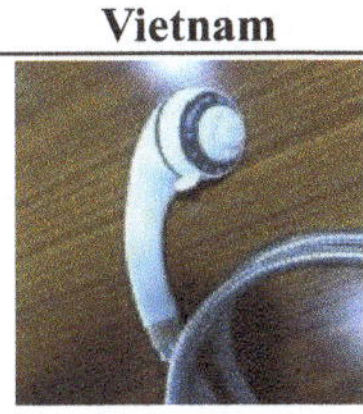	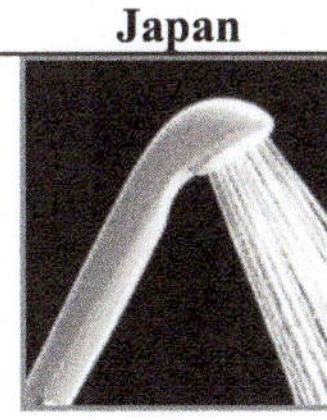	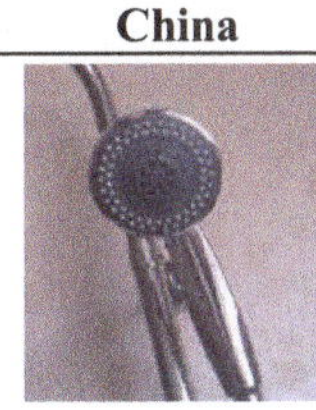	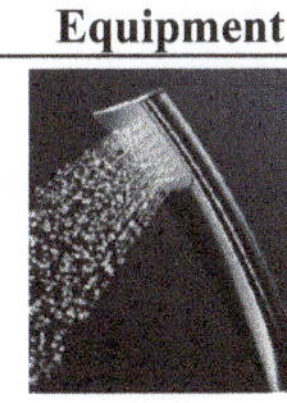
Performance [L/min] *	10.0	8.5	10.0	6.5

Note: * as an optimal flow rate.

Table 6. Shower usage model in Vietnam.

Showering Behavior	Hanoi City	Ho Chi Minh City
Number of showers per week	6.8	7.9
Shower duration time (min.)	10.1	7.6
Shower temperature (°C)	36.3	

Table 7. Water heater performance used in Vietnam.

Item	Instantaneous Electric Water Heater	Storage Electric Water Heater
Appearance		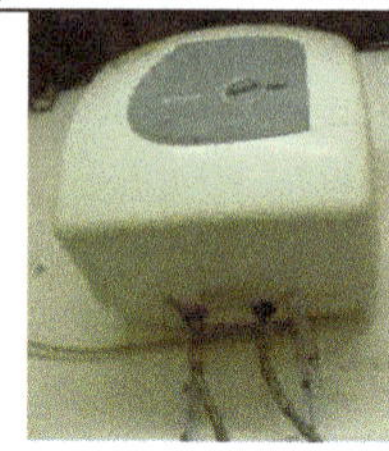
Power consumption (W)	4500	2500
Max. supply temperature (°C)	40	60
Storage capacity (L)	0	10–30

Also, based on interviews about shower use during the survey in Ho Chi Minh City, most respondents said that they could not return to taking cold showers once having started taking hot showers. It is easy to assume that, with future economic growth in Vietnam, water heaters for showers will become even more popular in the near future.

3.3. CO_2 Reduction Potential of Water Saving

The potential for reducing CO_2 emissions by popularizing water-saving equipment was evaluated. First, the CO_2 emission factor of water was determined based on the data for energy consumption rate, shown in Table 2, and CO_2 emission factors of used energies, such as electricity. The calculated result is shown in Table 8.

Table 8. CO_2 emission factor of water in Vietnam.

Process	Emission Factor (kg-CO_2/m^3)
Waterworks system	0.23
Sewer system	0.16
Water supply system in a house	0.32

Note: CO_2 emission factor of electricity: 0.576 kg-CO_2/kw·h.

The performance of water-saving toilets and showers to be introduced to Vietnam is shown in Tables 5 and 6. Although development of the water infrastructure is insufficient currently, in the future, it is expected that development of the water infrastructure will advance rapidly with the support of Official Development Assistance from Japan and other countries. It is also assumed that an improvement in living standards will rapidly increase energy consumption resulting from hot water supply for showers and so on. Accordingly, assuming a near-future society where water infrastructures have developed and hot showers are common, the potential for reducing CO_2 emissions by water saving was calculated.

Table 9 shows the results, along with the CO_2 reduction potential by water saving, as calculated for Japan and China.

Table 9. Contribution potential of water saving in Vietnam, Japan and China.

Item	Vietnam	Japan	China
Energy consumption rate of water (kWh/m^3)	0.66	0.85	1.32
CO_2 emission factor of electricity (kg-CO_2/kWh)	0.58	0.41	0.84
CO_2 emission factor of water (kg-CO_2/m^3)	0.39	0.44	1.11
CO_2 reduction potential by water saving (% ratio to whole emission)	8.8	1.0	1.8

Research that relates to city water consumption and energy has been carried out in advanced nations [3,15–17]. This research was influenced by the examination of water saving in cities where the waterworks and sewerage systems were already developed. According to these results, the widespread adoption of water-saving equipment has the potential to reduce CO_2 emissions by 1% in Japan and Taiwan. Contribution of water saving in Vietnam was estimated to reduce CO_2 emissions by 8%. Since the energy consumption of Vietnam is still small, this greatly influences the contribution of water saving.

The Vietnam National Green Growth Strategy was announced in September 2012, and actions against global warming are being promoted. It is possible that the popularization of water-saving equipment will become a policy that contributes to green growth.

4. Conclusions

So as to actively contribute to the fight against global warming, the Japanese government is using a joint crediting mechanism and promoting projects to leapfrog development in cities in developing countries so that they become low-carbon societies. As a water-infrastructure version of the green growth scenario to create low-carbon cities in developing countries conducted by the Japanese government, the authors evaluated the CO_2 reduction potential by the formation of a water-saving society. It appears that the formation of a water-saving society would prove effective in mitigating against the future increase in CO_2 emissions brought about by urbanization in Vietnam. When this idea is broadly accepted, and each developed country exercises its expertise in supporting developing countries, it is expected that the urban environment will become more environmentally friendly in many areas, and the reduction of carbon emissions will be accelerated.

Acknowledgments

This survey was conducted in Vietnam as a part of the "Global Warming Mitigation Technology Promotion Project (FY2012)" of the Ministry of Economy, Trade and Industry of Japan, and as a part of the "Project to Support the Large-Scale Formation of Joint Crediting Mechanism Programs to Realize Low Carbon Societies in Asia (FY2013)" of the Ministry of the Environment of Japan.

Author Contributions

The survey and analysis in this study were carried out by Takayuki Otani and Kanako Toyosada under the direction and supervision of Yasutoshi Shimizu. The first draft of the manuscript was prepared by Takayuki Otani, and later versions were revised and edited extensively by Yasutoshi Shimizu before publication. All authors discussed the results and commented on the manuscript at all stages.

Conflicts of Interest

The authors declare no conflict of interest

References

1. Ministry of Land, Infrastructure, Transport and Tourism of Japan, Water resources in Japan. Available online: http://www.mlit.go.jp/tochimizushigen/mizsei/water_resources/contents/current_state2.html (accessed on 31 December 2014).
2. Black, M.; King, J.; Lacey, C. *The Atlas of Water*; University of California Press: California, CA, USA, 2009.
3. Yasutoshi, S.; Kanako, T.; Kiyoshi, N. Prediction of CO_2 emissions associated with residential plumbing equipment (in Japanese). *J. Soc. Heat. Air-Cond. Sanit. Eng. Jpn.* **2010**, *163*, 11–18.
4. Yasutoshi, S.; Kanako, T.; Mari, Y.; Kyosuke, S. Creation of carbon credits by water saving. *Water* **2012**, *4*, 533–544.
5. Yasutoshi, S.; Satoshi, D.; Kanako, T. CO_2 emission factor for rainwater and reclaimed water used in buildings in Japan. *Water* **2013**, *5*, 394–404.
6. Yasutoshi, S.; Satoshi, D.; Kanako, T. CO_2 emission factor of water in Japan. *Water* **2012**, *4*, 759–769.
7. Japan Valve Manufacturers' Association: Monitoring Test Method of Showers for Bathing (Hot Water Saving B). Available online: http://www.j-valve.or.jp/suisen/setsuyu/setsuyu-monitor.pdf (accessed on 31 December 2014).
8. Ministry of Environment of Japan, Water Environmental Business in Asia. (In Japanese) Available online: http://www.env.go.jp/water/coop/asia_businessdownload/s_moe.pdf (accessed on 31 December 2014).
9. JETRO, Electricity Survey Report in Vietnam (In Japanese) Available online: http://www.jetro.go.jp/ext_images/jfile/report/07001271/vietnamelectricity2013-2.pdf (accessed on 31 December 2014).
10. General Statistics Office of Vietnam, Statistical Information. Available online: http://www.gso.gov.vn/default_en.aspx?tabid=491 (accessed on 31 December 2014).
11. JETRO, BPO Business Potential Survey Report in Vietnam. (In Japanese) Available online: http://www.jetro.go.jp/jfile/report/07000780/vnm_bop.pdf#search='%E3%83%9B%E3%83%BC%E3%83%81%E3%83%9F%E3%83%B3%E5%B8%82+%E5%B9%B3%E5%9D%87%E5%AE%B6%E6%97%8F%E4%BA%BA%E6%95%B0' (accessed on 31 December 2014).
12. Study on the Modeling of Toilet Usage and Bathing Habits in Vietnam. (In Japanese) Available online: http://www.shasej.org/ (accessed on 21 May 2015).

13. Kanako, T.; Yasutoshi, S.; Akihiko, I.; Kyosuke, S. Quantification of environmental impact reduction effect resulting from use of water-saving toilet bowls. *J. Soc. Heat. Air-Cond. Sanit. Eng. Apr.* **2013**, *193*, 1–8. (In Japanese)
14. Kanako, T.; Yasutoshi, S.; Satoshi, D.; Mari, Y.; Kyosuke, S. Evaluation of the potential of CO_2 emission reduction achieved by using water-efficient housing equipment in Dalian, China. In Proceedings of the 38th International Symposium CIB W062 on Water Supply and Drainage for Buildings, Edinburgh, Scotland, UK, 27–30 August 2012; pp. 587–598.
15. Cheng, C.L.; Liao, W.J.; Liu, Y.C.; Tseng, Y.C.; Chen, H.J.; Evaluation model of CO_2 emission for saving water strategy. In Proceedings of the 38th International Symposium CIB W062 on Water Supply and Drainage for Buildings, Edinburgh, Scotland, UK, 27–30 August 2012; pp. 537–548.
16. Wong, L.T.; Cheung, C.T.; Mui, K.W. Energy efficiency benchmarks of example roof-tank water supply system for high-rse low-cost housings in Hong Kong. In Proceedings of the 38th International Symposium CIB W062 on Water Supply and Drainage for Buildings, Nagano, Japan, 17–19 September 2013; pp. 349–359.
17. Matos, C.; Briga-Sá, A.; Pereira, S.; Silva-Afonso, A. Water and energy consumption in urban and rural households. In Proceedings of the 38th International Symposium CIB W062 on Water Supply and Drainage for Buildings, Nagano, Japan, 17–19 September 2013; pp. 209–221.

Permissions

All chapters in this book were first published in Water, by MDPI; hereby published with permission under the Creative Commons Attribution License or equivalent. Every chapter published in this book has been scrutinized by our experts. Their significance has been extensively debated. The topics covered herein carry significant findings which will fuel the growth of the discipline. They may even be implemented as practical applications or may be referred to as a beginning point for another development.

The contributors of this book come from diverse backgrounds, making this book a truly international effort. This book will bring forth new frontiers with its revolutionizing research information and detailed analysis of the nascent developments around the world.

We would like to thank all the contributing authors for lending their expertise to make the book truly unique. They have played a crucial role in the development of this book. Without their invaluable contributions this book wouldn't have been possible. They have made vital efforts to compile up to date information on the varied aspects of this subject to make this book a valuable addition to the collection of many professionals and students.

This book was conceptualized with the vision of imparting up-to-date information and advanced data in this field. To ensure the same, a matchless editorial board was set up. Every individual on the board went through rigorous rounds of assessment to prove their worth. After which they invested a large part of their time researching and compiling the most relevant data for our readers.

The editorial board has been involved in producing this book since its inception. They have spent rigorous hours researching and exploring the diverse topics which have resulted in the successful publishing of this book. They have passed on their knowledge of decades through this book. To expedite this challenging task, the publisher supported the team at every step. A small team of assistant editors was also appointed to further simplify the editing procedure and attain best results for the readers.

Apart from the editorial board, the designing team has also invested a significant amount of their time in understanding the subject and creating the most relevant covers. They scrutinized every image to scout for the most suitable representation of the subject and create an appropriate cover for the book.

The publishing team has been an ardent support to the editorial, designing and production team. Their endless efforts to recruit the best for this project, has resulted in the accomplishment of this book. They are a veteran in the field of academics and their pool of knowledge is as vast as their experience in printing. Their expertise and guidance has proved useful at every step. Their uncompromising quality standards have made this book an exceptional effort. Their encouragement from time to time has been an inspiration for everyone.

The publisher and the editorial board hope that this book will prove to be a valuable piece of knowledge for researchers, students, practitioners and scholars across the globe.

List of Contributors

Janesh Devkota
Dynamic Solutions International LLC, Edmonds, WA 98020, USA

Xing Fang
Department of Civil Engineering, Auburn University, Auburn, AL 36849-5337, USA

Bo Yang
College of Environment and Safety Engineering, Qingdao University of Science and Technology, Qingdao 266042, China

Xiao Kong
College of Environment and Safety Engineering, Qingdao University of Science and Technology, Qingdao 266042, China
Key Laboratory of Environmental Biotechnology, Research Center for Eco-Environmental Sciences, Chinese Academy of Sciences, Beijing 100085, China

Bingjian Cui
Key Laboratory of Environmental Biotechnology, Research Center for Eco-Environmental Sciences, Chinese Academy of Sciences, Beijing 100085, China
College of Resources and Environment, University of Chinese Academy of Sciences, Beijing 100049, China

Decai Jin
Key Laboratory of Environmental Biotechnology, Research Center for Eco-Environmental Sciences, Chinese Academy of Sciences, Beijing 100085, China
College of Resources and Environment, University of Chinese Academy of Sciences, Beijing 100049, China

Ye Deng
Key Laboratory of Environmental Biotechnology, Research Center for Eco-Environmental Sciences, Chinese Academy of Sciences, Beijing 100085, China
College of Resources and Environment, University of Chinese Academy of Sciences, Beijing 100049, China

Xuliang Zhuang
Key Laboratory of Environmental Biotechnology, Research Center for Eco-Environmental Sciences, Chinese Academy of Sciences, Beijing 100085, China
College of Resources and Environment, University of Chinese Academy of Sciences, Beijing 100049, China

Guoqiang Zhuang
Key Laboratory of Environmental Biotechnology, Research Center for Eco-Environmental Sciences, Chinese Academy of Sciences, Beijing 100085, China
College of Resources and Environment, University of Chinese Academy of Sciences, Beijing 100049, China

Zhihui Bai
Key Laboratory of Environmental Biotechnology, Research Center for Eco-Environmental Sciences, Chinese Academy of Sciences, Beijing 100085, China
College of Resources and Environment, University of Chinese Academy of Sciences, Beijing 100049, China

Monica Pinardi
Institute for Electromagnetic Sensing of the Environment, National Research Council, Via Bassini 15, Milan 20133, Italy
Department of Life Sciences, University of Parma, Viale Usberti 11/a, Parma 43124, Italy

Andrea Fenocchi
Department of Civil Engineering and Architecture, University of Pavia, Via Ferrata 3, Pavia 27100, Italy

Claudia Giardino
Institute for Electromagnetic Sensing of the Environment, National Research Council, Via Bassini 15, Milan 20133, Italy

Stefano Sibilla
Department of Civil Engineering and Architecture, University of Pavia, Via Ferrata 3, Pavia 27100, Italy

Marco Bartoli
Department of Life Sciences, University of Parma, Viale Usberti 11/a, Parma 43124, Italy

Mariano Bresciani
Institute for Electromagnetic Sensing of the Environment, National Research Council, Via Bassini 15, Milan 20133, Italy

Shaohua Liu
State Key Laboratory of Simulation and Regulation of Water Cycle in River Basin, China Institute of Water Resources and Hydropower Research (IWHR), Beijing 100038, China

Denghua Yan
State Key Laboratory of Simulation and Regulation of Water Cycle in River Basin, China Institute of Water Resources and Hydropower Research (IWHR), Beijing 100038, China

Jianhua Wang
State Key Laboratory of Simulation and Regulation of Water Cycle in River Basin, China Institute of Water Resources and Hydropower Research (IWHR), Beijing 100038, China

Baisha Weng
State Key Laboratory of Simulation and Regulation of Water Cycle in River Basin, China Institute of Water Resources and Hydropower Research (IWHR), Beijing 100038, China

Gang Wang
Beijing Municipal Research Institute of Environmental Protection, Beijing 100037, China

Meijian Yang
Beijing Municipal Research Institute of Environmental Protection, Beijing 100037, China

Seung Beom Seo
Department of Civil, Construction, and Environmental Engineering, North Carolina State University, Raleigh, NC 27606, USA

Young-Oh Kim
Department of Civil and Environmental Engineering, Seoul National University, 1 Gwanak-ro, Gwanak-gu, Seoul 151-741, Korea

Cho-Rong Kim
Korea Institute of Civil Engineering and Building Technology, 283 Goyangdae-ro, Ilsanseo-gu, Goyang-si, Gyeonggi-do 411-712, Korea

Jinggang Chu
School of Hydraulic Engineering, Dalian University of Technology, Dalian 116024, China

Yong Peng
School of Hydraulic Engineering, Dalian University of Technology, Dalian 116024, China

Wei Ding
School of Hydraulic Engineering, Dalian University of Technology, Dalian 116024, China

Yu Li
School of Hydraulic Engineering, Dalian University of Technology, Dalian 116024, China

James Horne
The Australian National University, Canberra ACT, 2600, Australia

Yu Yang
Department of Public Administration, Southeast University, Nanjing 210096, China

Lei Yin
School of Economics and Management, Southeast University, Nanjing 210096, China

Qingzong Zhang
School of Economics and Management, Southeast University, Nanjing 210096, China

Chiyuan Miao
Changjiang River Scientific Research Institute, Changjiang Water Resources Commission, Wuhan 430010, Hubei, China
State Key Laboratory of Earth Surface Processes and Resource Ecology, College of Global Change and Earth System Science, Beijing Normal University, Beijing 100875, China

Alistair G. L. Borthwick
School of Engineering, The University of Edinburgh, The King's Buildings, Edinburgh EH9 3JL, UK

Honghu Liu
Changjiang River Scientific Research Institute, Changjiang Water Resources Commission, Wuhan 430010, Hubei, China

Jigen Liu
Changjiang River Scientific Research Institute, Changjiang Water Resources Commission, Wuhan 430010, Hubei, China

Petri S. Juuti
International Environmental History Group, University of Tampere, Tampere FI-33014, Finland

Georgios P. Antoniou
Department of Architecture Engineering, NTUA, PhDc, Athens 10682, Greece

Walter Dragoni
Department of Physics and Geology, Perugia University, Perugia 06123, Italy

Fatma El-Gohary
Water Pollution Research Department, National Research Center, Bohouth Str. Dokki, Cairo 12622, Egypt

Giovanni De Feo
Department of Industrial Engineering, University of Salerno, via Ponte don Melillo, Fisciano 1-84084, Italy

Tapio S. Katko
Capacity Development in Water and Environmental Services, Tampere University of Technology, Tampere FI-33720, Finland

Riikka P. Rajala
International Environmental History Group, University of Tampere, Tampere FI-33014, Finland

Xiao Yun Zheng
Yunnan Academy of Social Sciences, 577 Huan Cheng West Road, Kunming City 650034, China

Renato Drusiani
Federutility, Rome 00192, Italy

Andreas N. Angelakis
Institute of Iraklion, National Foundation for Agricultural Research (N.AG.RE.F.), Iraklion 71307, Greece

Annalisa Di Piazza
National Research Council ISSIA—Institute of Intelligent Systems for Automation—UOS Palermo, Via Dante Alighieri 92, Palermo 90141, Italy

Francesco Lo Conti
Department of Civil Engineering, Environmental, Aerospace, of Materials University of Palermo, University of Palermo, Viale delle Scienze, Building 8, Palermo 90128, Italy

Francesco Viola
Department of Civil Engineering, Environmental, Aerospace, of Materials University of Palermo, University of Palermo, Viale delle Scienze, Building 8, Palermo 90128, Italy

Emanuele Eccel
Sustainable Agro-ecosystems and Bioresources Department, IASMA Research and Innovation Centre, Fondazione Edmund Mach, Via E. Mach 1, San Michele all'Adige (TN) 38010, Italy

Leonardo V. Noto
Department of Civil Engineering, Environmental, Aerospace, of Materials University of Palermo, University of Palermo, Viale delle Scienze, Building 8, Palermo 90128, Italy

Michael Ahlheim
Institute of Economics, University of Hohenheim, Stuttgart 70593, Germany

Oliver Frör
Institute for Environmental Sciences, University of Koblenz-Landau, Landau 76829, Germany

Jing Luo
Research Center for China's Borderland History and Geography, Chinese Academy of Social Sciences, Beijing 100732, China

Sonna Pelz
Institute of Economics, University of Hohenheim, Stuttgart 70593, Germany

Tong Jiang
National Climate Center, China Meteorological Administration, Beijing 100081, China
School of Geography and Remote Sensing/Collaborative Innovation Center on Forecast and Evaluation of Meteorological Disasters, Nanjing University of Information Science and Technology, Nanjing 210044, China

Takayuki Otani
ESG Promotion Department, TOTO Ltd., 2-1-1, Nakashima, Kokurakita-ku, Kitakyushu City 802-8601, Japan

Kanako Toyosada
Department of Environmental Science, International College of Arts and Science, Fukuoka Women's University, 1-1-1, Kasumigaoka, Higashi-ku, Fykuoka City 813-8529, Japan

Yasutoshi Shimizu
Water and Lifestyle Research Laboratory, 31-3, Munakata, Fukuoka 811-4141, Japan